FROM TOTAL QUALITY CONTROL TO LEAN SIX SIGMA

FROM TOTAL QUALITY CONTROL TO LEAN SIX SIGMA

by

Sanjay Borde

Wordpen Academics
Publishing Science That Matter

ISBN: 9789360847630

Published by : WORDPEN ACADEMICS
4736/23, Ansari Road, Darya Ganj,
New Delhi – 110 002
Phone: +91-011-35004336
E-mail: info@wordpenacademics.com

***Printed at* : Replika Press Pvt. Ltd.**

Exclusive Distributor: ***Bio-Green Books, New Delhi***

Preface

In the realm of quality management, the journey from Total Quality Control (TQC) to Lean Six Sigma represents a significant evolution in the way organizations approach process improvement, customer satisfaction, and operational excellence. Rooted in the principles of continuous improvement, waste reduction, and data-driven decision-making, both TQC and Lean Six Sigma methodologies have played instrumental roles in transforming industries, enhancing competitiveness, and driving organizational success. As we trace the evolution of quality management practices, we uncover valuable insights into the evolution of business excellence and the quest for continuous improvement in the modern era.

This book, titled "From Total Quality Control to Lean Six Sigma," serves as a comprehensive exploration of the evolution, principles, and applications of quality management methodologies over time. Drawing upon the collective wisdom of quality management experts, practitioners, and scholars, this volume offers insights, case studies, and practical guidance for organizations seeking to embrace the principles of TQC and Lean Six Sigma in their pursuit of excellence.

From the origins of Total Quality Control in post-war Japan to the integration of Lean principles and Six Sigma methodologies in contemporary business practices, "From Total Quality Control to Lean Six Sigma" traces the historical development and theoretical foundations of these quality management approaches. Through a blend of historical context, theoretical frameworks, and practical examples, readers will gain a deeper understanding of the underlying principles and key concepts that drive TQC and Lean Six Sigma initiatives.

Moreover, this book explores the practical applications and real-world implementations of TQC and Lean Six Sigma in various industries and organizational settings. By examining case studies, success stories, and lessons learned from leading companies, we highlight the transformative impact of quality management methodologies on business performance, customer satisfaction, and employee engagement.

As we navigate the complexities of quality management in the 21st century, it is our hope that "From Total Quality Control to Lean Six Sigma" will serve as a valuable resource for executives, managers, quality professionals, and aspiring practitioners seeking to harness the power of continuous improvement and operational excellence in their organizations. Whether you are embarking on your quality management journey or looking to elevate your organization's performance to new heights, this book invites you to explore the rich history and promising future of TQC and Lean Six Sigma.

Sanjay Borde

Contents

Preface v

1. **Artificial Intelligence Tools and Case Base Reasoning Approach for Improvement Business Process Performance** 3

   1. Introduction–2. The main targets, methods and contribution–3. Experimental research, areas for research and reasons for developing expert systems–3. Approach to developing expert system–4. Final considerations–References

2. **Improving 'Improvement' By Refocusing Learning: Experiences from An -Initially- Unsuccessful Six Sigma Project in Healthcare** 25

   1. Introduction–2. Theory and background–3 Method–4. The context: The Skaraborg Hospital Group and the Six Sigma quality programme–5 Results and analysis of the project–6 Integrating learning mechanisms into the SkaS quality programme–6 Conclusions–7 Further research–References

3. **Project Costs and Risks Estimation Regarding Quality Management System Implementation** 43

   1. Introduction–2. Projects costs management–2 Projects quality improvement through risks management process–3 Conclusions–References

4. **What Quality Management Allied to Information Can Do for Occupational Safety and Health** 69

   1. Introduction–3. The benefits of the fusion between quality management tools and knowledge management–4. The fusion between cycle PDCA + 5S = Organizational benefits?–5. The Internet as a source of data in occupational safety and health–6. Digital information access resources under the open archives Philosophy–7. OSH information management in São Paulo city, Brazil, under the professionals perception–8. Final considerations–References

5. Reducing Mirror Slippage of Nightstand with Plackett-Burman Doe and Ann Techniques **103**

1. Introduction–2. Application of Six Sigma–3. Application of ANN (Artificial Neural Network)–4. Conclusion–References

6. Redesigning the Service Process for Total Quality in Government Hospitals: Evidence From Kwara State **119**

1. Introduction–2. Problem statement–3. Methodology–4. Data analysis–5. Hypothesis testing–6. Recommendations–7. Conclusion–References

7. Some Applicable Methods to Analyze and Optimize System Processes in Quality Management **129**

1. Introduction–2. Review of system processes to reveal general engineering problems that are due to be solved by the mathematical modeling–3. Models and software tools to analyze information system processes–4. Models, software tools and methods to analyze and optimize system processes–5. Examples–6. Conclusion–References

8. Competence Education and Training for Quality **201**

1. Introduction–2. Term of quality–3. Education and training for quality–4. Competence of education and training for quality–5. Conclusion–References

9. The Integration of TQM and Six-Sigma **221**

1. Introduction–2. What is 'quality'–3. The evolution of quality management–4. The development and implementation of TQM–5. The development and implementation system of Six-Sigma program–6. Integrated model of TQM and Six Sigma–7. The development of a business excellence system–8. Case study–References

10. Qualitative and Quantitative Analysis of Six Sigma in Service Organizations **251**

1. Introduction–2. Background–3. Research methodology–4. The framework consolidation–5. Conclusion–References

11. Lean Six Sigma - Making it 'Business as Usual' **293**

1. Introduction–2. Lean Six Sigma – a brief step back in time–3. Approach–4. A final note–References

Index **303**

SECTION-1

QUALITY MANAGEMENT

1

Artificial Intelligence Tools and Case Base Reasoning Approach for Improvement Business Process Performance

1. Introduction

Contemporary and every day more perfect information achievement, becomes available for everybody, and simply, very quickly become a necessity. It is necessary that organizations use information technology as a tool for developing a sense of learning, acquire and use knowledge. Information tools should not be use like tools for automation of existing processes. There should be another aspect or already obsolete category. With this aspects, thinking and attitudes, it can be said that we living in the century of knowledge and that we have already overcome period of information technology which should be, simply, implemented like support in the way for achieving knowledge.

This informational environment has been recognized in the world and because of there are significant rising in the use of artificial intelligence tools. There is evidence that is a great number of eligible to use and easily available software for needs of the development of such as systems in the field of artificial intelligence. Also, in [1] states that investment and implementation of artificial intelligence show significant results, particularly in attempt of to get higher profit. The artificial intelligence, like the word itself says is the area that deals with the development of systems that mimic human intelligence and a man with tend to replace him in some activities based on knowledge. That is way for over viewing problem of human absence, cost of services, disinclination of people to provide knowledge and similar. Specified conditions, particularly from the standpoint of the necessities of knowledge, and also the fact that in area of research topic for the purposes of quality management systems, there are evident gap [2, 3-10, 11]. That facts justifying the author's striving to be in this research and accept to use artificial intelligence tools for developing systems oriented to knowledge. These views and attitudes were in agreement: that there is no correct programming software that has a strong base of knowledge that could assist in identification of a problem, that has not developed a single expert system that deals with the measurement, evaluation, corrective and preventive action to improve organizational performance and the like [12, 13-16, 10]. It is also an incentive to be based on such analogies create a foundation set up and entered the field of artificial intelligence in order to obtain knowledge as one of the most important factors for creating competitiveness in the market [17-19, 20].

Everything above can be understand like introduction for developing an research whit main aim for developing a system in the field of artificial intelligence that would be based on the analysis in the quality management system and that has given recommendations for achieving business excellence and improve the financial performance of the organization. The main parts and activities of that research stay in the basis of this chapter.

2. The main targets, methods and contribution

Based on the introduction and results of researching literature source and practice, in the scope of this research, it can be set up main targets, and that are:

- to find (regardless of size or type of organization) area in organization which have priority from the standpoint of improvement,
- to establish new concept of Degree of Readiness and Coefficient of Significance which can show intensity and type of action which should be provide in direction of achieving business excellence and
- to develop and testing in real condition an expert system for improvement business process performances even those of financial character base on analogy with human body function.

In this sense, it can be use science method for inductive and deductive way of deciding and concluding. First one was used for collecting, estimating and analyzing of experimental data, or to making general knowledge by using specific knowledge and particular facts. The second one was used for applying and checking specific conclusion in real condition.

Also, like science approaches it was used: analogy method, expert decision and "ex post facto" or previous case and facts.

Beside that, many other methods and tools were conducted like: knowledge discovery in data base, data mining, case base reasoning-CBR, object oriented programming, artificial intelligence tools, Analytic Hierarchy Process-AHP, expert choice, testing in real condition, Visual Basic and Select Query Language.

Through a detailed analysis of literature sources and software, it was found evident gap in applying artificial intelligence tools for improvement business process performances based on Quality Management System-QMS and especially in experience of other and case reasoning. In this research, analogy between human body function and process oriented organization were established, and areas in organization which is prior from the standpoint of improvement were identified. Two unique data bases and significant number of company and data, make original experimental value and bases for research. Also, new concept of Degree of Readiness and Coefficient of Significance for achieving business excellence stay in the basis of new expert system for achieving business excellence. By applying this expert system, especially on prior area, employees should drive they process performances to excellent condition, even those of financial character. Also, many actions for improvement with appropriate coefficients which show theirs intensity where found. This action should be understood also like preventive action for strengthening organizational condition to avoid some failure in the system. This expert system was tested in real conditions in one very successful organization which will be participant in competition for European Award for business excellence. This test and verification showed that the system could be useful and also the efficient and effective

3. Experimental research, areas for research and reasons for developing expert systems

The basic facts of this research are attempted to define two levels of experimental data. The first level of the data is related to quality management systems and nonconformities that have emerged. This is a basic level of data which reflects the situation in the quality management systems and identify critical places that are subject to improvement. The base of these data is unique and consists of the **1009** nonconformities (cases), identified in over than **350** organizations. If we know that in our area in the field of competent certification body has, approximately 500 certificates, then the number of 350 is about **70%** of the total number. That fact points out to the significance of sample for analysis.The term nonconformities refer to any non-conformance of requirements of ISO 9001, nonconformity non-fulfilment of a requirement [21]. During the external audits of quality management system, competent and trained auditors can identify several types of nonconformities (Figure 1). We are using most significant data from highest level of pyramid at which were collected at the level of many country like external estimation and evaluation of they performance and condition.Distribution of nonconformities depends on the rules that define the certification body itself. However, for the purposes of this research is used classification which is the most common in the literature, which is favour by the authoritative schools in the world in the field of management system and that is clearly recommended by European guidelines in the subject area, which is split into three levels. The first level is the disagreements that are evaluated as insignificant deviations from the standards and requirements which are interpreted as an oversight or random error. The other two categories are interpreted as nonconformities that represent a great deviation from the essential requirements, which are reflected in the frequent discrepancies in individual requirements, representing a deviation that brings into doubt the stability of the management system and threatening the operations of the organization.

Data base of nonconformities which is under consideration in this research contains only nonconformities in the domain of the other two categories, and that giving greater importance to this research and gives greater significance results.

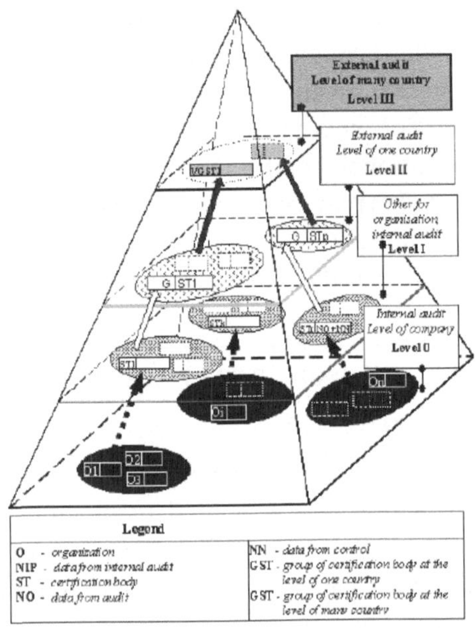

Figure 1. Data source (highest level of data significance)

Non-conformances are identified in accordance with the structure requirements defined in the ISO 9001 standard as follows:

- ❖ Quality management systems: 4.1 general requirements, 4.2 documentation requirements,
- ❖ Management responsibility (module 5): 5.1 management commitment, 5.2 customer focus, 5.3 quality policy, 5.4 planning, 5.5 responsibility, authority and communication, 5.6 management review,
- ❖ Resource management (module 6): 6.1 provision of resource, 6.2 human resources, 6.3 infrastructure, 6.4 work environment,
- ❖ Product realization (module 7): 7.1 Planning of product realization, 7.2 customer related processes, 7.3 design and development, 7.4 purchasing, 7.5 production and service provision, 7.6 control of monitoring and measuring devices,

- Measurement, analysis and improvement (module 8): 8.1 general, 8.2.1 customer satisfaction, 8.2.2 internal audit, 8.2.3 monitoring and measurement of processes, 8.2.4 monitoring and measurements of product, 8.3 control of nonconforming product, 8.4 analysis of data, 8.5 improvement.

Accordingly, for example in the field of 8.2.1 from the standpoint of the appearance of non- conformances organizations have a significant and frequent or large deviations in the sense that it does not follow the information about the observations of users, it did not define the methods for obtaining this information, they do not have strong communication with customers and similar. Or for example in the field of 8.2.3 with the observed aspect, organizations do not apply appropriate methods for monitoring and performance measurement processes, have not mechanisms for implementation of corrective measures in cases that have not achieved the planned performance of processes and the like.

This data will be used like the basis of CBR approach or approach where it is possible to make significant conclusion in the sense of main target of this research. This approach is shown in figure 2.

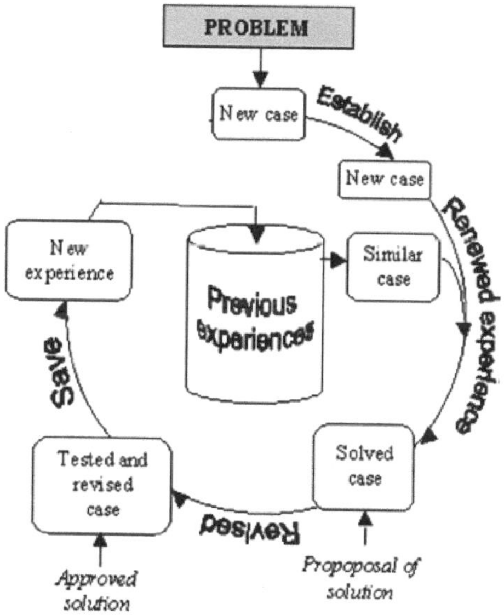

Figure 2. Case based approach

The second level of data consist data from evaluation organizations that participated in the competition for the quality award based on European Quality Award criteria. This database is unique, as well as in the previous case. Data were transferred in encoded form in order to secure the identity of the organization. Data were collected in 100% extent (34 organizations) and thus are significant and give a real picture of the situation in our organizations. These data are used for comparison with previous, basic level data. That is way for making improvement or exalt from basic level on the level of business excellence and way for making knowledge which reproduce expert system on his output. That is also comply with literature more existent attitude, and natural way that organization should first implement

Quality Management System and after that system which is based on Total Quality Management concept [22, 23-26, 27].

In order to show the current directions and trends in the field of development of software for quality, and to select under researched areas in the field of software quality, it was conducted a detailed review and analysis of a total of 143 software. All necessities information for that analysis are available in site (http://www.qualitymag.com) where are publish updated software items which are related to quality. The results of the analysis are shown in the figure 3.

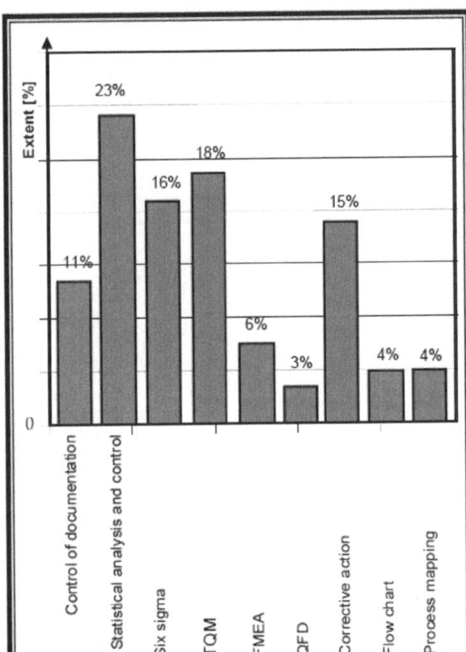

Figure 3. Results of analysis of existent software for quality

On the x axis diagrams are shown the software ability and orientation. Obviously is that the software in the field of quality is usually oriented to the control of documentation, statistical control and analysis, six sigma model, concept of total quality management, FMEA and QFD methodology, corrective action, flowchart and process mapping. However, there are specific tools for automation: the implementation of the quality management system documentation, description of information flow, implementation methods and techniques of quality, and more. Therefore, it can be concluded that there is no software that is based on the application of artificial intelligence tools in the sense of the definition of preventive actions for the purpose of improving the process. The greatest number of software is related to the application of statistical methods in the process of monitoring and improving quality. It is obviously that a large number of software is based on total quality management systems concept. The facts point out present approach which we develop in this research and also justify further research in this area. It is interesting that a large number of software are base on the corrective actions and on the other hand there is not any registered software that has application for output preventive action what is, of course, main recommendation of ISO 9000 series. This fact also gives stimulus in terms of development of software that emphasis to the prevention. That approach is unique in the field of software for quality and makes this research more significant.

Beside this analysis, in this research were analyzed huge amounts of available books in order to point out the justification of applying expert system. Expert systems are different from other artificial intelligence systems in that, they attempt to explicitly and unequivocally embody expertise and knowledge with the software [28]. Expert systems are also identified as one of the most commercial branches and in most number of projects used artificial intelligence tools [29, 30]. For example, it is estimated that in the first half of 21st century, even 75% of all legal documents be written with the assistance of expert systems [31]. Also expert systems will be of vital importance for measuring the quality of products and services [32-34]. Expert systems are an area of special importance with rise trends in modern business conditions [35, 36-38]. They have special significance in a highly developed countries where is actual knowledge based economy. This research highlight trends, significance and justification of developing and implementing expert systems.

Main idea and approach for developing expert system come from analogy between human body functions and process in some organization which was organized based on process modelling from ISO 9000 respect. This approach is present on figure 4.

This research tries to deal with perfection of functioning of the human body compare with a process modelling structures of the implemented quality management system. The challenge made in this way, tried to create a system that is universal for all sizes of organization, which incorporates a large number of gathered data, in fact a large number of experiences, in order to get a better image of the system status. This should be added to the primary goal which is to develop a model for improvement of management system, oriented to achieve BE according to show off how to maintain and improve the performance of the human body. However, the goal is also, to develop a system for measuring performance and capacity of each activity in the QMS, in order to obtain a true picture of the systems and capabilities in order to define the areas where improvements should be made, with clearly defined intensity of improvement. On the basis, thus established the analogy is made to compare elements of implemented QMS to the systems that have applied for Quality award for BE as a system with high performance.

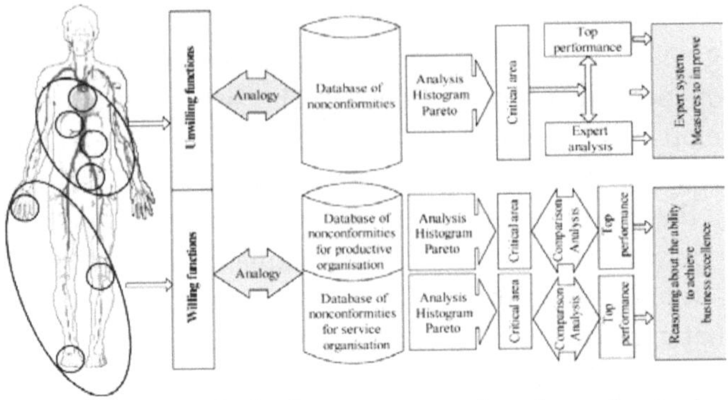

Figure 4. Analogy with the human organism in order to improve organizational performance

To establish the analogy between the process modulated organizational structure and the human organism, so as to create the system that is independent from organizational functions and based only on the process model, following division of man functions was made [39, 40]:

- willing and
- unwilling functions.

Willing functions (term "functions" is used in medical terminology, although it is equally correct, to use a term "activities" in view of ISO 9000 standard terminology. For reasons of consistent referencing and use of theories from the field of medicine, the author has chosen to use the term functions.) are those dependent on man's profession and performed by man's will. They are variable and dictated by a central control of the organism. For example, when a worker at the construction site lifts his hand, it is not the same as when a referee at the game lifts his hand and etc. Willing functions refer to functions of external motoric organs.

Second category is made of unwilling or automated functions and their use is given by their existence. There are functions that are same in all professions and all people (considering that they exist, i.e. that human body is in good health) and do not depend on the man will but are simply executed. For example, those are functions of secreting enzymes, hormones, heartbeats, and similar, like ordinary body functions, and functions that cannot be controlled [41, 42].

With such a ratio of functions in the human body, we can establish the analogy of the system with implemented quality management system. Analogy in term of willing function goes in direction to developed all data in to two category, production and service organization and make some analyses, which is not subject of this research.

In order to meet requirements of this research, only analogy in terms of unwilling functions has been considered. The idea is to use all nonconformities (undependable of organization type or size) and base on case base reasoning approach, make conclusion about readiness of systems to making some top form.

3. Approach to developing expert system

At the market today, we can find many tools for creating expert systems. These systems can be developed in a programmable environment through tools of type C + +, Visual Basic or some other programs which are related to development of expert systems. However, today are developed specialized tools for creating expert systems which allow a high degree of automation in process of developing expert systems. There are called expert system shells. From the standpoint of this research it was carried out choice of expert system shell from the aspect of next four criteria [43-45, Personal communication with group for consulting from London South Bank University, Business, Computing & Information Management, 2011):

- ❖ programmability,
- ❖ comprehensiveness,
- ❖ universality,
- ❖ price.

During the election, it was analyzed 58 shells. All information about shells are available on the Internet [2], and classified in a group of commercial shell. Detailed analyses were conducted separately for each tool through analyzing belonging site. For evaluation on the basis of the criteria it was adopted the scale of 1 to 5 where 1 is unsatisfactory grade. According to defined criteria as a most distinguished tool for the needs of the research was adopted ACQUIRE shell. That tool is non programming oriented and it has affordable price. This is a tool that supports the work of the Windows operational environment. It has possibility to develop all elements of expert system and supports forward, backward and combined chaining. For the presentation of knowledge it can be used production rules, the action table, or combined techniques. During a process of developing expert system, the role of engineer for knowledge took up first author, and the role of one expert took up second author. Also, as sources of knowledge were used following:

- ❖ experience from eleven prestigious organizations in the world of field of quality management systems, business excellence and organizational performance [46],
- ❖ guidelines from standards for improving organizational performance [47],
- ❖ best practices from auditing of ISO 9001 oriented system [48],
- ❖ experience and practice of organizations that participated in the competition for the Oscar of quality award [49],

- theory and principles of TQM [50],
- experiences that are listed in [51] and indicate the path to business excellence.

The expert systems are included and knowledge gained through many concrete practical projects of quality management systems implementation, and many training on that topic. That knowledge is next:

- knowledge that are specific to certain companies,
- knowledge derived from specific experiences and on specific way of solving problem,
- knowledge of those that are best for certain jobs and are passed special training,
- knowledge of those that is proven in practice for the specific job and similar.

For the purposes of this research, expert system was develop for modules 5 (management responsibility) and module 8 (measurement, analyses and improvement) of ISO 9001 standard. The reason for that is that these areas have the greatest importance in achieving business excellence [1] and therefore they should be considerate from the standpoint of improvement. Also, another reason is that module 8 has requirements that are oriented to the improvement and that is essence and priority.

The idea of this research is to make the integration of decision support systems (DSS) which is operate on first level of experimental data, and expert system. That is modern approach of integration a number of tools with the aim of acquiring a larger volume of better knowledge [52] and make system with higher level of intelligence. Today trends are integration expert systems and traditional decision support systems which as output give data and information [53].

Integration of expert systems and decision support system can be achieved in two ways [54]. The purpose of this research is to use model which is present on figure 5. based on the collection and analysis of data obtained at the output of the decision support system and it provide important information like one of inputs for expert system and its knowledge base. This is the model which is completely compatible with previous remarked analogy with human body. This two approach stay in base of this analogy integrative model for improvement business process performance.

For the purposes of this research, we developed a decision support system in the MS Access, Select Query Language and Visual Basic environment. This system is base on the first level of experimental data, and like one of outputs it gives results which are present on figure 6 (for module 8).

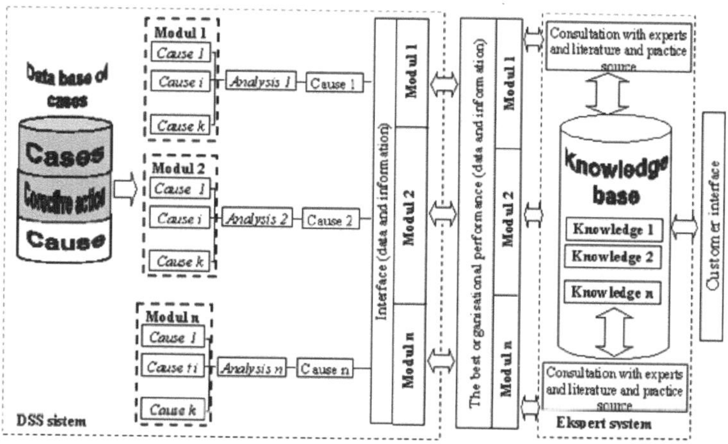

Figure 5. Integrative approach for merging expert system like separate part of DSS components

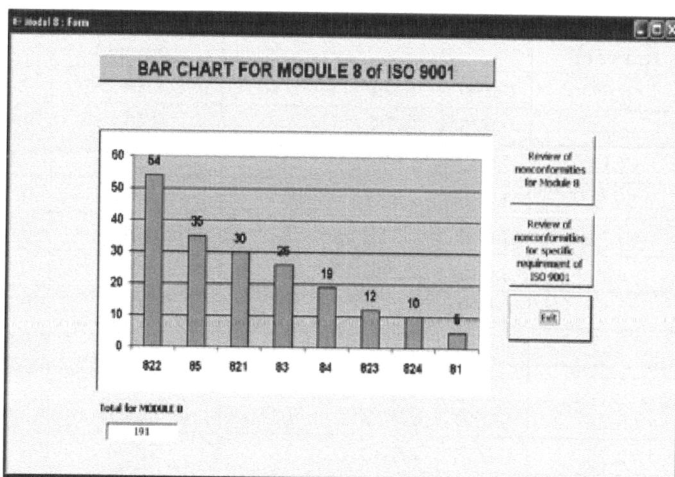

Figure 6. Results of DSS systems for module 8-measurement, analysis and

Applying Pareto method and rules of 70/30 it can be identified area which is crucial from the standpoint of improvement. Also, this system like support for making decision provides written presentation of nonconformities which can be shown as experience of other companies. That could be use like important data for the definition of knowledge in expert system. In addition, this system provides, and comparative analysis with the period of the four years before, which also has significance for the definition of knowledge in the expert system.

Connection between data from the first and data from the second level was achieved through the introduction of the concept of "Degree of readiness (Si)" in achieving business excellence, in accordance with the following expression:

$$S_i = N_z[\%] * K_z, \quad i=1,2,...,26 \tag{1}$$

where:

Si Degree of readiness for all type of organizations for all requests of ISO 9001

Nz Power of a standard clause in terms of percentage. Nz = *f* (number of nonconformities from experimental database)

Kz Coefficient of significance for achieving business excellence

That degree is applies to every single request of ISO 9001 and showing the willingness or the ability of organizations (both manufacturing and service sector) to attain business excellence in some areas. To find this degree, we are using method Analytic Hierarchy Process (AHP) and corresponding software Expert Choice. Results are shown in table 1.

Table 1. Review of the degree of readiness for all type organization in relation to every request of the standard

Requests	Organization (power)	Kz	Si	Requests	Organization (power)	Kz	Si
821	66.67	0.090	**6.00**	83	50	0.036	1.8
823	50	0.075	**3.75**	52	40	0.035	1.4
85	37.14	0.072	**2.674**	81	60	0.033	1.98
84	57.9	0.066	**3.821**	822	46.3	0.032	1.481
54	48.72	0.064	**3.118**	73	76.19	0.026	1.981
824	40	0.064	**2.56**	61	100	0.022	2.2
56	44	0.059	**2.596**	75	37.5	0.022	0.825
53	70	0.054	**3.78**	64	63.64	0.019	1.21
71	56.52	0.04	**2.26**	76	44.68	0.018	0.804
41	65.22	0.038	**2.478**	62	62.07	0.008	0.496
51	72.73	0.038	**2.763**	74	52.94	0.008	0.423
72	56.76	0.038	**2.157**	63	50	0.005	0.25
55	66.67	0.037	**2.467**	42	49.26	0.005	0.246

It is important to emphasize this because it was used and it is very important during definition of preventive measures in terms of defining their priorities and "power". Also, "power" of prevention was related with number of nonconformities in particular area. That means, larger number of nonconformities, or larger number of experience, make possibilities for defining more effective and efficient preventive action like output of expert system.

Through application of Pareto method, based on coefficient of significance following requests were identified as the most significant for achieving business excellence:

requests - 821, 823, 85, 84, 54, 824, 56, 53, 71, 41, 51, 72, 55

At the same time, this is important areas, and have high level of priority for improvement from the standpoint of achieving business excellence and it is very important for defining preventive action of expert system and intensity of that action. If we take a look at the list of "Coefficients of significance" for business excellence achieving, especially the most important ones and perform comparison with the list of variables and their significance in terms of: Business Process Reengineering (BPR), manufacturing strategy, benchmarking and performance measurement, being the result of the appreciated research [55] and [56] it may be found significant intercompatibility.

The concerned compatibility is especially reflected in the following variables, evaluated in the relative research as highly significant for the following four projects, *i.e.:* customer satisfaction, quality, employee satisfaction and personal growth, customer adaptability, identification of top managers with BPR goals, strong process orientation, results orientation, direct customer cooperation. On the other hand, the above mentioned four areas are considered as highly important for any market-oriented organization, thence it can be concluded that organizations by strengthening their capacities in areas of presented "Coefficients of significance" (especially the most important ones), are not only strengthened in terms of the business excellence achieving as per European Award model, but also in the stated four areas.

But some of these areas are much more important then other. Because that, the research was further elaborated in order to indicate most important area for improvement and area where should be focus attention and where should be provide very intensive action in order to achieve best organizational condition and results. This research was conduct from the standpoint of occurrence of nonconformities in all type of organisation regardless of they size or type (both for manufacturing and service organisation). Parallel the Pareto method (70/30) was carried out in that direction and based on that, it was identified next areas:

requests - 56, 75, 62, 822, 74, 76, 54, 72, 85, 821, 55, 63

Now we are search for common requests (area) that are most important and where should be oriented focus and where should be provide extensively action in terms of achieving business excellence regardless of type or size of organization. And they are:

1. 821 – customer satisfaction,
2. 72 - customer related processes,
3. 54 – planning,
4. 85 – continual improvement,
5. 56 – management review and
6. 55 – responsibility, authority and communication.

This area is most important for defining output of expert system and for defining intensity of action for improvement.

Objects were defined during the process of expert system developing. That were depend of problem which should be solved, base on ISO 9001 oriented check list an based on experience which can be find on DSS output. Base on results of DSS system, it is defined value of the object and relation between them. In that way, it is created decision tree, which is present on figure 7.

At the end, after starting the program, in a short time, system introduce user in a set of dialog boxes. One of them is shown on figure 8.

Depend on the answers, expert system produce user's report, like one which is presented on figure 9.

Data obtained from this report, user can use and implement knowledge that an expert system produces. However, users can improve performance of an organization in the field where such as performance are on lower level. Also, it can be improvement performances of other, non critical, area and can be reach level of business excellence.

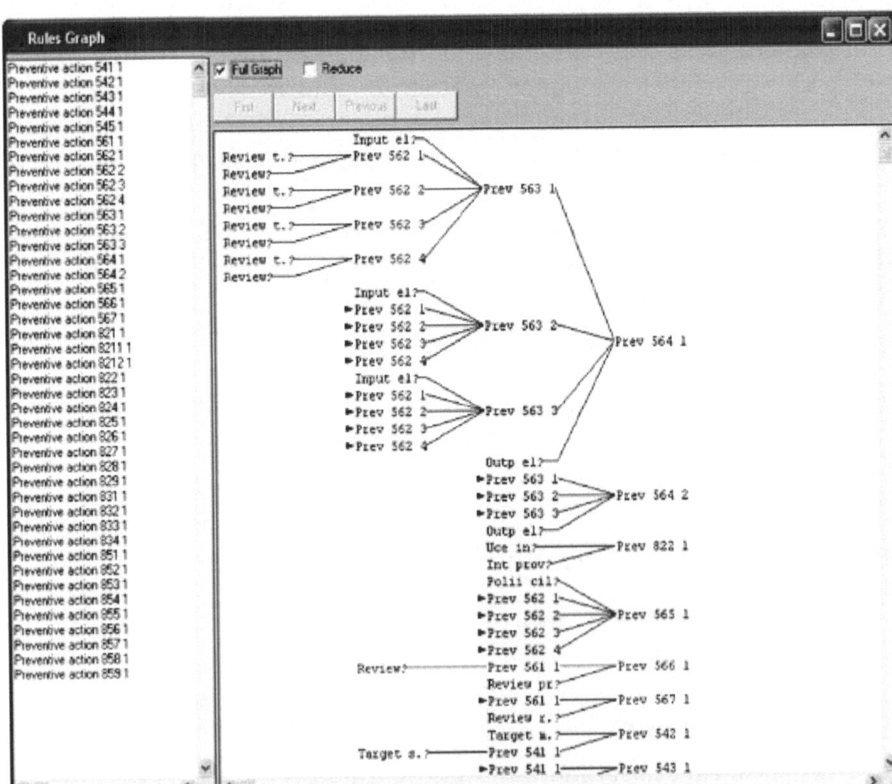

Figure 7. Decision tree

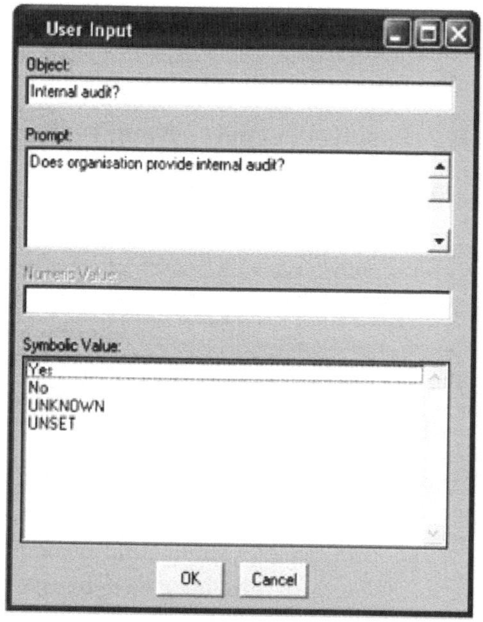

Figure 8. User's dialog box

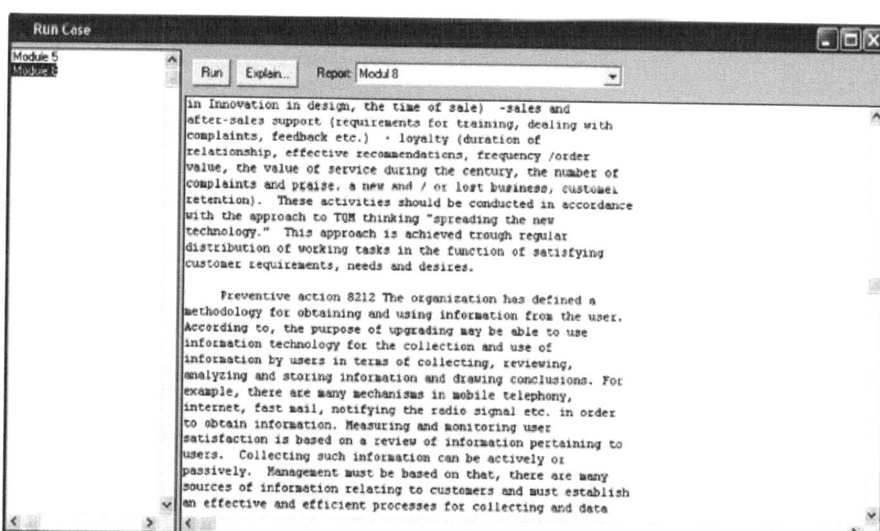

Figure 9. User's report

This expert system was developed in three iterative steps. Each of them resulted of the improvement, for example improvement of the definition of objects, set the input data, the relation between objects depending on the priorities of execution and more.

The expert system was implemented and tested in practical, real conditions in the organization that has a clear commitment to participate in the competition for the European Award for Business Excellence, also providing important measures in that direction. Evaluation was done on the basis of technical and ergonomic characteristics based on guidelines in standards ISO/IEC 9126/1:2001 for evaluation quality of software. The results are shown in Table 2.

Table 2. Results of expert system evaluation

	Category	**Awerage mark**	**Awerage mark**	**Total awerage mark**
Technical characteristic	*Fault of presented software*	8.2	8.7	9.2
	Benefit of new software	8.7		
	Influence on job organisation	9.2		
Ergonomic characteristic	*General ergonomic characteristic*	9.8.	9.7	
	System adaptability	9.6		

Figures showed significant high mark by categories, and thus the total amount. Software was evaluated positive in terms of technical characteristics and in terms of ergonomic. In this sense, product has small time of response, it is compatible with most used operating system, it has an excellent user's oriented interface, and it has easy data entry and a good view of the output, installation is simple and the software is very competitive. Also, in this sense, within the organization, it was carried out the reorganization of the priority areas from the viewpoint of improvement, implemented preventive measures for the potentially unstable areas and also applied the measures for the improvement (offered by this system) leading to business excellence achieving.

4. Final considerations

Nowadays, very small number (a few per cent) of the scientific research activities in area of quality management systems are based on topic of the collection and analysis of information with aim to improvement business results. That fact justify author's effort to make preventive actions for improvement business performances through establishing synergy between area of quality management and artificial intelligence like area which is strictly oriented on producing knowledge. Also, through analysis of the available software for quality management, it can be concluded that there are no any software from field of artificial intelligence that was developed for quality management systems improvement. That means that each further step in this direction brings positive scientific research results. The research point out necessity of making connection between more software solutions and tools in order to make the system with a higher level of intelligence. For this purpose, it is best to apply the integration of decision support systems and expert system. That is best world experience. With this approach it can be make system that producing knowledge and that is greatest resource which can make organization more competitive and can ensure improvement of organizations performances. Based on those facts in this research we developed unique analogy integrative approach which stays in the basis of model for improvement business process performance in the direction for achieving best organizational performances.

As the most important requests for achieving business excellences were identified requests which are mostly related to: measurement, analysis and improvement (module 8 - ISO 9001) and management responsibility (module 5-ISO 9001). The next area is most important for excellence organizational condition and at the same time area where should make very intensive action for improvement and strengthening: 821 – customer satisfaction, 72 - customer related processes, 54 – planning, 85 – continual improvement, 56 – management review and 55 – responsibility, authority and communication. It is interesting to highlight, that all activities and process which is related with customer and achieving his satisfaction and anticipation his needs, are in the focus and that should be direction and guidelines for all employees.

Also, it is shown that the strengthening, especially in these areas is used to lead to the significant progress in terms of: business process reengineering, manufacturing strategy, performance measurement and benchmarking, as very important aspects of market-oriented organization.

This research present interesting and useful results which should be use for defining measurement for improvement business performance in way for achieving business excellence. Those results are related with term of Degree of readiness which show part (every request) of ISO 9001 certified model and they ability for achieving top business form. Also, interesting results are present through values of Coefficient of significance. This two indexes show direction about area and intensity of action which should be provide to make best organisational condition.

In organizations that have specific information through database and information systems, it is necessary to develop systems that will assist staff in decision making. These systems provide data and output information on the basis of which, in accordance with the principle of decision making base on fact, the employees make business decisions that certainly contribute to improve organizational performance. However, in the today complex business condition, organization must make stride from level of data and information to level of knowledge. That is way for ensuring prestigious position on the market. That could be achieving through development expert system base on expert knowledge and base on output of decision support system.

This approach could be related with one modern approach, which calls case base reasoning. This approach is base on experience of other companies, and that approach could be use for defining preventive action. In this sense, it can be use a system that was developed in this work. That system was testing in real condition and proved to be very useful and that showed great level of efficiency and effectiveness for real business conditions. According to process of testing and estimation, users of the system were put ratings that are present in table 2. They indicate that this system can: make financial benefits, provide better organisation of job, stimulate all employees to improving own process, synchronise function in organisation, identify priority area for improvement, define intensity of action for improvement, stimulate preventive versus corrective action, encourage better involvement of new staff in to the activities, bring higher level of flexibility and other.

References

1. Vujovic A (2008). Improvement of business processes performances based on management systems by using artificial intelligence. PhD thesis, University of Montenegro.
2. Metaxios K S, Psarras J and Askounis D (2001). Expert systems in production planning and scheduling: A state-of-the-art survey. Journal of Intelligent Manufacturing, Vol. 13, No. 4, pp. 253-260.
3. Smith A (2001). The role of tacit and explicit knowledge in the workplace. Journal of Knowledge Management, Vol. 5, No. 4, pp. 11-21.
4. Lang C (2001). Managerial concerns in knowledge management. Journal of Knowledge Management, Vol. 5, No. 1, pp. 43-59.
5. Bollinger S and Smith D (2001). Managing organizational knowledge as a strategic asset. a. Journal of Knowledge Management, Vol. 5, No 1, pp. 8-18.
6. Clark T and Rollo C (2001). Corporate initiatives in knowledge management. Education and Training, Vol. 43, No 4, pp. 206-14.
7. Coff W (2002). Human capital, shared expertise, and the likelihood of impasse in corporate acquisitions. Journal of Management, Vol. 28, No. 1, pp.107-128.
8. Jacob M (2002). Implementing formal design review. Industrial Quality Control, Vol. 23, a. No. 8, pp. 398-404
9. Dommartin A (2003). ISO 9001:2000 and the EFQM excellence Model-The floor and the ceiling of the corporate house. ISO Management System, Vol. 3, No. 1, pp. 18-21.
10. Liao S (2005). Expert system methodologies and applications - a decade review from 1995 to 2004. Expert Systems with Applications, Vol. 28, No. 1, pp. 93-103.
11. Padhy N (2005). Artificial intelligence and Intelligent Systems. Oxford University Press: USA.
12. Kanji K G (1998). An innovative approach to make ISO 9000 standards more effective. a. Total Quality Management, Vol. 9, No. 1, pp. 67 - 78.
13. Najmi M and Kehoe F (2000). An integrated framework for post - ISO 9000 quality development. International Journal of Quality & Reliability Management, Vol. 17, No. 3, pp. 226 - 58.
14. Mohan R and Miller D (2004). Expert systems applications for productivity analysis. a. Industrial Management & Data Systems, Vol. 104, No. 9, pp. 776-785.

15. Elmuti D (2004). World-class standards for global competitiveness: an overview of ISO 9000. Industrial Management, Vol. 38, No. 5, pp. 5-9.
16. Srdoc A, Sluga A and Bratko I (2005). A quality management model based on the deep quality concept. International Journal of Quality & Reliability Management, Vol. 22, No. 3, pp. 278-302.
17. Shin M, Holden T and Schmidt A (2001). From knowledge theory to management practice: towards an integrated approach. Information Processing and Management, Vol. 37, No. 2, pp. 335–355.
18. Kreng V and Ming Tsai C (2003). The construct and application of knowledge diffusion model. Expert Systems with Applications, Vol. 25, No. 2, pp. 177-186.
19. Augier M and Knudsen T (2004). The architecture and design of thc knowledge organization. Journal of Knowledge Management, Vol. 8, No. 4, pp. 6-20.
20. Murray J and Chao M (2005). A cross-team framework of international knowledge acquisition on new product development capabilities and new product market performance. Journal of International Marketing, Vol. 13, No. 3, pp. 54-78.
21. ISO 9000:2005 (2005), "Quality management systems -- Fundamentals and vocabulary" International organisation for standardization, Geneva, 2005, 30 p.
22. Zhang Z (2000). Developing a model of quality management methods and evaluating their effects on business performance. Total Quality Management, Vol.11, No.1, pp. 129 37.
23. Sun H (2000). Total quality management, ISO 9000 certification and performance improvement. International Journal of Quality & Reliability Management, Vol.17, No.2, pp.168 - 79.
24. Escanciano C, Fernandez E and Vazquez C (2001). Influence of ISO 9000 certifcation on the progress of Spanish industry towards TQM. International Journal of Quality & Reliability Management, Vol.18, No.5, pp.481 - 94.
25. Gotzamani D and Tsiotras D (2002). The true motives behind ISO 9000 certification. Their effect on the overall certification benefits and their long term contribution towards TQM. International Journal of Quality & ReliabilityManagement, Vol.19, No.2, pp.151 - 69.
26. Tummala R and Tang L (2003). Strategic quality management, Malcom Baldrige and European quality awards and ISO 9000 certification: core concepts and comparative analysis. International Journal of Quality & Reliability Management, Vol. 13, No. 4, pp. 8-38.
27. Casadesus M and Karapetrovic S (2005). An Empirical Study of the Benefits and Costs of ISO 9001: 2000 Compared to ISO 9001/2/3: 1994. Total Quality Management, Vol. 16, No.1, pp.105–120.
28. Kosko B (1992). Neural networks and fuzzy systems: a dynamical systems approach to machine intelligence. Prentice Hall: New Jersey.
29. Bidgoli H (1997). Modern information system for managers. Department of Management, California State University: Bakersfield-California.

30. Welbank M (1983). A review of knowledge acquisition techniques for expert systems.
31. Ipswich: Martlesham Consultancy Services, Martlesham, United Kingdom. 49 p.
32. Paquin L C, Blanchard F and Thomasset C (1991). Loge–expert: from a legal expert system to an information system for non-lawyers. In Proceeding ICAIL '91 Proceedings of the 3rd international conference on Artificial intelligence and law in Oxford. St. Catherine's College: Oxford.
33. Eppinette M and Inman R A (1997). Expert systems and the implementation of quality customer service. Industrial Management &Data Systems, Vol. 97, No. 2, pp. 63–68.
34. Alavi, M. and Leidner D E (2001). Review: Knowledge Management and Knowledge Management Systems: Conceptual Foundations and Research Issues. MIS Quarterly, Vol. 25, No. 1, pp. 107-136.
35. Karhu K (2002). Expertise cycle – an advanced method for sharing expertise. Journal of Intellectual Capital, Vol. 3, No. 4, pp. 430-46.
36. Winn G, Gopalakrishnan B, Akladios M and Premkumar R (2005). Expert systems - what SH&E managers need to know about software verification and validation. Professional Safety, Vol. 50, No. 8, pp. 45-52.
37. Comesana J and Carlos J (2002). Creating an expert system for detailed scheduling. International Journal of Operations & Production Management, Vol. 22, No. 7, pp. 806- 819.
38. Giarratano J and Gary R (2004). Expert systems: Principles and Programming- 4th edition. PWS Publishing Company: Boston.
39. Fleet D, Peterson T and Van Fleet E (2005). Closing the performance feedback gap with expert systems. Academy of Management Executive, Vol. 19, No. 3, pp. 23-35.
40. Relja M and Klepac N (2002). From gene to therapy- huntingtonova korea. Biochemia medica, Vol. 12, No. 1-2, pp. 35-39.
41. Belak L, Gacina N and Radic T (2005). Food technology. In profesional material for students of High school for management in tourism in Sibenik: Sibenik.
42. Vilber K (2002). No limits-Popular psychology. Babun: Beograd.
43. Gilja I (2005). Prostate urinary incontinence. Medix, Vol. 11, No. 60/61, pp. 119-120.
44. Bielawski L and Lewand R (1988). Expert systems development - Building PC based applications. John Wiley & Sons, Hoboken: New Jersey.
45. Elder J and Abbott D (1998). A comparison of Leading Data Mining Tools. In Knowledge Discovery and Data Mining proceedings of the Fourth international Conference in New York, Elder Reasearch: New York.
46. Goebel M and Gruenwald L (1999). A surveyof data mining and knowledge discovery software tools. SIGKDD Explorations, Vol. 1. No.1, pp. 20-33.

47. Oakland J (2001). Total Organizational Excellence - Achieving world class performance. a. Elsevier Butterworth – Heinemann: Oxford.

48. MEST ISO 9004:2009 (2009), "Managing for the sustained success of an organization. A quality management approach", International Organization for Standardization, Geneva, 2009, p. 54.

49. Tricker R (2006). Best Practice - ISO 9001:2000 - The Quality Management Process. Van Haren Publishing: Zaltbommel-Netherlands.

50. FQCE (2004), "Course for auditors based on Oscar for quality award", Fond for Quality Culture Excellence, Beograd, 2004, p.217.

51. Perovic M (2003). Management – informatics – quality, Kragujevac. CIM Center-Faculty of Mechanical Engineering: Kragujevac.

52. Rakovic R (2006). Quality toward business excellence. Energoprojekt – Ingraf: Beograd. [52] Aamodt, A., and Plaza, E. (2004). Case-Based Reasoning: Foundational Issues, a. Methodological Variations, and System. AI Communications - IOS Press, Vol. 7, No. 1, pp. 39 - 59.

53. Wen W, Wang W and Wang C (2005). A knowledge - based intelligent decision support system for national defense budget planning. Expert Systems with Applications, Vol. 28, No. 1, pp. 55–66.

54. Turban E and Watson H (2008). Integrating expert systems and decision support systems. Management Information Systems Quarterly, Vol. 10, No. 2, pp. 121-136

55. Vujica-Herzog N, Polajnar A and Tonchia S (2007). Development and validation of business process reengineering (BPR) variables: a survey research in Slovenian companies. International Journal of Production Research., Vol. 45, No. 24, pp. 5811- 5834.

56. Vujica-Herzog N, Tonchia S and Polajnar A (2009). Linkages between manufacturing strategy, benchmarking, performance measurement and business process reengineering. Computers & Industrial Engineering, Vol. 57, No. 3, pp. 963-975.

2

Improving 'Improvement' By Refocusing Learning: Experiences from An -Initially- Unsuccessful Six Sigma Project in Healthcare

1. Introduction

The Skaraborg Hospital Group (SkaS) is the first hospital group in the Nordic Countries that has added Six Sigma on a large scale to its quality programme to improve care processes (Lifvergren et al. 2010). Unlike many change efforts in the healthcare sector that are neither successful nor sustainable (Chakravorty 2009; Øvretveit 2009, 1997; Thor et al. 2007; Zimmermann and Weiss 2005), the success rate of improvement projects in the programme in this period was 75%, in some respects due to lessons learned from this particular project.

Still, the high success rate of the programme might be surprising, given the fact that the presumed success of planned or programmatic change has been seriously questioned in a number of articles and books (Alvesson and Svenningson 2007; Beer et al. 2000; Beer et al. 1990; Dawson 2003; Duck 1993; Kotter 1995; Schaffer et al. 1992; Strebel 1996). It is argued that organizations are not rational entities where people do as they are told and follow the latest strategic 'n-step model' – on the contrary, organizational change is to a high degree seen as contextual and processual, unpredictable and beyond the realms of detailed plans (Alvesson and Svenningson 2007; Dawson 2003; Stacey 2007). The culture and history of the actual organization define what strategies for change are possible. What may work in one organization might be impossible to carry out in another. In other words, improvement strategies seem to be notoriously difficult to transfer between organizations.

Change and improvement is about learning and apparently, organizations seem to have difficulties to learn. Furthermore, daily problem solving activities may inhibit organizational learning (Tucker et al. 2002). It is difficult for organizations to recognize and capitalize on the learning opportunities posed by operational failures (Tucker 2004) and the how-aspect of learning is vital in this respect (Tucker et al. 2007). Creating arenas for learning in a non-punitive climate is thus critical and the role of the managers is essential in this respect (Tucker et al. 2007; 2003). Consequently, we see learning as a crucial perspective in change and improvement programmes. A better understanding on how learning can be facilitated in organizations is thus essential.

In this chapter, we describe how an enhanced focus on learning through 'learning mechanisms' (Docherty and Shani 2008; Shani and Docherty 2008, 2003) has contributed to the high project success rate of 75% in the Six Sigma programme at SkaS. We present the analysis of a traditional Six Sigma project that failed initially, but eventually led to an enhanced approach emphasizing learning. This entailed a refocusing on actively planning for learning within and between projects – 'learning by design'– involving the integration of cognitive, structural and procedural learning mechanisms (Docherty and Shani 2008; Shani and Docherty 2008, 2003). The ensuring success from utilizing learning mechanisms inspired us a) to redesign the Six Sigma roadmap –DMAIC, incorporating an 'L' for 'learning mechanisms' – DMAICL, b) to establish permanent arenas for learning between organizational units and, c) to institutionalize parallel learning networks consisting of specially educated improvement managers that support and facilitate local improvement projects. We suggest that learning mechanisms can provide a useful framework to the how- aspects of learning (Tucker et al. 2007) when designing organizational change initiatives that leave room for the cultural and historical contexts inherent in every organization.

We will first, however, give a brief overview of Six Sigma before moving on to the theoretical underpinnings of this chapter – cognitive, structural and procedural learning mechanisms. The concept of learning mechanisms is explored in some detail, connecting theories of organizational learning to learning mechanisms, thus elucidating the application of the mechanisms as a way to enhance organizational learning. These theories are then positioned in relation to theories of individual learning and of improvement cycles in quality improvement.

The context of the project is then described in some detail; SkaS, the Six Sigma quality programme, and the actual emergency ward (EW). We then describe the actual improvement project and its initially failed results before moving on to the project analysis using an action research approach. We then present how lessons learned from the analyses were used to integrate learning mechanisms in the Six Sigma programme, thus contributing to its high project success rate. In particular, we present how the analysis contributed to a successful re-take on the project.

We conclude with some proposals that might be valuable to other healthcare organizations facing the difficulties of larger change initiatives and, finally, provide some suggestions for further research.

2. Theory and background

2.1 Six Sigma

There are many definitions of Six Sigma in the literature. Antony et al. (2007) defines Six Sigma as "a process-focused data driven methodology aimed at near elimination of defects in all processes which are critical to customers" (p. 242). According to Harry and Schroeder (2000) "Six Sigma is a disciplined method of using extremely rigorous data- gathering and statistical analysis to pinpoint sources of errors and ways of eliminating them" (p. 23). Recent research also points to the parallel organizational structure that supports improvements within Six Sigma (Schroeder et al. 2008; Zu et al. 2008). Based on case study data and literature, Schroeder et al. (2008) more specifically define Six Sigma as "an organized, parallel-meso structure to reduce variation in organizational processes by using improvement specialists, a structured method, and performance metrics with the aim of achieving strategic objectives" (p. 540). This definition also captures some of the elements that distinguish Six Sigma from TQM —the role structure and the structured improvement procedure (Zu et al. 2008). The role structure is often referred to as the 'belt system' and could be seen as a way to standardize the improvement competences in an organization. The black belt role signifies a co-worker with advanced improvement knowledge, working fulltime as an improvement expert. The structured improvement procedure – DMAIC (Define-Measure-Analyze-Improve-Control) –is used to solve quality problems of greater complexity and with unknown root causes (Schroeder et al. 2008). The Define phase identifies the process or product that needs improvement, while the Measure phase identifies and measures the characteristics of the process/product that are critical to customer satisfaction. The Analyze phase evaluates the current operation of the process to determine the potential sources of variation for critical performance parameters. Improved process/product characteristics are designed and implemented and cost/benefit analyses are carried out in the Improvement phase and, finally, the solutions are documented and monitored via statistical process control methods in the Control phase (Dahlgaard and Dahlgaard-Park 2006; Schroeder et al. 2008). Iterations of the procedure are sometimes necessary but also desirable for successful project completion. Significant for this and other descriptions of the DMAIC roadmap is the instrumental approach oriented towards tools and procedures (see e.g. Antony et al. 2007; Dahlgaard and Dahlgaard-Park 2006; Schroeder et al. 2008; Zu et al. 2008). However, the how-aspects of learning in the improvement cycles are seldom explored or described (Antony et al. 2007; Dahlgaard and Dahlgaard-Park 2006; Schroeder et al. 2008).

2.2 Using learning mechanisms to enhance organizational learning

Unquestionably, organizational learning has been described, defined and studied in many ways and from different theoretical angles (e.g. Argyris 1999; Argyris and Schön 1978; Crossan et al. 1999; Dixon 1999; Friedman et al. 2001; Garvin 2000; Hedberg 1981; Senge 1990; Weick 1995). Many psychologists maintain that only people can learn, though organizational theorists refer to 'organizational learning' by attributing the term to observable changes in the structures, procedures and formal frameworks of the organization, expressed in such documents as policies, strategies and value statements, when theses changes can be clearly related to preceding events and developments in the organization.

Many studies have shown that learning at work, like learning in formal educational settings, is a matter of design and not evolution (Docherty and Shani 2008; Ellström 2006, 2001; Fenwick 2003; Shani and Docherty 2008, 2003). That is, it is a matter of organizing the workplace, not only for production, but also for supporting learning at work. Most studies of learning at work focus on individual workers. Crossan et al. (1999) provide a '4 I' framework that links individual learning (Insight), through networks of collective or group learning (Interpretation and Integration) until it meets a senior management group whose decisions make important changes in the organization (Institutionalization), that is termed 'organizational learning'. Shani and Docherty (2008, 2003) use the term 'learning mechanisms' for the preconditions that are designed to promote and facilitate individual, collective and organizational learning. They use three main categories; cognitive, structural and procedural. *Cognitive mechanisms* are concepts, values and frameworks expressed in the values, strategy and policies of the organization and, ideally, underpin the practice-based learning processes at different organizational levels. *Structural mechanisms* are organizational infrastructures that encourage practice-based learning. An example would be lateral structures that enable learning of new practices across various organizational units. Finally, *procedural mechanisms* concern the routines, methods, and tools that support and promote learning, e.g. the introduction and, eventually, the institutionalization of a new problem-solving method. Learning mechanisms in practice may include more than one of these components, e.g. both structural and procedural (see e.g. Lifvergren et al. 2009 for an application of learning mechanisms in healthcare). In other words, learning mechanisms aim to encourage individual and collective learning eventually leading to organizational learning.

Thus, *individual* learning is a prerequisite for organizational learning. Without doubt, individuals can learn and learning takes place in iterative action/reflection cycles (or loops). Moreover, researchers who maintain that organizations can learn relate this directly to human learning, i.e. the learning of organizational members (Argyris and Schön 1978; Huzzard and Wenglén 2007; Kolb 1984; Shani and Docherty 2003).

Argyris and Schön (1978) take their departure from the concept of 'single – and double loop learning', where the former refers to our adaption of activities without questioning the 'a priori' – our taken-for-granted assumptions. Consequently, the latter signifies the alteration of our preconceptions in order to act or behave in new ways (ibid. 1978; but also Argyris 2001; Huzzard and Wenglén 2007).

Kolb (1984) pictures learning in an iterating four-phase cycle (or, rather, spiral), where learning is depicted as the interplay between theoretical knowledge that leads to activities (experiments), generating new experiences. These experiences further inform reflection, leading to new knowledge.

2.3 Learning cycles in continual improvement

Beyond doubt, there is a close connection between theories of learning and the improvement cycles of quality improvement. At the core of every quality programme, including Six Sigma, lies the concept of Continual Improvement, CI, in which learning cycles (or loops) should be used in every problem solving process (Bergman and Klefsjö 2010; Bergman and Mauleon 2007).

Already in the 1930s, Walter Shewhart proposed that mass production could be seen as constituting "a continuing and self-corrective method for making the most efficient use of raw and fabricated materials" (Shewhart 1939, p. 45). By repeating the steps of specification –production – inspection in a continuous spiral, a circular path representing 'the idealized state' could be reached. Deming (1986), inspired by Shewhart, proclaimed that the management should construct "an organization to guide continual improvement of quality", in which a four-step cycle, the 'Shewhart-cycle', should be utilized (p. 88). In other writings by Deming, this cycle is referred to as the PDSA-cycle (Plan, Do, Study, Act), see e.g. Deming 1994, where 'Act' also signifies reflection and learning. Similarly, Joseph Juran highlighted the importance of quality improvement, meaning "the organized creation of beneficial change" (1989, p. 28). All improvement should take place "project by project", where a project is defined as a "problem scheduled for solution..." (ibid., p. 35), and in which recurrent learning cycles should be applied. In Japan, the concept of CI, partly inspired by Juran and Deming (see e.g. Bergman and Klefsjö 2010), has been deeply ingrained in quality initiatives since the 1960s. Imai elucidated 'kaizen,' signifying "ongoing improvement involving everyone, including both managers and workers" (1986, p. 3) using the continuation of the Deming wheel: "Japanese executives thus recast the Deming wheel and called it the PDCA wheel (Plan, Do, Check, Act), to be applied in all phases and situations" (ibid., p. 60). According to Imai, the concept of Kaizen has been the most important and distinguishing feature of the Japanese quality movement. The DMAIC roadmap of Six Sigma shares the same origin from Shewhart and can be seen as an extension of the PDSA cycle and an enhanced version thereof, often used in the Japanese improvement descriptions, the QC-story (Bergman and Klefsjö 2010, Smith 1990). Evidently, Shewhart as well as Deming brought forward the importance of learning in the iterating PDSA cycles of today's CI, emphasizing the importance of action as well as reflection on the action (Bergman and Mauleon 2009, 2007).

3 Method

3.1 Action research

In this project, an action research approach has been used. Action research could be described as an orientation to inquiry where the intention to improve the studied system is achieved by designing iterative action-reflection loops involving both the researchers and the practitioners in the workplaces involved in the projects. The research question usually stems from problems that need to be solved in the studied organization. In action research projects, researchers and co-workers share a participative community, in which all the members are equally important in generating actionable knowledge. Co-workers are thus considered to be co-researchers in the inquiry process. The purpose of action research projects is mainly twofold; to generate actionable knowledge that help to solve the local problem, but also to contribute to the body of generalized knowledge (Bradbury and Reason 2008). Two project workshops were used in this research, see section 5, where a co- generative model inspired by Greenwood and Levin (2007, p. 93) and Lewin (1948, p. 143-152) was used.

Emanating from the action research framework already described, a co-generative dialogue starts out from a distinct problem definition where outsiders, in this case the project mentor, the development director and an insider through mutual reflection and learning try to solve the problem. The solutions are formulated and tested using iterative reflection-action loops to further enhance the creation of opportunities for learning and reflection.

4. The context: The Skaraborg Hospital Group and the Six Sigma quality programme

4.1 SkaS

The Skaraborg Hospital Group, (SkaS), is situated in the Western Region of Sweden and serves a population of 260 000 citizens. The group consists of the hospitals in four towns, Lidköping, Skövde, Mariestad and Falköping. The services offered by SkaS include acute and planned care in a large number of specialties. In total there are more than 700 beds and around 4500 employees at SkaS. There are two emergency wards (EW) in two separate hospitals at SkaS. Each ward is responsible for all aspects of acute care in its constituency.

SkaS have a long tradition of quality development using different types of quality improvement approaches, such as TQM, organizational audits, small scale improvement cycles, the Collaborative Breakthrough Series (IHI 2003). Still, in 2005 it was unclear if the many improvement efforts contributed to the realization of the overall quality strategy. In many cases, poor formulation of project goals made it difficult to assess whether the improvement initiatives had failed or succeeded. Furthermore, the economic outcomes from different improvement efforts were not measured. Drawing on these experiences and inspired by a pilot Six Sigma project in 2005 (Lifvergren et al. 2010) the senior management team decided to add Six Sigma to the SkaS quality methods tool box. Six Sigma would contribute to the quality strategy by systematically searching for and reducing unwanted variation in critical healthcare processes, and by sustaining an even flow in the processes. More than 50 black belts have been trained at SkaS in the period from 2005 to 2010. Half of them now work as fulltime internal consultants leading various improvement efforts at SkaS.

SkaS also initiated an action research collaboration with Chalmers University of Technology in 2006 to explore how Six Sigma can be embedded in a healthcare setting and to improve the DMAIC-roadmap to better correspond to healthcare process improvement.

4.2 The initial project at the emergency ward

From a patient's perspective, long patient waiting times at emergency wards (EW) are unacceptable. Studies have shown that the mean patient Length of Stay (LoS) at an EW correlates to increased morbidity and mortality (Sibbritt and Isbister 2006). At one of the EWs at SkaS, the LoS was increasing during 2005. An analysis revealed that about 16 000 patients were treated that year. The average LoS at the EW during the first six months was hours. Furthermore, the variation in LoS was also significant. Nearly 10% of patients had a LoS of five hours or more, and almost 20% had a LoS of more than four hours.

To address the problem the owner of the emergency process at the EW - the manager of the surgical clinic - decided to start an improvement project in the spring of 2006, aiming to decrease the mean LoS by 20 minutes, thereby increasing patient satisfaction and safety, improving working environment and improving resource utilization. A reason for this initiative was that LoS at EWs was a topic that appeared frequently in the national patient safety discourse. The project group consisted of interested co-workers at the EW and was led by two internal black belts. A steering committee consisting of the medical and surgical clinical managers was established. The first line managers responsible for the different clinics in the emergency department followed the project.

The daily operations of the EW are admittedly complex. About 16 000 cases pass through the department each year, and each patient is unique. Some patients must receive immediate treatment in the EW, while others' treatment is less pressing. The inflow of patients varies from week to week, depending on such factors as the weather (e.g. slipperiness in the streets), epidemics (e.g. influenza) in the population and healthcare articles in newspapers. The EW is also heavily dependent on a well-functioning collaboration with other units –primarily the x-ray department and the laboratory unit –to achieve an even flow through the department. The complex operations sometimes lead to increased LoS, which is worrying, tiresome and potentially dangerous to the patients. A high inflow of patients also contributes to a stressful working environment. In addition, increased LoS put a higher demand on the resources at hand. When there is an accumulation of patients due to different bottlenecks, the tail of the patient flow has to be handled late at nights at a higher cost.

The EW is organized under the surgical clinic; nurses and assistant nurses are employed at the EW whereas the doctors responsible for the EW come from the medical and the surgical departments following a scheme for emergency duty. There are two on-duty lines; the primary doctor on duty (usually a resident) works together with front line staff at the EW, whereas the secondary doctor (a senior physician) is on standby duty, always reachable by phone and obliged to appear within 20 minutes at the ward if called for.

The DMAIC roadmap of Six Sigma was used to assess the emergency process in order to detect root causes explaining the long waiting times. Several tools and methods were used; process analysis of the patient flow, e.g. how the inpatient clinics responded to a request to admit a patient; analyses of different lead times in the process, e.g. patient in need of x-ray.

Interviewing members of the project group, the information flow in the departments was also analyzed. The most important reasons for prolonged LoS were:

- Patients that should be admitted had to wait too long for the doctor's examination;
- the waiting times for patients in need of x-ray were too long;
- patients with fractures had to wait too long for pain-relieving treatment;
- the communication between doctors and other co-workers at the EW was poor;
- new residents were not introduced to the procedures used at the EW and;
- there were no clear rules for when the secondary doctor on-call should be contacted.

With these root causes in mind, several improvements were suggested and implemented, e.g. nurses should be allowed to remit the patient for x-ray in case of suspected hip fractures and they should also be permitted to give pain-relieving treatment to these patients without consulting a doctor. A common routine for the improved communication between different categories of staff was created. In addition, a mandatory introduction program for intern doctors was developed in which important routines at the emergency ward were taught. The proposed solutions were shared with co-workers including the physicians at regular work place meetings. The results of the proposed solutions were monitored using control charts, continuously assessing the overall LoS. Random inspections were also used to make sure that the proposed solutions were implemented.

5 Results and analysis of the project

5.1 Initial results show no improvement

Surprisingly, the LoS at the EWwere not affected at all but appeared to increase during the first three months after implementation of the suggested solutions by the initial Six Sigma project. It was the only one of eight on-going projects during 2006 that did not produce any positive results (Lifvergren et al. 2010). In order to learn from the initial failure, a deeper analysis of the project was carried out to reveal the causes of the failure and to improve the conditions for future projects. The development director (Svante Lifvergren) initiated the analysis. Two workshop dialogues, inspired by the co-generative model, were carried out. The purpose of the dialogues was to reveal the reasons to why the project had not succeeded so far. In the first workshop the development director, the supervisor of the Six Sigma program and one of the project managers participated. The second workshop also included the clinical manager and the assistant clinical manager at the surgical department, the other project manager and the manager at the EW. The results of the dialogues were also discussed with the outsider researchers, in this case Bo Bergman. Several plausible reasons explaining the failure of the project could be agreed upon (see figure 1).

The causes could be categorized into two groups; 'failure of implementation' and 'insufficient analysis'. These groups where then subdivided according to the figure and the relations between the different subgroups were visualized using arrows, thus showing the believed cause and effect relations between the subgroups. Each subgroup was further investigated using '5-why' in repetitive root cause analyses, depicted below (Table 1).

Table 1. Root cause analyses in the different subgroups

Subgroup(s)	Root cause analysis	Probable root causes
Poor project information at the EW and Lack of commitment	Lack of commitment among the ward manager, the physicians and the co workers »» (due to) poor knowledge of the usefulness of the project »» poor information about the project »» the information about the project from management was insufficient »» the management did not realize that the information had not reached all co workers »» poor management knowledge of the importance of project communication and how this should be accomplished /visual engagement from management in project was lacking	1. Management knowledge of the importance of project communication and how this should be accomplished was lacking 2. Poor management knowledge of the importance of physically being involved and showing engagement in the project
Poor support for the local project group	The project group lacked authority »» strong informal leaders didn't commit to/support the project (co- workers at the EW as well as physicians) »» management was not able to convince key personnel about the importance of the project »» management did not realize the importance of recruiting key personnel to the project group or to communicate to informal leaders about the project »» the project managers also lacked this knowledge »» not enough focus on project stakeholder issues early on in the Six Sigma education at SkaS	3. Not enough focus on critical project stakeholder issues early on in the Six Sigma education
Other methods not exploited	Project managers lacked knowledge of and experience from other methods and concepts, e.g. lean, discrete simulation, Design for Six Sigma etc. »» to learn the DMAIC-roadmap was time consuming »» project managers lacked time to study other methods »» the education was too compressed and did not contain other methods	4. The Six Sigma education was too compressed and did not contain other methods as well
True root causes not found	Data and risk analyses insufficient »» no actual root cause analysis from data »» insufficient amount of data »» project scope too large »» not enough time to gather data »» the project mentor did not give enough support to the project managers in helping them delimiting the scope of the project but also in suggesting alternative methods »» poor communication between mentor and project managers and inexperienced mentor	5. The project mentor did not give enough support to the project managers in helping them delimiting the scope of the project but also in suggesting alternative methods 6. Poor communication between mentor and project managers 7. Inexperienced project mentor and project managers

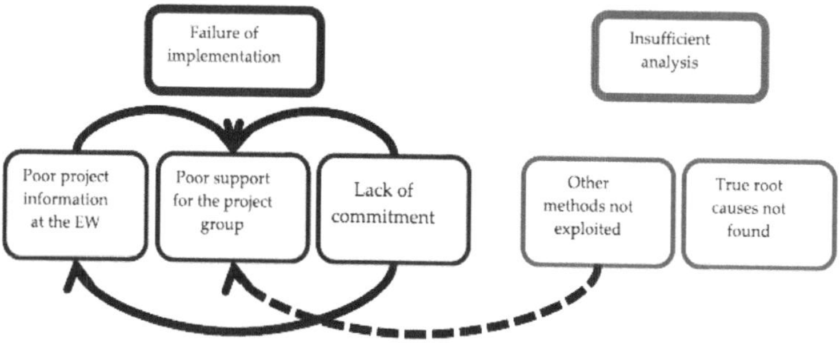

Figure 1. An affinity diagram from the workshops displaying probable causes of the, initially, failed project

6 Integrating learning mechanisms into the SkaS quality programme

Although only this project initially failed in its impact on the operational units concerned during 2006, we believe that it was the most successful through its impact on the hospitals' development strategy and procedures, especially regarding the design and integration of learning mechanisms – cognitive, structural and procedural – into the ongoing quality program at SkaS. The lessons learned were used to redesign the Six Sigma solving process – cognitive and procedural mechanisms. Moreover, the causes to the failure so far have been shared in parallel networks –a structural mechanism. Finally, templates to be used in future projects also have been designed to prevent the mistakes to reappear in the SkaS quality programme (Table 2). This also led to a second, successful project in the EW.

6.1 Cognitive learning mechanisms

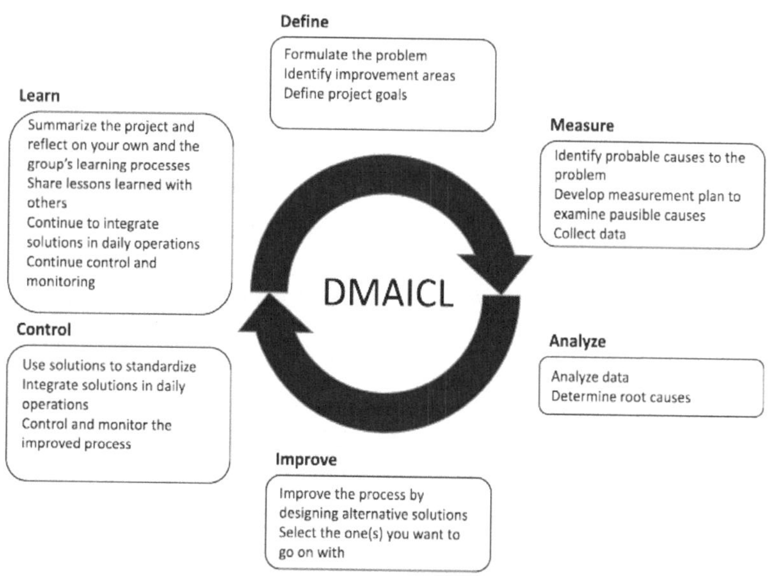

Figure 2. The improved DMAICL-cycle currently used at SkaS

Table 2. The integration of lessons learned from the failed project into the SkaS quality programme

Root causes of the failed project	Solutions integrated into the SkaS quality programme	Type of learning mechanism
1 Management knowledge of the importance of project communication and how this should be accomplished was lacking	Feedback to top management and a revision of the SkaS quality system highlighting the importance of management involvement Revised templates for the problem solving procedure	Cognitive, structural and procedural
2 Poor management knowledge of the importance of physically being involved and showing engagement in the project	Same as above	Cognitive, structural and procedural
3 Not enough focus on critical project stakeholder issues early on in the Six Sigma education	A stakeholder template was incorporated into the 'DMAICL' roadmap The importance of stakeholder involvement was elucidated in the Six Sigma education	Cognitive and procedural
4 The Six Sigma education was too compressed and did not contain other methods as well	Revision of the education; Lean and Design for Six Sigma were added to the Six Sigma education and the education was prolonged	Cognitive
5 The project mentor did not give enough support to the project managers in helping them delimiting the scope of the project but also in suggesting alternative methods	A revised problem solving procedure –DMAICL – was established Revised templates for delimiting projects	Cognitive and procedural
6 Poor communication between mentor and project managers	Accelerating learning through the establishment of parallel learning structures at SkaS Revised templates for project communication	Structural and procedural
7 Inexperienced project mentor and project managers	Same as above	Structural and procedural

Cognitive mechanisms provide language, concepts, models, values and theories for thinking, reasoning and understanding learning issues. Some examples would be models and approaches for improvement, company value statements but also strategy documents (Shani & Docherty 2008, 2003; Docherty & Shani, 2008). In this case, the analysis of the failed project revealed both the absence of reflection during the project, and also the negative impact of no reflective feedback being shared with and among co-workers at the actual workplace – the how-aspect of learning (Tucker et al. 2007). The lessons learned inspired us to redesign a) the SkaS quality system elucidating the importance of management commitment, and b) a revised Six Sigma roadmap –DMAIC, incorporating an 'L' for learning and reflection –DMAICL (see figure 2). In the Learning phase, the project manager and the members of the project group conjointly reflect on the project process in order to 'improve the improvement processes'. Moreover, the sixth phase adds important time to the delivery of the solutions in the daily operations of management; this was indicated by earlier Six Sigma project experiences. The DMAICL roadmap has been used in every black belt and green belt project at SkaS since 2006 and could thus be seen as an institutionalization of a learning mechanism throughout the organization (Crossan et al. 1999). The importance of iterations of the DMAICL- cycle has also been highlighted at SkaS.

5.2 Structural learning mechanisms

Structural mechanisms concern organizational, technical and physical infrastructures that enhance learning, e.g. different feedback and communication channels, arenas/forums and networks for dialogue, but also specific learning structures such as parallel learning structures.

As a result from the analysis of the actual project but also drawing from other parallel project experiences at SkaS (Lifvergren et al. 2010, 2008), horizontal permanent arenas for learning have been established. In these forums, project managers and quality coordinators from different organizational units meet every month to learn from improvement 'successes' and 'failures'. Improvement efforts are monitored and analyzed in order to learn how to improve the 'project process' itself. From these network activities, important learning is spread throughout the hospital; e.g. project groups and project mentors can learn from each other. Also, sharing the 'L' from every project inspires reflection and second loop learning between projects. Moreover, an intranet project database displaying concluded as well as ongoing and future improvement projects has been established.

A parallel learning structure has been instituted –integrating the selection and training of operational personnel to conduct the Six Sigma projects in operational units. Many return to their units, while others after further training, become internal consultants (cf. Bushe and Shani 1991).

5.3 Procedural learning mechanisms

Procedural mechanisms pertain to the rules, routines, procedures and methods that can be institutionalized in the organization to promote and support learning, e.g. assessment methods and standards (Docherty and Shani 2008; Shani and Docherty 2008). In this particular case, the cognitive and procedural mechanisms overlap, where roadmap templates to be used in every larger improvement project support the new cognitive model DMAICL (figure 2).

5.4 A second and successful retake on the project using learning mechanisms

The problem with long LoS persisted, so a new improvement project was initiated in 2008. Learning from the root cause analyses of the failed project, the clinical manager was deeply involved in the project, supporting it and requesting regular feedback on its progress; the project was subdivided in to several subprojects. Even more emphasis was put on continual and regular project communication to involve all co-workers at the EW. All these efforts are examples of *procedural* learning mechanisms. This resulted in several improvement suggestions from the front line staff. Moreover, expert knowledge on flow theory in daily operations was brought to the project – a *cognitive* learning mechanism. Much effort was also invested in involving and motivating the physicians. Finally, an improved DMAICL roadmap – signifying cognitive and procedural mechanisms –was followed (see figure 2). As a result of all these efforts, the project managed to reduce mean LoS at the EW by 20 minutes during 2008, an improvement that has been sustainable during 2009 and 2010.

6 Conclusions

In this chapter, we have described how a deeper analysis of a project that initially failed its client has led to emphasizing learning and integrating learning mechanisms into the SkaS Sigma programme, thus contributing to the present project success rate of 75% (Lifvergren et al. 2010).

In every improvement programme, the concept of Continual Improvement (CI) plays a vital role. At the core of CI, we find the learning cycles – PDSA, PDCA, DMAIC – critical to joint sensemaking (Weick 1995) and learning that creates actionable knowledge (Bradbury and Reason 2001, 2008). This particular project has disclosed how the importance of learning has been played down in the DMAIC roadmap in favor of more instrumentally emphasized problem solving techniques, e.g. templates, project charters and statistical analyses (Anthony et al. 2007; Schroeder et al. 2008). The experiences presented here have led to the addition of an 'L' in the roadmap, DMAICL, thus highlighting the original intentions of the learning cycles that somehow got lost on the way (Deming 1986; Shewhart 1939). We further propose that the 'L' might signify cognitive and procedural learning mechanisms, the intention of which is to invoke second loop learning within and between project groups and operational units.

Moreover, through the use of learning mechanisms, a design approach to institutionalization can be adopted (Crossan et al. 1999; Shani and Docherty 2003), including the contextual, historical and cultural factors present in every organization, while reducing the unpredictability of organizational change.

To openly show and analyze your failures adds knowledge and enhances a reflexive approach and a non-punitive culture in the organization. In that respect, the initially unsuccessful project could actually be considered to be the most successful project in SkaS Six Sigma programme, contributing to the how-aspects of learning (Tucker et al. 2007).

7 Further research

The challenges facing healthcare calls for sustainable changes, necessitating long term approaches. The integration of learning mechanisms in the change efforts taking place at SkaS but also in other healthcare organizations of the Western Region in Sweden will be followed continuously. How learning mechanisms are interpreted and adopted in other healthcare systems given their unique culture and history is also a question that deserves further investigation. Moreover, can learning mechanism be adopted to alleviate the conflicts that often emerge between vertical, hierarchical management structures *and* improvement projects that seek to solve problems pertaining to the value creating horizontal patient processes (Hellstrom et al. 2010)?

References

1. Argyris C (2001) On organizational learning. 2nd edition, Blackwell Publishing, MA
2. Argyris C, Schön D (1978) Organizational learning– a theory of action perspective. 2nd edition. Addison Wesley Publishing, MA
3. Alvesson M, Sveningsson S (2007) Changing organizational culture: cultural change work in progress. Routledge, New York, NY
4. Alvesson M, Sköldberg K (2000) Reflexive methodology - new vistas for qualitative research. Sage, London
5. Beer M, Nohria N (2000) Cracking the code of change. Harv Bus Rev May-June: 133-139
6. Beer M, Elsenstat RA, Spector B (1990) Why change programs don't produce change. Harv Bus Rev Nov-Dec: 158-166
7. Bergman B, Klefsjö B (2010) Quality from customer needs to customer satisfaction.3rd edition, Studentlitteratur, Lund
8. Bergman B, Mauleon C (2009) Exploring the epistemological origins of Shewhart's and Deming's theory of quality. Influences from Lewis conceptualistic pragmatism. Int J Qual Serv Sci 2:160-171
9. Bergman B, Mauleon C (2007) Ständiga förbättringar och dess rötter i pragmatismen ('Continuous improvements and its roots in pragmatism). In: Elg M, Gauthtereau Witell L (eds) Att lyckas med förbättringsarbete ('To succeed with improvements'). Studentlitteratur, Lund
10. Boaden R, Harvey G, Moxham C, Proudlove N (2008) Quality improvement: theory and practice in healthcare. NHS Institute for Innovation and Improvement, Coventry
11. Bradbury H, Reason P (2008) The Sage handbook of action research. 2nd edition, Sage, London
12. Bushe GR, Shani AB (1991) Parallel learning structures: increasing innovation in bureaucracies. Addison-Wesley Publishing, Reading, MA

13. Cederquist J, Hjortendal Hellman E (2005) Iakttagelser om landsting ('Observations on counties'). Finansdepartementet, Stockholm
14. Chakravorty SS (2009) Six sigma failures: an escalation model. Oper Manag Res 2:44-55. doi:10.1007/s12063-009-0020-8
15. Crossan MM, Lane HW, White RE (1999) An organizational learning framework: from intuition to institution. Acad Manag Rev 24: 522–537
16. Dahlgaard JJ, Dahlgaard-Park SM (2006) Lean production, six sigma quality, TQM and company culture. TQM Mag 18:263–281
17. Dawson P (2003) Understanding organizational change– the contemporary experience of people at work. Sage, London
18. Deming WE (1994) The new economics. 2nd edition. MIT Press, Cambridge, MA Deming WE (1986) Out of the crisis. MIT Press, Cambridge, MA
19. Dixon N (1999) The organizational learning cycle: how we can learn collectively. 2nd edition. McGraw-Hill, London
20. Docherty P, Shani AB (2008) Learning mechanisms as means and ends in collaborative management research. In: Shani AB, Mohrman SA, Pasmore WA, Stymne B, Adler N (eds) Handbook of Collaborative Management Research. Sage, Thousand Oaks, CA
21. Duck JD (1993) Managing change: the art of balancing. Harv Bus Rev, Nov-Dec: 109- 118.
22. Ellström PE (2006) The meaning and role of reflection in informal learning at work. In: Boud D, Cressey P, Docherty P (eds) Productive reflection at work: learning for changing organizations. Routledge, London
23. Ellström PE (2001) Integrating learning and work: conceptual issues and critical conditions. Hum Resour Dev Q 12: 4
24. Eriksson N (2005) Friska vindar i sjukvården ('Fresh breezes in healthcare'). Dissertation. Förvaltningshögskolan, Gothenburg University, Gothenburg
25. Fenwick TJ (2003) Professional growth plans: possibilities and limitations of an organization-wide employee development strategy. Hum Resour Dev Q 14: 59– 77.
26. Friedman VJ, Lipshitz R, Overmeer W (2001) Creating conditions for organizational learning. In: Dierkes M, Berthion Antal A, Child J, Nonaka I (eds) Handbook of organizational learning and knowledge. Oxford University Press, Oxford
27. Garvin DA (2000) Learning in action: a guide for putting the learning organization to work. Harvard Business School Press, Boston, MA
28. Greenwood DJ, Levin M (2007) Introduction to action research. Sage, London
29. Harry M, Schroeder R (2000) Six sigma— the breakthrough management strategy. Doubleday, New York
30. Hedberg BLT (1981) How organizations learn and unlearn. In: Nystrom PC, Starbuck WH (eds) Handbook of organizational design, vol. 1. Adapting organizations to their environments. Oxford University Press, Oxford

31. Hyer NL, Wemmerlöv U, Morris J, Shtub A (2003) Cellular manufacturing in health care: the case of a trauma unit. In: Spina J et al. (eds) One world? One view of OM? The Challenges of integrating research & practice. EurOMA–POMS Conference, Como, Italy, 16–18 June, Vol I: 913–917
32. Imai M (1986) Kaizen –the key to Japan's competitive success. Random House Business Division, New York
33. Institue for Healthcare Improvement (2003) The Colloborative Breakthrough Series. IHI, Boston: IHI.
34. Juran J (1989) On leadership for quality –an executive handbook. The Free Press, New York
35. Kolb D (1984) Experiential learning: Experience as the source of learning and development. Prentice-Hall, New York
36. Kotter J (1995) Leading change: why transformation efforts fail. Harv Bus Rev, Mar-Apr: 59- 67
37. Lewin K (1948) Resolving social conflicts & field theory in social science. American Psychological Association, Washington DC
38. Lifvergren S, Gremyr I, Hellström A, Bergman B (2010) Lessons from Sweden's first large- scale implementation of six sigma in healthcare. Oper Manag Res, DOI 10.1007/s12063- 010-0038-y
39. Lifvergren S, Huzzard T, Docherty P (2009) A development coalition for sustainability in healthcare. In: Docherty P, Kira M, Shani AB (eds) Creating sustainable work systems: developing social sustainability. 2nd edition. Routledge, London
40. Lifvergren S, Chakhunashvili A, Bergman B. Docherty P (2008) Online statistical monitoring of critical patient data increases patient safety. PICMET-conference, Proceedings, Capetown
41. Lindgard Larsen M, Gertsen F, Johansen J (2003) Applying lean thinking in hospitals exploring implementation difficulties. Paper presented at the 3rd International conference on the management of healthcare and medical technology, Warwick
42. OECD. (2007). OECD Health data: Statistics and indicators for 30 Countries. OECD, Paris
43. Olsson J, Kammerlind P, Thor J, and Elgh M (2003): 'Surveying improvement activities in health care on a national level –the Swedish internal collaborative strategy and its challenges', Qual Manag Healthc 12: 202-216
44. Øvretveit J (2009) Does improving quality save money? A review of evidence of which improvements to quality reduce costs to health service providers. The Health Foundation, London
45. Øvretveit J (1997) A comparison of hospital quality programmes: lessons for other services. Int J Serv Ind Manag 3:220–235
46. Schroeder RG, Linderman K, Liedtke C, Choo AS (2008) Six sigma: definition and underlying theory. J Oper Manag 26: 536–554

47. Schaffer R, Thomson H (1992) Successful change programs begin with results. Harv Bus Rev, Jan-Feb: 80-89.
48. Senge P (1990) The fifth discipline: the art and practice of the learning organization. Broadway Business, London
49. Shani AB, Docherty P (2008) Learning by design. Key mechanisms in organization development. In: Cummings TG (ed) Handbook of organization development. Sage, Thousand Oaks, CA
50. Shani AB, Docherty P (2003) Learning by design: building sustainable organizations. Blackwell, Oxford
51. Shewhart W (1939) Statistical method from the viewpoint of quality control. Dover Publications, New York
52. Sibbritt D, Isbister G (2006). Emergency department performance indicators that encompass the patient journey. Qual Manag Healthc 15: 27-38
53. Smith B (1990) The Motorola story. Lecture by Bill Smith, Senior Quality Manager at Motorola at World Class Quality Event, Västerås, Sweden
54. Spear S (2005) Fixing healthcare from the inside, today. Harv Bus Rev, Sept: 1-20
55. Stacey RD (2007) Strategic management and organizational dynamics. The challenge of complexity. 5th edition. Prentice Hall, Harlow, England
56. Strebel P (1996) Why do employees resist change? Harv Bus Rev, May-June: 86-92
57. Thor J, Herrlin B, Wittlöv K, Øvretveit J, Brommels M (2007) Evolution and outcomes of a quality improvement program. Appended paper in Thor J (2007) Getting going on getting better. Dissertation. Department of Learning, Informatics, Management and Ethics, Karolinska Institutet, Stockholm
58. Tucker AL, Nembhard I, Edmondson AC (2007) Implementing new practices: An empirical study of organizational learning in hospital intensive care units. Manag Sci 6, June: 894- 907
59. Tucker AL (2004) The impact of operational failures on hospital nurses and their patients. J Oper Manag 22: April
60. Tucker AL, Edmondson AC (2003) Why hospitals don't learn from failures: organizational and psychological dynamics that inhibit system change. Calif Manag Rev 45, no 2 (winter)
61. Tucker AL, Edmondson AC, Spear S (2002) When problem solving prevents organizational learning. J Org Change Manag 15: 2
62. Weick KE (1995) Sensemaking in organizations. Sage, London
63. Zimmerman JP, Weiss J (2005) Six Sigma's seven deadly sins. Quality, January: 62–67. http://www.qualitymag.com/
64. Zu X, Fredendall L D, Douglas T J (2008) The evolving theory of quality management: the role of six sigma. J Oper Manag 26: 630-650

3

Project Costs and Risks Estimation Regarding Quality Management System Implementation

1. Introduction

Project management is the art — because it requires the skills, tact and finesse to manage people, and science because it demands an in-depth knowledge of an assortment of technical tools, of managing relatively short-term efforts, having finite beginning and ending points, usually with a specific budget, and it must meet or exceed customers' needs and expectations (Duicu et al., 2011). This involves balancing competing demands among:

- Scope, time, cost and quality;
- Stakeholders with different needs and expectations;
- Identified requirements and expectations.

All projects share the same characteristic - the design ideas and transform them into new activities and achievements. Elements of risk and uncertainty always present show that the activities and tasks necessary to implement the projects may never be planned with absolute accuracy when very complex projects, the very possibility of their successful completion can sometimes be called into question. Project management follows the processes and guidelines established by the PMI, (2004).

Project management uses a set of principles, rules, expertise, methods and tools for planning, necessary to start the deployment and successful completion of a project. In addition, project management is a system based on: financial resources, human and time.

Within each phases of project development, there are many processes, which must be completed before a project can move into the next phase. Project Management Institute suggests that the five process control groups should be used to define these processes within each phase of a project for a successful implementation. Based on the classification of each project, different combinations of processes should be used to successfully complete the project. Some factors included in this measurement of classification include complexity of scope, risk, size, period, institutional experience, and access to resources, maturity, and industry and application area. The figure 1 provides an overview of the process groups that will be implemented in any phase of a project (ASU, 2012).

One of the key elements of the competitiveness is quality. Quality management is an essential component of the project management along with other processes. Growth and continuous improvement of performance of a project depends heavily on how to ensure proper management of quality. Quality of project management not only refers to time and budget, but to specification and quality requirements.

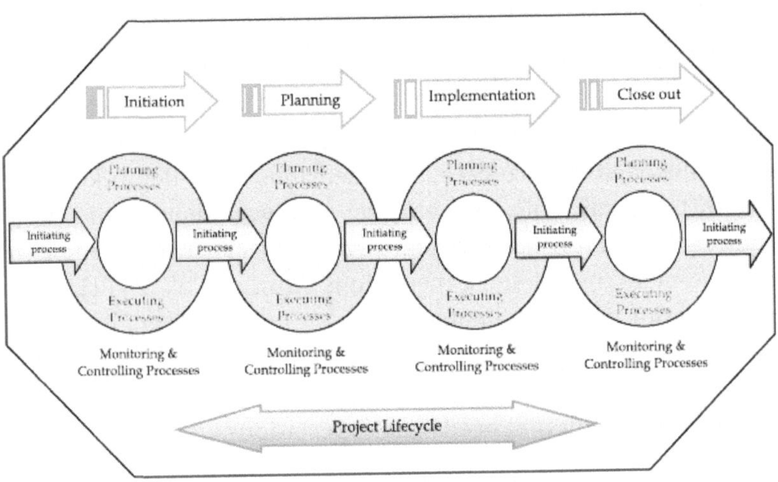

Figure 1. Project process groups

Project quality management consists of processes to ensure that the project will meet the requirements defined and planned, that quality planning processes, quality assurance and quality control (PMD, 2008).

Project quality management includes all management activities that will ensure the quality policy, objectives, and responsibilities and fulfill them through planning and improving quality through quality assurance and quality control. The project quality management processes are specified in figure 2.

Project quality management is the process required that ensures that the project meets requirements and expectations of the beneficiary involved in the project consists of: identification of relevant quality levels for the project and how to meet them, planned activities implemented quality system intended to ensure that the project will be within the parameters of quality planning, monitoring results of project activities and assessing their quality standards, ways to eliminate the causes which led to unsuccessful and continuous improvement (Nedelcu & Dumitrascu, 2010).

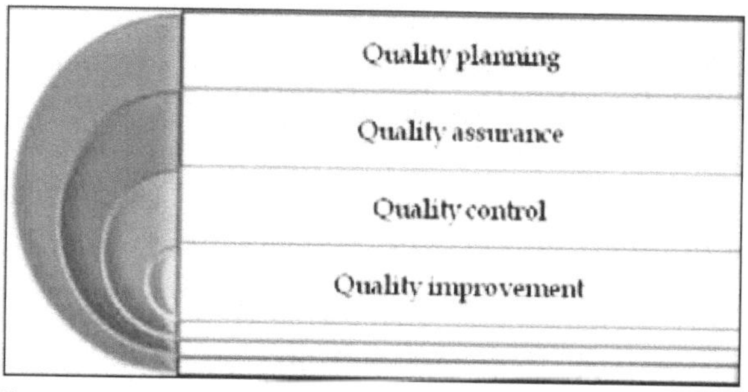

Figure 2. Project quality management processes

In figure 3 is illustrated an approach model of the quality management system in projects.

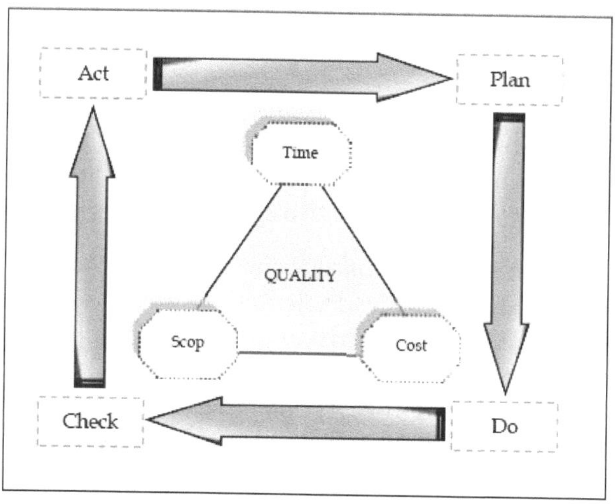

Figure 3. Project quality management

Quality management is a continuous process that starts and ends with the project. It is part of every project management processes from the moment the project initiates to the final steps in the project closure phase.

The purpose of quality management system is:

- To exclude possible errors in planning, coordination and all other phases;
- To ensure a controlled way that qualitative requirements on processes, approaches and products are respected at all stages;
- To find faults / errors as soon as possible, remove them and determine measures to avoid repeated mistakes;

- Check permanent measures to ensure quality and efficiency, the need to initiate corrective action;
- To determine and initiate corrective action / preventive measures.

2. Projects costs management

1.1 Theoretical aspects

Project cost management includes the processes involved in planning, estimating, budgeting, and controlling costs so that the project can be completed within the approved budget. The related knowledge area processes are (PMI, 2004):

- Cost estimating – developing an approximation of the costs of the resources needed to complete project activities;
- Cost budgeting – aggregating the estimated costs of individual activities or work packages to establish a cost baseline;
- Cost control – influencing the factors that create cost variances and controlling changes to the project budget.

According to the American Association of Cost Engineers, cost engineering is defined as that area of engineering practice where engineering judgment and experience are utilized in the application of scientific principles and techniques to the problem of cost estimation, cost control and profitability.

Project cost management is primarily concerned with the cost of the resources needed to complete schedule activities. Project cost management should also consider the effect of project decisions on the cost of using, maintaining, and supporting the product, service, or result of the project. Life-cycle costing, together with value engineering techniques, can improve decision-making and is used to reduce cost and execution time and to improve the quality and performance of the project deliverable. On some projects, especially ones of smaller scope, cost estimating and cost budgeting are so tightly linked that they are viewed as a single process that can be performed by a single person over a relatively short period of time. These processes are presented here as distinct processes because the tools and techniques for each are different. The ability to influence cost is greatest at the early stages of the project, and this is why early scope definition is critical (PMI, 2004).

The dependence between project cost management processes and project phases are detailed in table 1.

Table 1. Processes and phases of project cost management

Process	Project phase	Key deliverables
Estimate costs	Planning	Activity cost estimates, Basis of estimates
Determine budget	Planning	Cost performance baseline
Control costs	Monitoring and controlling	Work performance measurements

Project cost estimates are a key component of the planning process and provide a basis for key decisions. Cost estimate represents a prediction of quantities, cost, and/or price of resources required by the scope of an asset investment option, activity, or project. As a prediction, an estimate must address risks and uncertainties. Estimates are used primarily as inputs for budgeting, cost or value analysis, decision making in business, asset and project planning, or for project cost and schedule control processes. Cost estimates are determined using experience and calculating and forecasting the future cost of resources, methods, and management within a scheduled time frame (ISO, 2010).

An activity cost estimate is a quantitative assessment of the likely costs of the resources required to complete schedule activities. This type of estimate can be presented in summary form or in detail. Costs are estimated for all resources that are applied to the activity cost estimate. This includes, but is not limited to, labor, materials, equipment, services, facilities, information technology, and special categories such as an inflation allowance or cost contingency reserve.

The level of quality can be assessed with costs. It is normal for an organization to strive for the high quality products and services, but this procedure should not result in expenses that may cause, in turn, its bankruptcy. Thus, it is necessary to develop a budget for improving the quality and compared with expected profit. It is also necessary to establish special measures for sub-suppliers quality assurance, quality system continuously monitored, providing feedback information.

Data quality control is essential to ensure the integrity of results from quality improvements projects. Feasible methods are available and important to help to ensure that stakeholder's decisions are based on accurate data.

1.2 Costs estimation methods - modeling and simulation

Estimating the cost of project is one of the most crucial tasks for project managers. The main factors that are typically estimated at the beginning of a development project are: cost, size, schedule, quality, people resources, effort, resources, maintenance costs, and complexity. Cost estimation tools, or model-based estimation techniques use data collected from past projects combined with mathematical formulae to estimate project cost. They usually require factors such as the system size as inputs into the model.

The major software cost and schedule estimation techniques can be grouped and classified as regression-based models, learning-oriented models, expert based approaches and finally composite-Bayesian methods (Keaveney & Conboy, 2011). Therefore, there is a natural erroneous tendency associated with any form of estimation primarily because "an estimate is a probabilistic assessment of a future condition" and accuracy can therefore rarely be expected in the estimation process (Stamelos & Angelis, 2001). The causes of inaccurate estimates in development projects were grouped into four categories by Lederer & Prasad (1995), namely methodology, politics, user communication and management control.

A project simulation uses a model that translates the uncertainties specified at a detailed level of the project into their potential impact on project objectives. Simulations are typically performed using the Monte-Carlo technique.

In Monte-Carlo method calculations are repeated several times using the same, deterministic model of a physical phenomenon, but each time for different, randomly selected values of particular arguments, from among uncertainty range given a priori.

In a simulation, the project model is computed many times (iterated), with the input values randomized from a probability distribution function (e.g., cost of project elements or duration of schedule activities) chosen for each iteration from the probability distributions of each variable. A probability distribution (e.g., total cost or completion date) is calculated.

The Monte-Carlo technique is a device for modeling and simulating processes that involve chance variable. Monte-Carlo simulation requires hundreds or thousands of iterations. Each sample yields one possible outcome for the variable(s) of interest. By studying the distributions of results, we can see the range of possible outcomes and the most likely results. Using simulation, a deterministic value can become a stochastic variable. We can then study the impact of changes in the variable on the rest of the spreadsheet.

One of the most used distributions is the triangular because the input data can be obtained very easily and it does not require laborious investigations. Recent popularity of the triangular distribution can be attributed to its use in Monte Carlo simulation modeling and its use in standard uncertainty analysis software. The triangular distribution is also found in cases where two uniformly distributed errors with the same mean and bounding limits are combined linearly (Castrup, 2009).

Uncertainties may be modeled by the distribution where Johnson and Kotz (1999) discuss the asymmetric triangular distribution. Suppose that:

$$x_i = \begin{cases} a + \sqrt{z_i \cdot (b-a) \cdot (m-a)}, a < x_i \le m \\ b - \sqrt{(1-z_i) \cdot (b-a) \cdot (b-m)}, m < x_i \le b \end{cases}, \quad (1)$$

where $a\hat{}$ is lower estimate, $m\hat{}$ is most likely estimate value and $b\hat{}$ is maximum estimate value.

The mean and standard deviation are given by:

$$\mu = \frac{a+m+b}{3}, \quad (2)$$

$$\sigma = \sqrt{\frac{a^2 + m^2 + b^2 - am - ab - mb}{18}}. \qquad (3)$$

The distribution emerges in numerous papers (Mohan et. al, 2007; Keefer & Verdini, 1993) and the probability density function for asymmetric three-parameter is given by:

$$f(x) = \begin{cases} \frac{2(x-a)}{(b-a)\cdot(m-a)}, a < x \leq m \\ \frac{2(b-x)}{(b-a)\cdot(b-m)}, m < x < b \\ 0, elsewhere \end{cases}. \qquad (4)$$

Cumulative distribution function is defined by:

$$F(x) = \begin{cases} 0, x \leq a \\ \frac{(x-a)^2}{(b-a)\cdot(m-a)}, a < x \leq m \\ 1 - \frac{(b-x)}{(b-a)\cdot(b-m)}, m < x < b \\ 1, x \geq b \end{cases}. \qquad (5)$$

Cumulative distributions functions are usually presented graphically in the form of ogives, where we plot the cumulative frequencies at the class boundaries. The resulting points are connected by means of straight lines, as shown in chapter 2.3.

If the relative frequency is plotted on normal probability graph paper the ogive will be a straight line for a normally distributed random variable. The normal probability graph paper is a useful device for checking whether the observations come from a normally distributed population, but such a device is approximate. One usually rejects normality when remarkable departure from linearity is quite evident (Gibra, 1973).

1.3 Case study regarding project cost estimation of the quality management system implementation

Implementation of a quality management system (QMS) takes a lot of time and effort, whereas the top management wants benefits on a short period of time. The purpose of this case study is to provide the applicative aspects of project costs estimation of a QMS implementation.

The specific objectives achieved by this project are:

- Correct identification of processes within the organization;
- Development of specific documents of the QMS (quality manual, documented procedures of quality management system, specific documents);
- Implementation and certification of QMS according to ISO 9001:2008 accredited by a recognized certification organization in order to increase the performances of the own organization.

Quality management system implementation aims:

- Ongoing activities to meet customers' expectations;
- Products comply with applicable standards or specifications;
- Products offered at competitive prices;
- Products obtained in terms of profit.

The profit can be obtained by applying a rigorous prediction of specific costs. There plays a significant role the application of efficient and effective cost management. The project must be within allocated budget. Competitiveness of products and services is linked to quality. To achieve the objectives is not sufficient to implement the various techniques and methods, but quality must be acknowledged and implemented in all fields and at all levels. All these are possible only by introducing and supporting the documentation of quality management system.

Quality management system will assist by:

- Managing costs and risks;
- Increasing effectiveness and productivity;
- Identifying improvement opportunities;
- Increasing customer satisfaction.

A well-managed quality system will have an impact on:

- Customer loyalty and repeat business;
- Market share and industry reputation;
- Operational efficiencies;
- Flexibility and ability to respond to market opportunities;
- Effective and efficient use of resources;
- Cost reductions and competitive advantages;

- ❖ Participation and motivation of human resources;
- ❖ Control on all processes within organization.

Project quality evaluation represents an essential component of project success. The implementing process of a quality management system in an organization may be considered as a project. The general objective of this project is implementation, maintaining, certification and continuous improvement of the quality management system in organization in accordance with ISO 9001:2008. Therefore, in figure 4 are detailed the phases within a project's lifecycle and it will be considered complete and operational.

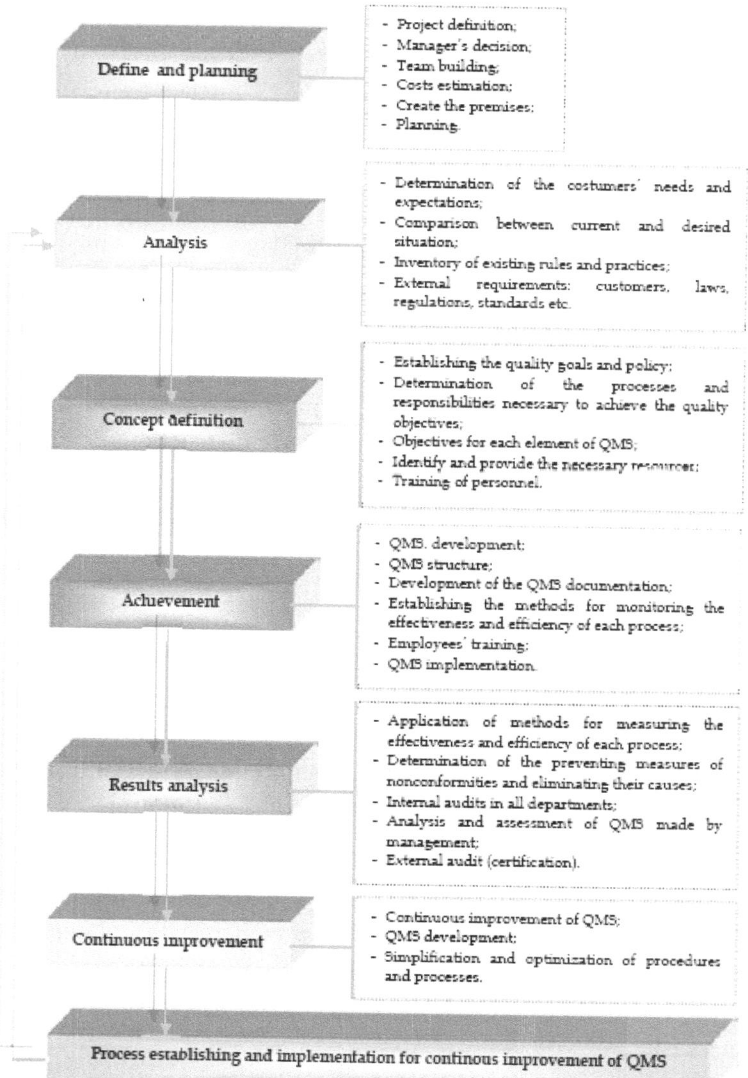

Figure 4. Project phases of a quality management system implementation

The applicative researches focuses on the relevant activities related to the QMS implementation and it covers the following stages:

- Diagnostic audit according to ISO 9001:2008 requirements;
- Establishing the quality policy and goals;
- Establishment of quality management program;
- Training of top management team;
- Initial evaluation, processes planning and identification relevant to quality system;
- Development of the quality management system documents;
- Approval, multiplication and dissemination of quality management system documents;
- Implementation of quality management system;
- Documentation for certification of quality management system;
- Preliminary discussion with selected accredited (optional);
- Internal quality audit;
- Transmission the documents of quality management system to certification organization (quality manual, procedures, work instructions);
- Reviewing the documentations by certification organization;
- Audit plan;
- Certification audit;
- Surveillance audit (annual);
- Recertification audit (three years).

For the most important tasks that imply substantial costs for QMS implementation process it was performed the simulated researches of the project activities costs using triangular distribution. The analyzed activities are presented in figure 5.

To estimate the cost of an activity we need to know how long it can run, that human or material resources will be involved, raw materials and materials used in the execution of activities. Consequently, the estimation process in a project must start from the estimation times. To do this, we can assess the effort required to execute each task.

Considering the steps of quality management system implementation described above, the Monte-Carlo simulation program is detailed in figure 6.

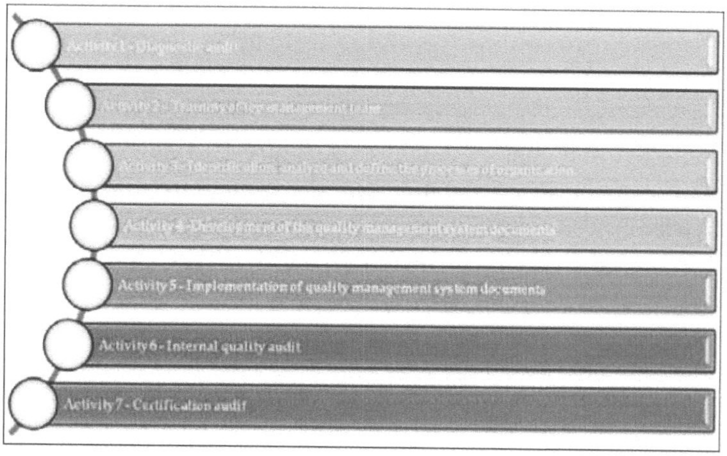

Figure 5. Main stages of quality management system implementation specific to costs simulation process

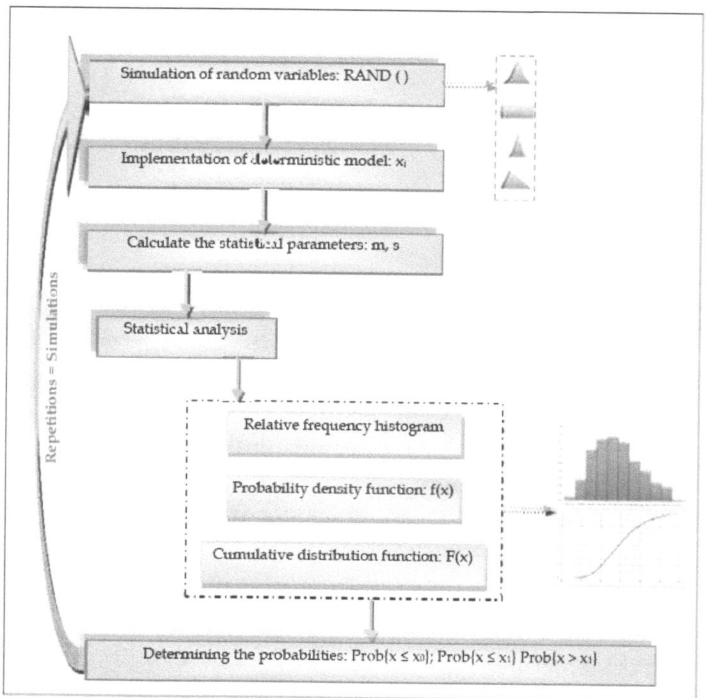

Figure 6. Monte-Carlo simulation program

Excel is not optimized for statistics, so other vendors have created add-ins offering more features. They not only calculate probabilities, but also permit Monte Carlo simulation to draw repeated samples from a distribution.

Consider an outcome, such as the task cost and we want to simulate what the actual costs might be if we know the lowest cost, the highest and the most likely.

As a result, the simulation of specific costs for analyzed activities are represented by relative frequency histograms (see figure 7, figure 9, figure 11, figure 13, figure 15, figure 17, figure 19). If we run a stochastically simulation on these tasks costs, it can construct a relative cumulative frequency graph (ogive chart) for these data and looks like in figure 8, figure 10, figure 12, figure 14, figure 16, figure 18, figure 20). Cumulative distribution curve indicates the probability that we will complete this project tasks in less cost than the deterministically predicted. Each time the simulation is executed, the cell will be updated to show a random value drawn from the specified distribution.

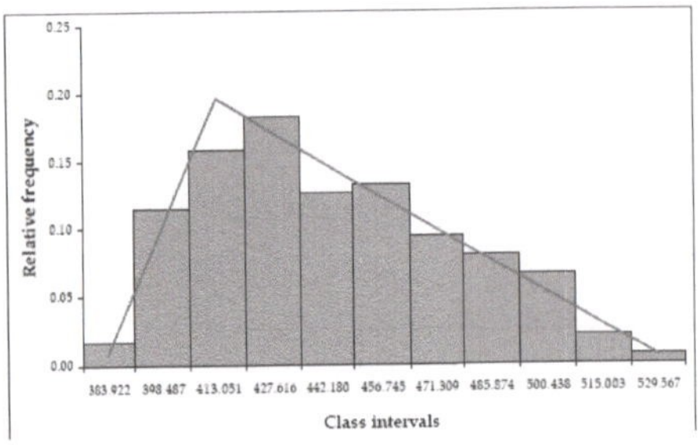

Figure 7. Relative frequency histogram for task 1

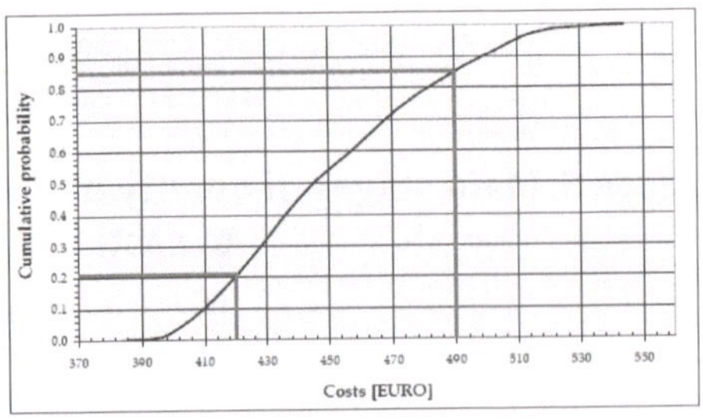

Figure 8. Cumulative frequency distribution for task 1

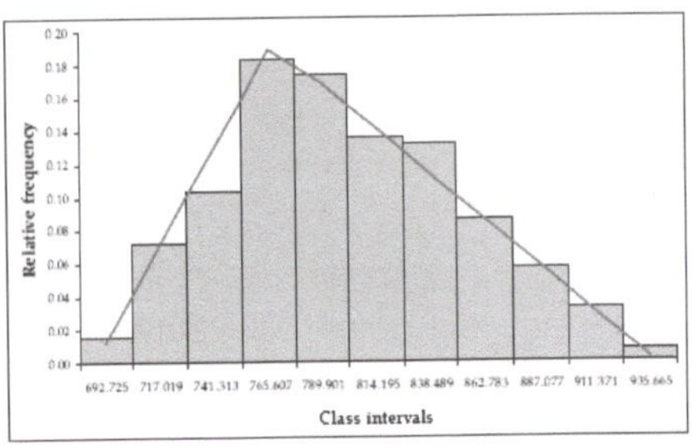

Figure 9. Relative frequency histogram for task 2

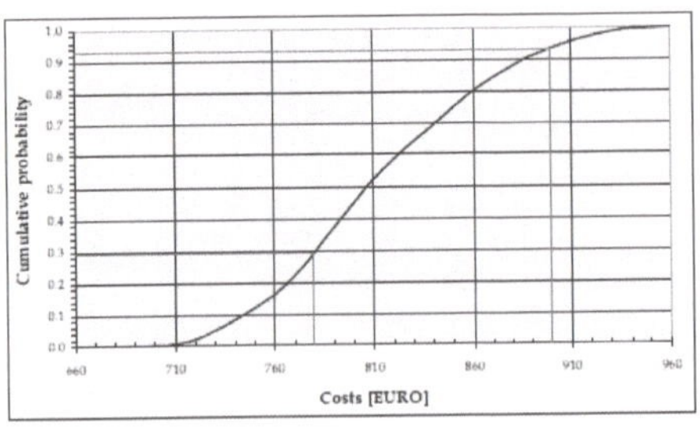

Figure 10. Cumulative frequency distribution for task 2

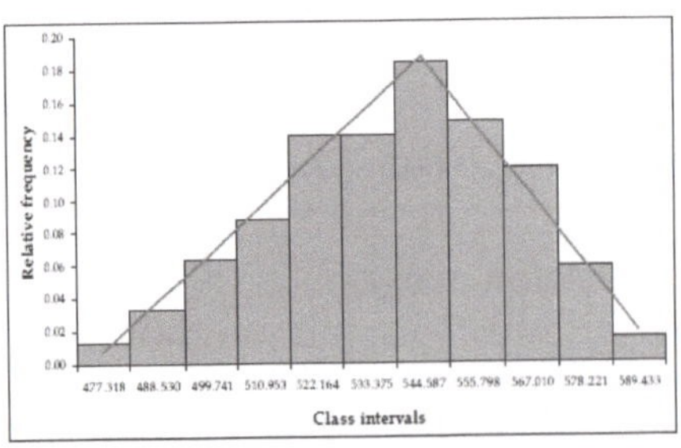

Figure 11. Relative frequency histogram for task 3

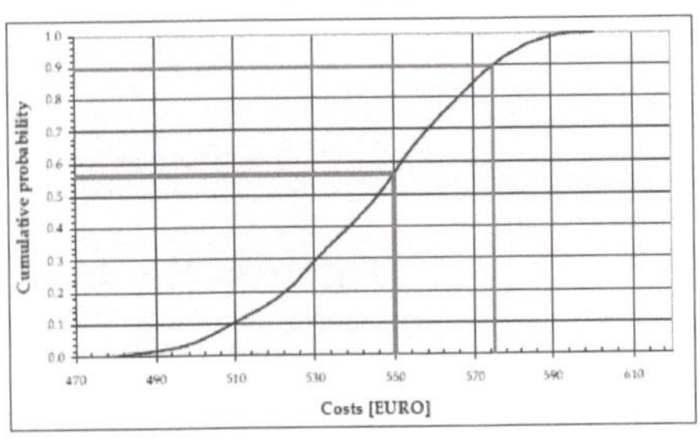

Figure 12. Cumulative frequency distribution for task 3

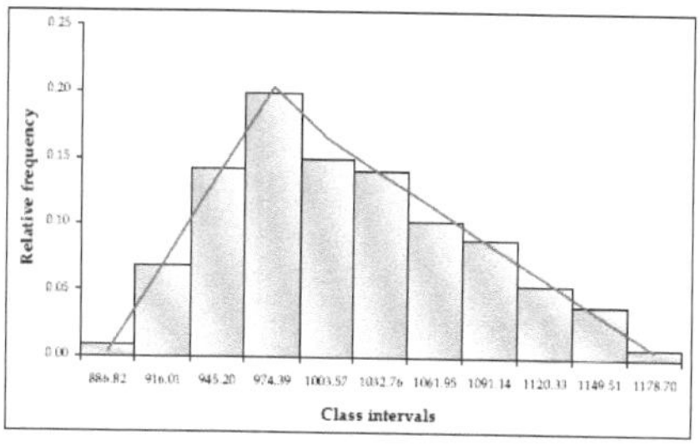

Figure 13. Relative frequency histogram for task 4

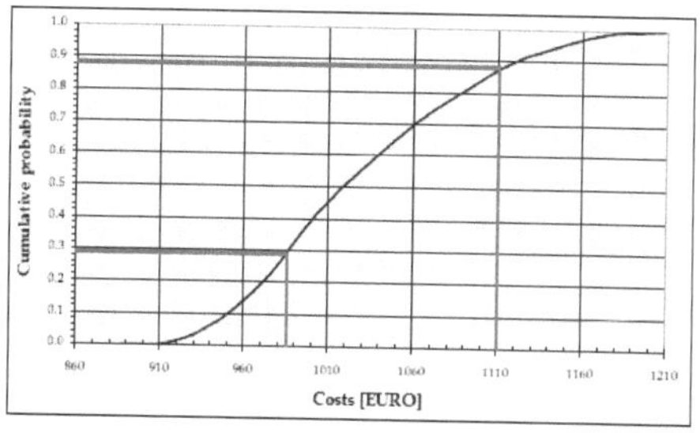

Figure 14. Cumulative frequency distribution for task 4

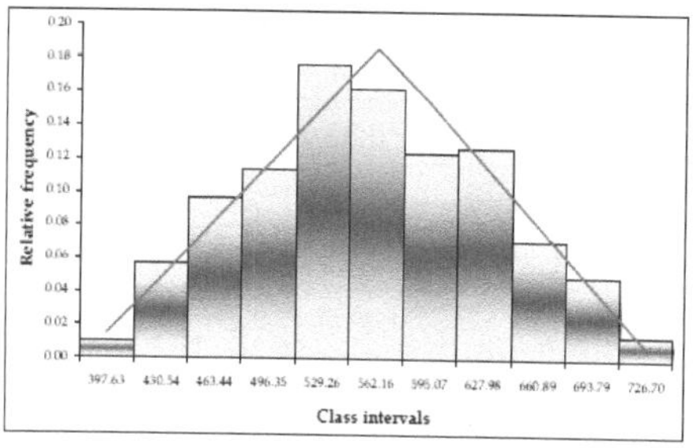

Figure 15. Relative frequency histogram for task 5

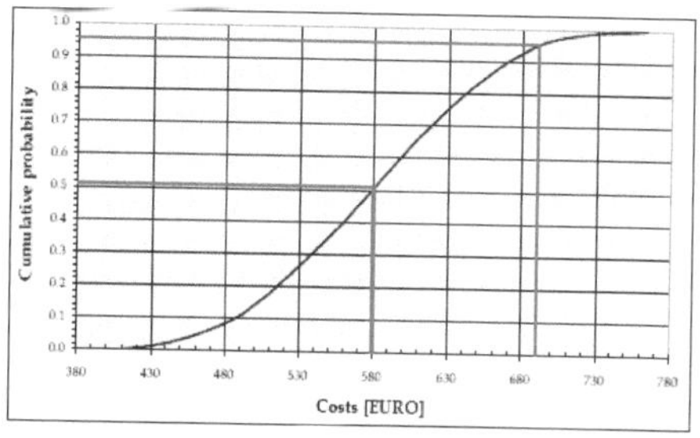

Figure 16. Cumulative frequency distribution for task 5

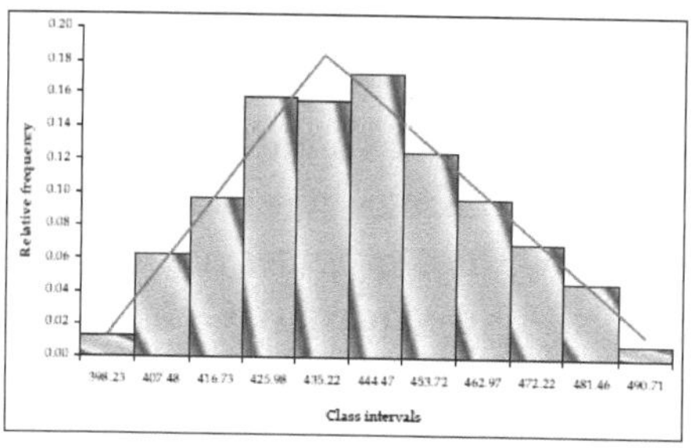

Figure 17. Relative frequency histogram for task 6

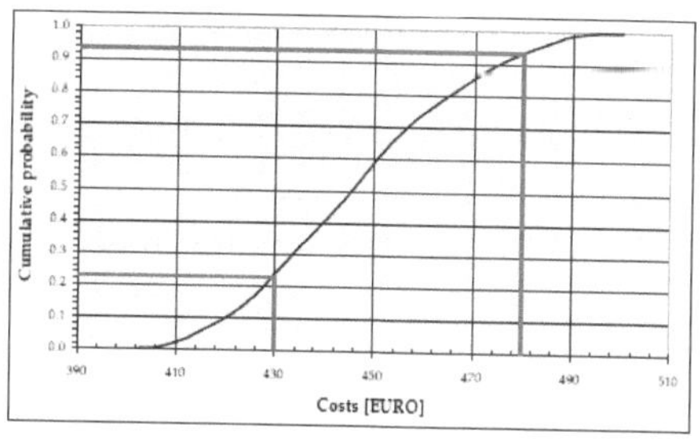

Figure 18. Cumulative frequency distribution for task 6

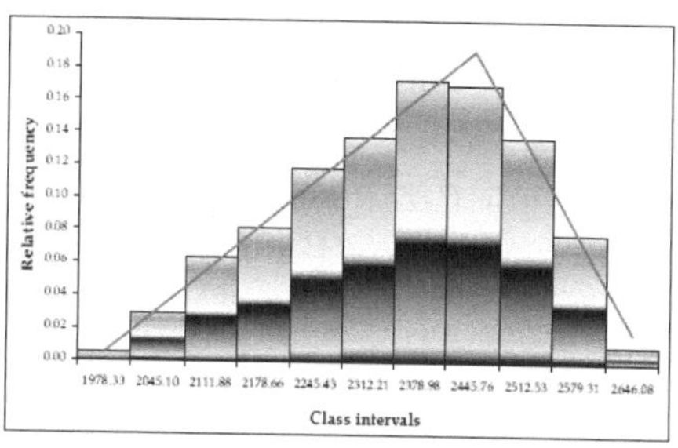

Figure 19. Relative frequency histogram for task 7

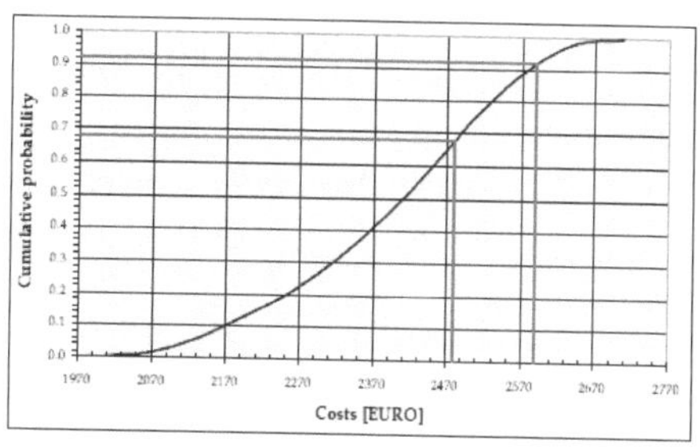

Figure 20. Cumulative frequency distribution for task 7

In table 2 are detailed the input data and the results of simulation process. It can be observed that the probability to exceed the maximum allocated costs of the tasks is smaller than 14%. Taking into account the contingency funds for the overall project tasks (approximately 10-20%), we will be able to accomplish the project activities with the planned costs and time.

Table 2. Simulation process results

Tasks	Minimum	Likely	Maximum	Prob{x ≤ x0}	Prob{x ≤ x1}	Prob{x > x1}
Activity 1	390	420	550	0.212	0.863	0.137
Activity 2	700	780	950	0.292	0.937	0.063
Activity 3	400	550	600	0.561	0.891	0.109
Activity 4	900	985	1200	0.283	0.864	0.136
Activity 5	400	580	750	0.514	0.956	0.044
Activity 6	400	440	500	0.226	0.926	0.074
Activity 7	2000	2480	2700	0.679	0.923	0.077

Analyzing the relative frequency histograms, it can be seen the standard asymmetric triangular distribution. For activities where the distribution is skewed to the left, we might finish the tasks ahead of schedule and for the distribution is skewed to the right, the project implies supplementary costs.

2 Projects quality improvement through risks management process

Management of risk is an integral part of good business practice and quality management. Learning how to manage risk effectively enables managers to improve outcomes by identifying and analyzing the wider range of issues and providing a systematic way to make informed decisions. A structured risk management approach also enhances and encourages the identification of greater opportunities for continuous improvement through innovation (AUSAID, 2006).

Project risk management offers a great opportunity to improve project performance dramatically. The amount of work needed to implement the project risk management process is considerably less than that devoted to other project variables such as cost, schedule, or quality, yet the benefits are equally great. The greatest challenge to implementing a project risk management lies in changing corporate culture. However, once this is done, and risk management becomes routine, it will add greatly to the probability of project success (Naughton, 2012).

To be competitive, an organization must be proactive in managing the risks to successful achievement of the cost and schedule objectives for its projects. The risks management implementation can improve the quality of supplied products and processes through the identification, monitoring and control of risks.

2.1 Concepts, principles and components of risks management

Project risk represents uncertain events or situations that potentially can adversely affect a project as planned, usually in terms of cost, schedule, and/or product quality.

Project risk management includes the processes concerned with conducting risk management planning, identification, analysis, responses, and monitoring and control on a project; most of these processes are updated throughout the project (PMI, 2004). It involves processes, tools, and techniques that will help the project manager to maximize the likelihood and consequences of positive events and minimize the probability and consequences of adverse events. In figure 21 it is shown the processes of project risk management.

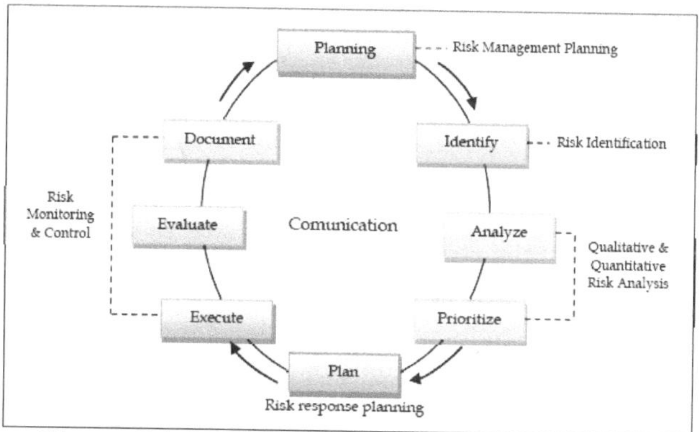

Figure 21. Project risk management process

Risks are prioritized according to their potential implications for meeting the project's objectives. A risk matrix is used to combine likelihood and impact ratings values to obtain a risk score. The risk score may be used to aid decision making and help in deciding what action to take in view of the overall risk. How the risk score is derived can be seen from the sample risk matrix shown in table 3. The organization can define as many risk levels as it believe are necessary. In our case the matrix presents tree domains: high, moderate and low risks.

Table 3. Risk assessment with risk matrix

	Probability				
0.90	0.05	0.09	0.18	0.36	0.72
0.70	0.04	0.07	0.14	0.28	0.56
0.50	0.03	0.05	0.10	0.20	0.40
0.30	0.02	0.03	0.06	0.12	0.24
0.10	0.01	0.01	0.02	0.04	0.08
	0.05	0.10	0.20	0.40	0.80
			Impact		

2.2 Risks estimation of quality management system implementation

Risk management for implementing a QMS consists of dealing with big and small objections, coming from people all over the organization, and can, in general, not be quantified in the same way as the risks in a production process (CERCO, 2000).

Risk management is a facet of quality, using basic techniques of analysis and measurement to ensure that risks are properly identified, classified, and managed.

The main objectives of this case study are:

- Identification and definition of risks categories;
- Establishing the criteria (factors) of risks analysis and their levels of assessment;
- Determining the risk score;
- Risk ranking in three categories: high, medium or small level;
- Risks prioritization and implement the corrective or preventive actions.

2.2.1 Risks Identification

It is evaluate the potential risks to the opportunity, to be able to build a project plan that maximizes the probability of project success. Risk identification is generally done as part of a feasibility study, at the beginning of the active project work, and at each new phase of a large project.

The project team considers:

- **Risks:** what might go wrong;
- **Opportunities:** better methods of achieving the project's purpose and need;
- **Triggers:** symptoms and warning signs that indicate whether each risk is likely to occur.

After the assessment, risk categories that influence the project development of the quality management system implementation are detailed in table 4.

Table 4. Identified risks categories

Risks categories	**Risk description**
Quality risk	Risk that influence product quality or service supplied
Project management risk	Risk which influence development of the project due to the planning process.
Internal risk	Risk influencing the project development caused internal factors of the organization
External risk	Risk affecting the whole system caused by the changes of external entities (importers, laws, authorizations, etc.)
Financial risk	Risk that affects the allocated budget

2.2.2 Qualitative analysis

In this step are assessed the impacts of identified risks (table 5).

Table 5. Qualitative analysis of risks

No.	Risks categories	Risk impact (I)	Risk impact evaluation
R1	Choosing a consulting organization that does not know / not to meet the requirements of field / organization	0.8	Major impact (catastrophic) on the project, such as deviations more than 25% from the project scope, schedule or budget
R2	Exceeding the budget	0.4	Significant impact on the project, such as deviations of 10-25% from the project scope, schedule or budget
R3	Exceeding allotted	0.4	Significant impact on the project, such as deviations of 10-25% from the project scope, schedule or budget
R4	Wrong identification of processes	0.4	Significant impact on the project, such as deviations of 10-25% from the project scope, schedule or budget
R5	Omission of the stages in the description of processes	0.2	Measurable impact on the project, such as deviations of 5-10% from the project scope, schedule or budget
R6	There are not evaluated all departments	0.2	Measurable impact on the project, such as deviations of 5-10% from the project scope, schedule or budget
R7	Incomplete planning of processes / products	0.4	Significant impact on the project, such as deviations of 10-25% from the project scope, schedule or budget
R8	Documentation is incomplete	0.1	Minor impact on the project, such as deviations less than 5% from the project scope, schedule or budget
R9	Evaluation is not real	0.1	Minor impact on the project, such as deviations less than 5% from the project scope, schedule or budget
R10	Documentation is imposed without discussed	0.4	Significant impact on the project, such as deviations of 10-25% from the project scope, schedule or budget
R11	Do not have "client" in center of processes	0.2	Measurable impact on the project, such as deviations of 5-10% from the project scope, schedule or budget
R12	Do not reflect the commitment for customer satisfaction	0.2	Measurable impact on the project, such as deviations of 5- 0% from the project scope, schedule or budget
R13	Establishment of unrealistic targets	0.4	Significant impact on the project, such as deviations of 10-25% from the project scope, schedule or budget
R14	It does not comply with annual audit plan	0.1	Minor impact on the project, such as deviations less than 5% from the project scope, schedule or budget
R15	Audits weren't done	0.4	Significant impact on the project, such as deviations of 10-25% from the project scope, schedule or budget

No.	Risks categories	Risk impact (I)	Risk impact evaluation
R16	Audits only formally	0.4	Significant impact on the project, such as deviations of 10-25% from the project scope, schedule or budget
R17	Unsolved of the prepared nonconformities report	0.05	Insignificant impact on the project. It is not possible to quantify the impact, which is extremely low.
R18	Decisions weren't fulfilled	0.1	Minor impact on the project, such as deviations less than 5% from the project scope, schedule or budget
R19	Resumption of process if it is found that implementation was not effective	0.8	Major impact (catastrophic) on the project, such as deviations more than 25% from the project scope, schedule or budget
R20	Audit report to be prepared superficially	0.1	Minor impact on the project, such as deviations less than 5% from the project scope, schedule or budget
R21	Incomplete documentation and the certification process extension	0.4	Significant impact on the project, such as deviations of 10-25% from the project scope, schedule or budget
R22	Delay to transmission of the audit plan	0.1	Minor impact on the project, such as deviations less than 5% from the project scope, schedule or budget
R23	Identification of major nonconformities	0.4	Significant impact on the project, such as deviations of 10-25% from the project scope, schedule or budget
R24	Isn't taken certification after audit	0.8	Major impact (catastrophic) on the project, such as deviations more than 25% from the project scope, schedule or budget

2.2.3 Quantitative analysis

This process aims to analyze numerically the probability of each risk and consequences on project objectives (table 6).

Table 6. Quantitative analysis of risks

No.	Risks categories	Likelihood (P)	Impact consequences
R1	Choosing a consulting organization that does not know / not to meet the requirements of field / organization	0.1	Certification is not obtained
R2	Exceeding the budget	0.9	More expenses than was anticipated
R3	Exceeding allotted	0.5	Financial losses and low productivity
R4	Wrong identification of processes	0.3	Delay to finish the project
R5	Omission of the stages in the description of processes	0.5	Delay to finish the project
R6	There are not evaluated all departments	0.5	There are not included all aspects
R7	Incomplete planning of processes / products	0.9	Delay to finish the project

No.	Risks categories	Likelihood (P)	Impact consequences
R8	Documentation is incomplete	0.9	Delay to finish the project
R9	Evaluation is not real	0.7	Measures plans are not effective
R10	Documentation is imposed without discussed	0.7	Delay to finish the project
R11	Do not have "client" in center of processes	0.3	Delay to finish the project
R12	Do not reflect the commitment for customer satisfaction	0.3	Delay to finish the project
R13	Establishment of unrealistic targets	0.3	Delay to finish the project
R14	It does not comply with annual audit plan	0.7	Delay to finish the project
R15	Audits weren't done	0.1	Delay to finish the project / Certification is not obtained
R16	Audits only formally	0.1	Delay to finish the project / Certification is not obtained
R17	Unsolved of the prepared nonconformities report	0.3	Resumption of internal audits
R18	Decisions weren't fulfilled	0.3	Delay to finish the project
R19	Resumption of process if it is found that implementation was not effective	0.1	Delay to finish the project / Certificate wasn't taken at scheduled time
R20	Audit report to be prepared superficially	0.7	Certification is not obtained
R21	Incomplete documentation and the certification process extension	0.5	Delay to finish the project
R22	Delay to transmission of the audit plan	0.3	Delay to finish the project
R23	Identification of major nonconformities	0.5	Delay to finish the project
R24	Isn't taken certification after audit	0.3	Delay to finish the project

2.2.4 Monitoring and control of risks

This stage supposes keeping track of identified risks, monitoring residual risks, and identifying new risks.

In table 7 are quantify the identified risks based on the risk matrix specified in table 3.

Table 7. Risks score

No.	Risks categories	Risk score (I x P)
R1	Choosing a consulting organization that does not know / not to meet the requirements of field / organization	0.08
R2	Exceeding the budget	0.36
R3	Exceeding allotted	0.20
R4	Wrong identification of processes	0.12
R5	Omission of the stages in the description of processes	0.10
R6	There are not evaluated all departments	0.10
R7	Incomplete planning of processes / products	0.36
R8	Documentation is incomplete	0.09
R9	Evaluation is not real	0.07
R10	Documentation is imposed without discussed	0.28
R11	Do not have "client" in center of processes	0.06
R12	Do not reflect the commitment for customer satisfaction	0.06

No.	Risks categories	Risk score (I x P)
R13	Establishment of unrealistic targets	0.12
R14	It does not comply with annual audit plan	0.07
R15	Audits weren't done	0.04
R16	Audits only formally	0.04
R17	Unsolved of the prepared nonconformities report	0.015
R18	Decisions weren't fulfilled	0.03
R19	Resumption of process if it is found that implementation was not effective	0.08
R20	Audit report to be prepared superficially	0.07
R21	Incomplete documentation and the certification process extension	0.20
R22	Delay to transmission of the audit plan	0.03
R23	Identification of major nonconformities	0.20
R24	Isn't taken certification after audit	0.24

The identified risks are categorized by degrees of probability and impact, and they are distributed into high, moderate, and low risk categories.

After making analysis were results twenty-four risks for all stages of project development of a quality management system implementation. The most important risks that must be treated it can be seen in figure 22.

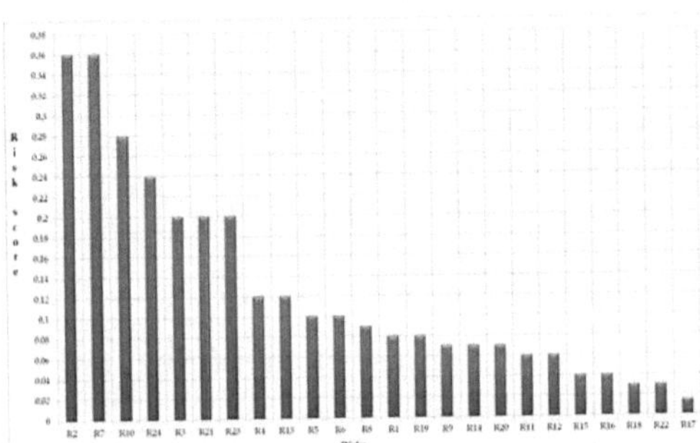

Figure 22. Risks prioritization

Also, implementing the monitoring and control plans, the risks situated in the high domain can be eliminated or minimized to a medium or low level.

Nothing can be controlled which cannot be measured. In a project there are three things which can always be measured - the schedule, the cost, and the users' satisfaction. If the project meets all these three criteria, it is right to consider it a successful project. If it meets two of them and comes close to the third it is probably successful. Few projects which are very late and very much over budget are considered to be successful and they meet the users' requirements. Therefore, the essence of risk management is the avoidance of anything which extends the schedule, increases the costs, or impairs the users' satisfaction with the product of the project.

3 Conclusions

Customer satisfaction provides valuable information for saving resources by tuning those processes and its quality aspects. Therefore the measurement of customer's satisfaction should be seen as an essential part of business management and should be carefully designed as one of the key measures of the success of an organization.

Monitoring quality costs is essential when implementing a quality management system as this gives relevant information about the balance between efforts and investments in quality to reduce non quality and what remaining non quality still costs. However it is difficult to identify all the relevant elements and the quality indicators are necessary because they provide a concrete tool to measure how the QMS is improving the organization's efficiency and effectiveness (CERCO, 2000).

The case study regarding costs estimation can be extending to the stochastically modeling of the total project schedules. There is uncertainty surrounding any estimate of how long a task will take. If it is a short, simple task we have done before, the uncertainty range may be so small that it can safely be ignored. With large, complex tasks, the uncertainty becomes very significant, especially if we're depending on factors beyond our control.

Anyone who is serious about realistically forecasting project schedules, anticipating potential trouble spots, and taking action to mitigate against likely problems – in other words, truly managing major projects, rather than just monitoring them – should be using Monte-Carlo simulation software to plan and analyze projects stochastically. It is the best way to avoid late and over budget problems. Simulation has many advantages for risk assessment. Among them are:

- ❖ Simulation forces us to state our assumptions clearly;
- ❖ Simulation helps us visualize the implications of our assumptions;
- ❖ Simulation reveals the potential variation in the variables;
- ❖ Simulation quantifies risk (the probability of a given event).

One of the key measures of the resilience of any project is its ability to reach completion on time and on budget, regardless of the turbulent and uncertain environment it may operate within. Cost estimation and tracking are therefore paramount when developing a system.

The magnitude of each costs components depends on the size of the organization, the nature of the project as well as the organization management, among many considerations, and the owner is interested in achieving the lowest possible overall project cost.

Budget estimate must be adopted early enough for planning long term financing of the facility. Consequently, the detailed estimate is often used as the budget estimate since it is sufficient definitive to reflect the project scope and is available long before the engineer's estimate. As the work progresses, the budgeted cost must be revised periodically to reflect the estimated cost to completion. A revised estimated cost is necessary either because of change orders initiated by the owner or due to unexpected cost overruns or savings (Dumitrascu et al., 2010).

When we plan a project, we are attempting to predict what will happen in the future what we will produce, by when, for how much money. Prediction is always fraught with uncertainty. Our estimates are usually wrong because we cannot accurately predict the future. This is why we need risk assessment - it helps us to manage the uncertainty of the future (Naughton, 2012).

The risks of implementing and maintaining a QMS are well known. Although they cannot necessarily be eliminated they can be managed, and their impact reduced.

Based on information, obtained risks score and analyzed the case study regarding the risk management applied to a quality management system implementation, we can draw the following conclusions:

- Identified risks have a significant level of appearance and a relatively small impact;
- It has been identified risks, and the appearance causes on which will be applied corrective or contingent actions;
- Risk management plan ensure the prevention and treatment of risks arising for each stage of implementation;
- Risk management plan can be rebuilt according to other risks identified in the project. Also, the case studies can be extended to costs assessment of project risks.

It is important to determine quantitatively the impact of event risk and determine the risks that cause harm (or benefit) for the project. So, it can build the sensitivity. They are used to visualize possible effects on the project (the best results and lower) the different unknown variables than their initial values. Variable sensitivity is modeled as uncertain value, while all other variables are held at baseline (stable). Tornado diagram is a variation of sensitivity analysis where the variable with the greatest impact is positioned on top of the chart, followed by other variables, in descending order of impact. It can be used for sensitivity analysis of project objectives (cost, time, quality and risk) (Barsan- Pipu, 2003).

References

1. ASU, (2012). *Project Lifecycle*, Angelo State University, IT Project Office, Available from: http://www.angelo.edu/services/project_management/Project_Lifecycle.php
2. AUSAID, (January 2006). Performance Review and Audit (AUDIT) Section, In: *Risk Management Guide*, Australian Government, Available from: http://www.ausaid.gov.au/publications/pdf/Risk_Management_Guide-final_7-3-06.pdf

3. Barsan-Pipu, N. & Popescu, I. (2003). *Risk Management. Concepts. Methods. Applications.* (in Romanian), Transilvania University Publishing House, Brasov, ISBN 973-635-180-7
4. Castrup, H. (January 2009). *Error Distribution Variances and Other Statistics,* Integrated Sciences Group, 1/15/2009, Available from: http://www.isgmax.com/Articles_Papers/Error%20Distributions%20and%20Other%20Statistics.pdf
5. CERCO, (August 2000). *Handbook for Implementing a Quality Management System in a National Mapping Agency*, CERCO Working Group on Quality, August 18, Available from: http://www.eurogeographics.org/sites/default/files/handbook_V1.pdf
6. Duicu, S., Dumitrascu, A.-E. & Lepadatescu, B. (2011). Project Quality Evaluation - An Essential Component of Project Success, *Proceedings of WSEAS International Conference - Mathematics and Computers in Biology, Business and Acoustics*, ISBN 978-960-474-293-6, pp. 139-142, Brasov, Romania, April 11-13, 2011
7. Dumitrascu, A.-E., Nedelcu, S., Lepadatescu, B. (2010). Project Cost Estimation of the Quality Management System Implementation, *Proceedings of 7th WSEAS International Conference on Engineering Education*, ISBN 978-960-474-202-8, pp. 267-270, Corfu Island, Greece, July 22-24, 2010
8. Gibra, I.N. (1973). *Probability and Statistical Inference for Scientists and Engineers*, Prentice-Hall Publishing House, Englewood Cliffs, New Jersy, ISBN 0-13-711622-5
9. Hesse, R. (May 2000). *Triangle Distribution: Mathematica Link for Excel*, Graziadia Graduate School of Business, Pepperdine University, p. 12-14, Available from: http://www.decisionsciences.org/decisionline/Vol31/31_3/31_3clas.pdf
10. Isken, M. (June 2003). Distributions, Excel Functions, and Simulation, June 11, Available from:http://www.sba.oakland.edu/faculty/isken/hcm540/Session03_DMUncertainty/Dis tributionsSimulationExcelFunctions.pdf
11. ISO, (2008) Quality management systems - Requirements, ISO 9001:2008, ISO technical committee, Available from: http://www.iso.org/iso/iso_9001_2008
12. ISO, (September 2010). *Attachment D to Planning Procedure 4*, September 17, Available from: http://www.iso-ne.com/rules_proceds/isone_plan/pp4_0_attachment_d.pdf
13. Johnson, N.L. & Kotz, S. (1999). Non-Smooth Sailing or Triangular Distributions Revisited After Some 50 Years. In: *Journal of the Royal Statistical Society Series D - The Statistician*, No. 48, pp. 179–187
14. Johnson, R. (n.d.). *Probability and Statistics for Engineers*, Department of Statistics, University of Wisconsin, Madison, Available from: http://www.stat.wisc.edu/courses/st311- rich/Ch1-2.pdf
15. Keaveney, S. & Conboy, K. (2011). *Cost Estimation in Agile Development Projects*, National University of Ireland, Galway, Ireland
16. Keefer, D.L. & Verdini, W.A. (1993). Better Estimation of PERT Activity Time Parameters. In:

17. *Management Science*, Vol. 39, No. 9, pp. 1086–1091

18. LHS, (n.d.). *AP Statistics Solutions to Packet 1, Exploring Data: Displaying Distributions with Graphs Describing Distributions with Numbers*, Lakeridge High School, Available from:http://lhs.loswego.k12.or.us/z- abbottd/Stats/Homework%20Solution%20Packets/soln1.pdf

19. Lederer, A.L. & Prasad, J. (1995). Perceptual Congruence and Information Systems Cost Estimating, *ACM SIGCPR Conference on Supporting Teams, Groups, and Learning Inside and Outside the IS Function Reinventing IS*, Nashville, Tennessee.

20. Lock, D. (2000). *Project Management* (in Romanian), Codecs Publishing House, Bucharest, ISBN 973-8060-25-7

21. Mangino, J. (2011). Chapter 8 - Quality Assurance and Quality Control (QA/QC), In: *IPCC Good Practice Guidance and Uncertainty Management in National Greenhouse Gas Inventories*, C.M. Lòpez Cabrera and L.A. Meyer, (Ed.), pp. 8.1-8.17, Available from: http://www.ipcc-nggip.iges.or.jp/public/gp/english/8_QA-QC.pdf

22. McDonald, R. (1997). Normal-Ogive Multidimensional Model, In: *Handbook of Modern Item- Response Theory*, W. Van der Linden and R. Hambleton, (Ed.), pp. 257–269. New York: Springer.

23. Mohan, S., Gopalakrishnan, M., Balasubramanian, H. & Chandrashekar, A. (2007). A Lognormal Approximation of Activity Duration in PERT Using Two Time Estimates, In: *Journal of the Operational Research Society*, No. 58, pp. 827–831

24. Naughton, E. (n.d.). *Improving Project Performance through Risk Management*, Institute of Project Management, Available from: http://www.projectmanagement.ie/images/assets/pdf/

25. Nedelcu A. & Dumitrascu D. (2010). Some Aspects Regarding Quality Management in Industrial Projects, *Proceedings of 7th WSEAS International Conference on Engineering Education (EDUCATION '10), International Conference on Education and Educational Technologies*, pp. 263-266, ISBN: 978-960-474-202-8, Corfu, Greece, July 22-24, 2010

26. Niskanen, S. (May 2009). *Development of an Open Source Model Rocket Simulation Software*, Master's thesis, Helsinki University of Technology, Faculty of Information and Natural Sciences, 05.20.2009, Available from: http://openrocket.sourceforge.net/thesis.pdf

27. PDPM, (December 2007). Chapter 20 - Project Development Cost Estimates, In: *Project Development Procedures Manual*, 12/15/2007, Available from: http://www.dot.ca.gov/hq/oppd/pdpm/chap_pdf/chapt20.pdf

28. PMD, (2008). *Project Management for Development Organizations - A methodology to manage development projects for international humanitarian assistance and relief organizations*, Available from: www.pm4dev.com

29. PMI, (2004). *A Guide to the Project Management Body of Knowledge - Third Edition*, Project Management Institute, an American National Standard ANSI/PMI 99-001-2004, Newtown Square, USA

30. PMI, (2008). *A Guide to the Project Management. Body of Knowledge - Fourth Edition*, Project Management Institute, Available from: http://pmi.org
31. Powell, T.C. (1995). Total Quality Management into the Projects, In: *Strategic Management Journal*, Vol. 16, No. 1, pp. 15-37, John Wiley & Sons
32. Stamelos, I. & Angelis, L. (2001). Managing Uncertainty in Project Portfolio Cost Estimation, In: *Information and Software Technology*, No. 43, pp. 759-768
33. Wideman, R.M. (October 2002). *Project Quality Management*, Project Management Consulting, AEW Services, October 2002, Available from: www.maxwideman.com WKMOG, (June 2008). *Report of the Workshop on Maturity Ogive Estimation for Stock*
34. *Assessment (WKMOG)*, 3-6 June, ICES WKMOG report 2008, pp. 33-72, Lisbon, Portugal

4

What Quality Management Allied to Information Can Do for Occupational Safety and Health

1. Introduction

What quality management and scientific information can do for an organization will hardly depend on how it is applied in the workplace. Excellence seeking is an inherent practice to most organizations regardless its branch of activity. Loss prevention and market gains are usually presented as a justification for that. In consequence, quality management is closely linked to capital gain, and that is why enterprises apply high investments in this field. What is not explicit in this scenery is that all of this is also linked to scientific information and its correct dissemination. On the other hand, it is not a secret that market share, competitive advantage and profit are directly proportional to how informed one organization's leaders are about their business. In this context, it is pertinent to say that information is money, especially considering its high costs. It is also not a secret that information generates information. It means that the more widespread, more refined information gets, according to the spiral of knowledge that will be discussed ahead. So, the major input that can culminate in findings that may contribute for Science advancement is scientific information. However, two of several potential impediments to accessing and exploiting this vast field are resource scarceness and high costs of information access, and the scenery is the same in occupational safety and health (OSH) field. The barriers faced by OSH professionals during the information retrieval process, especially those derived from scientific experiments, generate the need for reflection on ways of promoting the intellectual capital turnover among professionals and this encourages the development of research instruments that enable the access to scientific information by the OSH community.

On this way, the Internet contributes significantly, whereas the virtual space favors fluidity, agility and possible reduction of communication noise between researchers, due to the possibility of real time communication, and that promotes the knowledge growing and also the improvement of security practices in the workplace. The significant increase in the volume of publications disseminated in the Internet without any criteria of quality and reliability of data is the result of the internet popularization, which contributed to the appearance of barriers in the selection of relevance information, and in certification of the legitimacy of presented data, particularly in OSH, in which information is scarce in relation to other areas of human knowledge.

An alternative to minimize the impacts caused by noise in information flowing is the creation of specialized virtual environments, called open archives (OA), which enable direction of the information to a target audience, promote break of bureaucratic practices in processes of documents publishing, provide interoperability between systems and communication in the scientific community besides promoting multidimensional ways of accessing studies made available by such routes.

Therein, the utilization of digital repositories of information under the philosophy of information spreading applied to the open archives initiative may be considered as an alternative to promote the intellectual capital turning as well as scientific communication among researchers. The initiative in question is presented as a strategy to support management of health and safety at work, together with quality management tools PDCA and 5S.

Among the barriers to information access it is important to highlight the linguistic obstacles that may difficult or even make inaccessible the consult to relevant studies, generally published in expressive international journals. These factors justify the promotion of technical and scientific exchange in order to form a professional experiences exchange network, that may assist OSH professionals in achieving satisfactory results on reversing the current degrading situation of working conditions, that are still found in several branches of professional acting.

Information is crucial to the effectiveness of OSH management system, however, in many situations, the information necessary for effective interventions are not available, restricted, for example, by temporal contingency. [...] The evaluation of the OSH management system is a systematic examination of its functionality and effectiveness in order to produce consistent information that may assist in decision making processes for the conception of new strategies to promote the improvement of OSH performance [1].

Engineering, Medicine, Law. All of these and the other fields of human knowledge exist because of information. It is also true that professional acting in any area offer occupational hazards to workers. Some more intense, some less so, but the fact is that occupational safety and health must be the basis of any professional activity, be at the office, or in heavy industry, so that the worker physical and mental integrity is always preserved.

Basically, two kind of information concern to OSH: prevention to accidents and occupational diseases and rehabilitation of occupationally sick or injured workers. The first one has proactive characteristics, while the second one seems to be reactive. An active posture allows the company to preview and prevent negative events, like an occupational accident, for example, and by means of this, save money or decrease spending, considering the expenses that the organization may no longer employ in professional rehabilitation. This means that the costs used in the prevention of occupational accidents and diseases should be treated as investments rather than as expenses.

Once the organization have identified the risks and opportunities in a pro-active way, the manager enables the tools to prevent or mitigate risks, maximize opportunities and reduce or eradicate harmful events to the employee and to the organization as a whole in consequence.

The burden of adopting a reactive posture is to expose the company to a totally vulnerable situation, in which it may not have enough time to any reaction in an emergency situation, staying at the mercy of events.

Scientific and technical activities are the fountain from where flow technical and scientific knowledge that will be transformed, after registered, into technical and scientific information. But, on reverse, these activities only exist, and become real against this information. Information is the Science sap. Without information, Science can not develop and live. Without information research would be useless and there would not be the knowledge [2].

Manage means to carry on business affairs, finance and human resources administration, what also implies to watch out for occupational health and safety. The purpose is common to all: to lead a particular activity or a group of them, to the achievement of one or more targets, which may vary according to the organization goals. Therefore, a range of compliances should be reached, and in case it does not occur, losses certainly will come, in most situations, with dimensions and impacts may vary from case to case. These factors highlight the importance of management and know-how to take advantage of all available resources, always keeping in mind the organization goals to perform tasks with the best coefficient of cost-effectiveness relationship. Management has the function to "build organization that work" [3], and to work, all sectors must maintain a synergy, so, one single action has the strength to move a whole, like a set of gears. By the way, this is the recipe of success which is unknown by so many companies. It is important that workers feel important to the organization and that their work has a real value to the company. As a returning, the worker would receive determination, and motivation, not only merely driven by obligation or interest (or duty) of economic nature. He will feel pleasure to work.

In the industry, machines and equipments receive a special treatment, with periodic reviews and replacement of components. The same way, the employee also requires some care from the organization, like the workplace adjustment and maintenance, for example. After all, the worker is the one who holds the most valuable asset of the organizations: the intellectual capital. The employee is the "keeper and guardian" of the organizational wealth, which is the knowledge. So, the same way a machine demands periodic reviews, small (or not so small) repairs, and even "vacations", the worker demands satisfactory work conditions as well, so it will be made the most of his potential efficiency. Psychological sphere certainly is very important in an occupational safety and health management plan. Considering that most part of the day of an employee is spent in his professional activities, the manager should also value the maintenance of a healthy work environment, both in physical and mental aspects. Many practices may be adopted in this sense: a good interpersonal relationship, relaxation techniques, profits sharing, regular meetings with the teamwork to identify their needs, and also the encouragement of reading and sports practicing are some valid measures for achieving a healthy work environment.

The adoption of such measures, linked do scientific communication may assist OSH professionals in developing management tools that support the employee in a large-scale way, providing better working conditions and health, both physically and mentally. In this case, information has the role to complement the knowledge of professionals who are responsible for zealous for workers' health and safety, besides giving wide publicity to the good practices adopted in this regard. Trainings, technical visits, lectures, simulations and other forms of professional improvement are also widely successfully applied. Added to appropriate information flowing, and also the valorization and respect for the worker, which may be made visible even in supplying adequate individual and collective protection equipments, the workplace may become a place that provides pleasant and well-being that may also contribute to occupational accidents and diseases prevention.

Occupational accidents, as well as any other accidents are unwanted events, unforeseen and which are generally accompanied by physical and / or material damages. Avoiding accidents and occupational diseases is the major concern of OSH professionals, and also a duty for any citizen. Appearing of occupational diseases, due to noise, high temperatures, radiation, exposure to chemical substances, and so on, is one of the greatest concern to hygienists [4]. An effective prevention program results in declining of occupational accidents and diseases rates and also declining of repair and consequence(s) of a possible accident [5].

This way, information sharing is crucial. A seemingly routine situation for one organization may be considered a risk task for one company which may already have had the unpleasant experience of an occupational accident as a result of that activity. Information like this should be as most widespread as possible (even though the company's name is preserved) to prevent other similar accidents from happening in other plants, or other organizations. However, the fear of intellectual overcoming, the organizational culture or even the fear of exposing thoughts to academic criticism sometimes keep the OSH professionals from starting an occupational hazards prevention discussion with other professionals.

Generating wealth and improvements from intangible assets is precisely the task of knowledge management, which is defined as a "new way of working, a new organizational culture, in which the environment and values allow to generate the necessary motivation for learning, sharing or transfer and application of knowledge [6]." In proportion to the knowledge becomes an essential and strategic asset, organizational success depends more and more on increasing the company's ability to produce, gather, store and disseminating knowledge [7]. Thus, it falls to the company to promote knowledge sharing among the work team, and consequently the intellectual capital turning within the limits of their own plant. This point by itself is already a plausible justification for building an institutional repository of scientific information. Managing knowledge in organizations is to provide conditions for this knowledge to be constantly produced, codified and shared across the enterprise. Facilitate the interactive knowledge flowing in the organization, adding value to information and distributing them correspond to the role of knowledge management, transforming knowledge into competitive advantage [8].

However, for knowledge to be effectively converted into competitive advantage, it must be subject to application and transmission in the company. This conversion takes place through the transformation of tacit knowledge into explicit knowledge and tacit knowledge again and so successively until the results are tangible enough to satisfy, or preferably exceed the organization needs and expectations.

Tacit knowledge is acquired individually by each person, through courses, readings, lectures, or even the professional experience. The access to this knowledge depends on some factors, including the potential ability of explanation of the person who holds it. Another factor is the "ways of doing things": each person develops their own way of working, aiming to make it easier and more comfortable. However, such procedures and techniques generally are not found in books, journals, films, or any other type of informational support, because this kind of knowledge only can be acquired through professional practice and experience.

On the other hand, explicit knowledge is available everywhere, to whom it may concern. So, explicit knowledge is more perceptible than tacit knowledge in an organizational environment. That is the case of books, journals, newspapers, magazines and so on. Explicit knowledge is what one can transmit in a formal and systematic language. It is the knowledge that can be registered in books, guides, or be transmitted by electronic mail or hardcopy [9]. In fact, information in general corresponds to explicit knowledge. Implicit knowledge is a more recent knowledge that serves to describe a knowledge that, though it has not been documented, it is likely to be. It is the knowledge one keeps, and is able of transmitting be in a more or less assisted way. Indeed this is the knowledge that can be made explicit, but has not been yet. For example: the way between one's home and his workplace is registered anywhere, but this person can draw a map or explain someone how to get there. This knowledge is implicit: it is not registered, but it can be, if there is one's will to do so. On counterpart, tacit knowledge the one that everyone keep, and have no conscience of that. It is personal, acquired through practice, experience, mistakes and successes, it is difficult to be formulated and transmitted in a formal way. A cake recipe (explicit knowledge), for example, can generate totally different results, depending on the experience and sensibility of the person who makes it. That happens because of the act of making a cake also involves a tacit knowledge, which is personal.

The conversion of tacit knowledge into explicit and vice versa, as in a cycle, promotes besides the human capital circulation and socialization, also the opportunity of renewal, acquisition of new knowledge, and promotes the improvement of the whole set of information already gathered. This information exchange cycle provides an intellectual capital spin, which encourage the socialization of generated knowledge, as well as the investigation and the reflection towards new knowledge creation [10]. The authors propose a model of knowledge conversion, what became known as knowledge spiral.

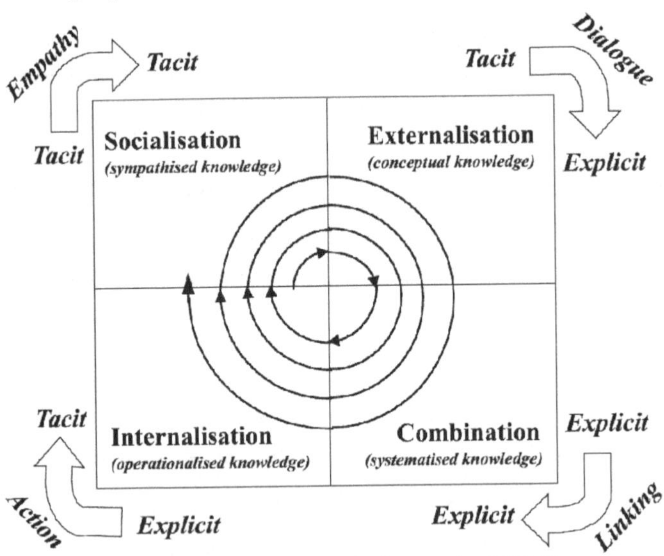

Figure 1. The knowledge spiral [10]

According to the model proposed by the authors, the process of acquiring knowledge passes through four stages:

- Socialization (tacit knowledge to tacit knowledge):

Observation and also the practice are ways of transmitting knowledge through socialization. In this process, there is much experience exchange, what is the big shot to knowledge acquisition. Trainings, brainstorming, and even one's own professional experience can be mentioned as examples of knowledge socialization.

❖ Externalization (tacit knowledge to explicit knowledge):

Externalization is the highest point of knowledge transmission. It is when occurs the encoding of tacit knowledge into explicit knowledge, so that other professionals can use such information, and convert them back into tacit knowledge. One of the tacit knowledge characteristics is precisely the verbalization high difficult degree. Scientific papers publicizing is a way to facilitate the transmission of this kind of knowledge.

Combination (explicit knowledge to explicit knowledge):

New ideas, new question and projects are designed in this stage. Explicit knowledge is confronted with another explicit knowledge. The result of that is the development of new information. Combination is very present in meetings, discussions, be in person, or by Internet, phone call, or even during a report, or even any other document reading.

❖ Internalization (explicit knowledge to tacit knowledge):

The conversion of explicit knowledge again into tacit knowledge occurs during internalization process. This is when acquired information is absorbed and right after, converted into knowledge.

Each time the cycle is completed, it should be restarted. Each restart is a higher degree of information complexity, specificity and consistency. Scope is also amplified as well, and may start with a single person and after expand across boundary of people, departments, sectors, units and companies. And information gets more and more improved and specific, until its holder becomes an expert in such question. To enable creation of organizational knowledge, tacit knowledge accumulated needs to be socialized with other members of the organization, and that gets another spiral of knowledge started again, as in a cycle with no end [8].

knowledge creation is divided in five main phases:

❖ The sharing of tacit knowledge;

❖ The creation of concepts;

❖ The justification of the concepts;

❖ The construction of an archetype and;

❖ The spread of interactive knowledge [10].

At the phase of tacit knowledge sharing, occurs the socialization, when information is shared through out the company. The creation of concepts corresponds to the phase in which there is conversion of tacit knowledge into explicit knowledge, as a new concept (externalization). In the third phase, justification of the concepts, the company evaluates and decides whether the new concept proceeds or not, and if there is continuity, archetypes are constructed and may generate new prototypes of products services and so on. The last phase, which corresponds to the interactive dissemination of knowledge, is responsible for systematic sharing of applicable information, in extra organizational sphere. This phase is extremely important for OSH management once this field of study focus is to promote welfare and protect worker's physical and mental integrity, while performing their work tasks. The capital gain and the competitive advantage originated from this process, although are both desired and very well seen, are consequences of the knowledge spiral cycle completion. The interaction of one company not only with the other ones, but also with the academic community, may bring great benefits in the sense of human capital acquiring, whereas active professionals in the labor market not always have the opportunity to research and publish new knowledge. So, the scientific community which traditionally carries out this task has much to contribute to the dissemination of research results, and by consequence, to the OSH management inside the companies, reaching, this way, until the shop floor, which can be considered the apple of OSH's eyes.

Any shop floor workers' tacit and explicit knowledge that the company makes use is very healthy to the organization as a whole. Even though experts develop great theories, the more accurate knowledge is held by the employee, which is the one who knows each single part of the processes in depth. As time passes by, working activity becomes an automatic process for a person, due to one's high intimacy level with the activity in question, and this relationship creates bonds that only the familiarity and the professional practice can establish. The "best and more efficient way of doing things" only can be indicated by someone who realizes it frequently.

Facing this, the interaction between the company and the employee is indispensable for the organizational learning to put foward tangible and satisfactory results, though it is also required the employee engagement, in order to be open to new experiences, new information and so on.

Relations with other organizations in the ambit of OSH can increase the spin of the spiral of knowledge, what can bring benefits for all involved in this process, because in contradiction to mathematical laws, knowledge gets multiplied when shared, and may become applicable to other areas like management risk, for example. Once there is a failure that may lead to an occupational accident, it is not enough only to fix it, but it is also necessary to warn the whole working team about the occurrence, for it not to be recurrence, avoiding any occupational accident, regardless its proportions. From the moment that there is the transmission of this kind of information, it becomes part of the knowledge of each single person of the working team, and that makes the likelihood of repeating the same incidence decrease. Through knowledge, companies become more effective and efficient in the use of their scarce resources. On the other hand, without knowledge they become less efficient and effective using resources, and fail, at last [7].

One of the factors that may contribute to accident preventions in an industrial plant is adopting a good quality management system. In contrast, quality is not put in practice without access to information. The efficiency is nothing more than the top of a set of information, arising very often from unwanted incidents, which are responsible for the generation of lessons learned, applied to a specific process, providing satisfactory results in order to improve products and processes. Considering these points it is interesting to stimulate investments in quality management and knowledge management, which is responsible for bringing up the tacit knowledge and experience of the workers, who usually hide valuable data for management in general. This fusion, in fact can bring great benefits to the organization in its various aspects.

3. The benefits of the fusion between quality management tools and knowledge management

A quality management tool is a methodology to improve processes using as less financial resources as possible, optimizing the working processes, increasing production, and also profit, by rational use of space, time and material and human resources.

Quality management tools are widely applied in organizational processes, and generally provide positive results. After all, quality is a crucial factor for optimization of processes, resources using, and the entire operation of an organization. The knowledge management benefits may be enhanced through the use of the resources suggested by instruments of quality management.

3.1. Organizational knowledge management and PDCA cycle

Organizational learning is a prerequisite for an organization existence. Or it is an organization that learns, or it will be condemned to failure soon. Considering the specificity of OSH field, the professional responsibility of monitoring the evolution in the industry becomes even greater and must be observed by each one involved in this knowledge area. However, for a company to learn it is necessary its employees learn first. In this sense, it is interesting to create policies to incentive the human capital acquisition aiming the organizational growth and improvement of processes in a larger sphere. The policy to stimulate education and multiply knowledge is important for the organization and in most cases, spending on research and development are usually recovered at short term, through improvements achieved by this process. This justifies high investments that developed countries apply in research development. So, a permanent cycle of knowledge acquirement provides to the organization the potential to achieve better results, exceeding expectations and achieving goals. The tasks improvement and staff expertise are merely consequences in this scenery.

One of the tools applied in this sense is the PDCA cycle, defined as method of managing for processes or systems. That's the way of achieving the goals assigned to the products or enterprise systems [11]. The word goal is the result of a junction between meta + hodos, that means a path to the goal. Thus, it is possible to affirm that the PDCA cycle is a way to achieve a goal [11, 12].

As the organization itself, all knowledge in any field of knowledge is like a living organism, constantly in change. Advances in Science and technology has created an scenery in which there is a need to create new ways of doing work, new strategies, new work tools and/or adapt existing ones to the new reality imposed by XXI Century. Each one of these items should function orderly among themselves, like a gear. So, all the processes of the larger organism, the organization as a whole, are able to flow correctly. Therefore, it is necessary to domain of some essential data for the proper functioning of this system. These data are often not found in books, or in the scientific literature, but in each employee's own experience, the tacit knowledge. In this sense, knowledge management assumes the role of contributing for each employee's intellectual qualifications to be always aligned to the mission and vision of the company which must include the promotion of safe work, and therefore, the occupational accident prevention.

Dissociating theory from practice is impossible, whatever is the activity about to be developed. Theory, due to hard studies, is the main responsible for the facilities and conveniences that can be observed specially on factory floor, as the machine protection and tasks automation, for example. Therefore, as business management, management in OSH knowledge should also be part of a continuous improvement process, like PDCA.

Conceived by Walter A. Shewhart in the 1930s at Bell Laboratories in the United States [12] and applied two decades later by the quality expert W. Edwards Deming (that's why PDCA cycle is also known as Deming Cycle) in statistics and sampling methods using, the PDCA method is recognized as a process of improvement and controlling that should be familiar to the employees of an organization to be effective [13, 14] CENTRO DE TECNOLOGIA DE EDIFICAÇÕES, 1994). Basically the PDCA cycle can be defined as

A method that aims to control and achieve effective and reliable results in the activities of an organization. It is an efficient way to provide processes improvement. It standardizes quality control information, avoids logical errors in its analysis and makes information easier to understand. It also may be used to facilitate the transition to a management style directed to continuous improvement [15].

The procedure proposed by PDCA implies constant evaluation of the entire system, enabling the early detection of possible failures or improvement points. This ideology makes necessary to conduct detailed audits at all stages of work. Each time the proposed cycle (plan, do, check and act) is completed, occurs the continuous improvement and then, improvement of knowledge of those involved with the activity in question, by consequence. In fact, the PDCA cycle is a total quality management method, which may (and should) also include the prevention of occupational accidents and diseases.

Whereas the world is constantly evolving, even activities performed satisfactorily are subject to improvements that must be pursued as incessant searching process for an unreachable perfection. That is called continuous improvement. Whenever it does not occur, the consequences are certain: occupational accidents are potentialized along with their indices of severity, losses and damages.

Occupational accidents do not happen by chance. They are just a materialization of a sequence of failures, whether human or not (although machine or mechanical failures also can occur due to human errors) which in most cases could have been avoided.

It is common for us to accept as true when the newspaper disclose, even before the experts report conclusion, that the origin of a fire was a short circuit in electrical installations. It is such an often cause, which examples are innumerable, that not even is necessary to wait a long time for the confirmation that was foreseen by journalists from interviewing experts and testimonials of the users of the hit building. [...] What is not always demonstrated for the population is that the referred short circuit was not a surprise occasioned by an uncontrollable phenomenon, but the last breath of an installation that was agonizing. Fires, originated by electrical causes start, general rule, in overloaded circuits, without proper protection and maintenance, and handled by curious laymen covered up by "economic professionals" that seek by "the lowest price" the technical solutions for problems that should have been the cause for specialized diagnosis [16].

Failures, that not always culminate in accidents, but in near-misses, may be product of several factors, among which the lack (or insufficiency) of safety audits and knowledge domain by operators responsible for the activity that has any kind of inconsistency. The set of failures, combined with temporal factors may culminate in industrial accidents. Therefore, improvement must always be pursued by the organization, regardless its maturity level or any other factor.

The cycle proposed by PDCA requires constant evaluation of the whole system, in order to detect any possible failure, in anticipation of the occurrence of inconvenient and undesirable situations. Under this ideology, it is made necessary to conduct accurate audits at each single stage of the work. So, it will always be found new (and better) ways of doing the same things in a more efficient and safety way. Following the cycle Plan, Do, Check and Act (PDCA) is a way of promoting continuous improvement and thus, improves also the knowledge of everyone involved with the activity in question. Moreover, PDCA cycle is a method of total quality management, what also includes occupational safety and health management.

So, PDCA cycle is basically composed by four phases [12]:

P ⇒ Plan

D ⇒ Do

C ⇒ Check

A ⇒ Act

PDCA – P: Plan

The planning phase comprises the step in which all activities are outlined having the company's mission and vision as a point guide, besides other documents that may concern, like standards for example. Planning should point all the parameters of the activity, how it should be executed, financial, material and human needed resources, deadlines and all the information about the characteristics of the planned action, as in a project itself.

It is also in planning phase that are often identified needs hidden by the routine or even by convenience. Goals and methods of reaching them are all defined in this phase. Basically planning phase consists on the development and delineation of the project scope itself. That is why this is the proper moment to identify organizational weaknesses and strengths and develop strategies to convert weaknesses into strengths and strengths into points of excellence.

PDCA – D: Do

To do is to put into practice all that was planned in planning phase (P). It is like preparing a cake (do phase) following the instructions of a recipe (planning phase). This is the moment for training and educating employees, besides implementing all practices according to the project scope. Simultaneously, it can be a pre-audit, when positive and negative points of the plan should be identified and registered as a way of ensuring that all objectives previously established will be achieved in its fullness. This information also will be useful during the phase of checking the implemented actions.

PDCA – C: Check

Checking is the phase which most involves indicators and performance metrics, and comprises the comparison between planned and achieved results. Audits, processes analysis, evaluations, and satisfaction surveys are very common at this stage of the PDCA cycle. No inconsistency founded means the satisfactory following of the cycle. Otherwise, it is time to start the next phase: Act.

PDCA – A: Act

If the results are as expected, nothing should be done until the moment that any inconsistency is detected, when it is necessary to perform corrective measures to bring the project back to the scope initially established. However, besides implementing corrective actions, acting phase also may be in order to standardize processes that worked well, as a manner of registering information that can be used as stepping stone in similar situations. This phase corresponds to the PDCA cycle completion.

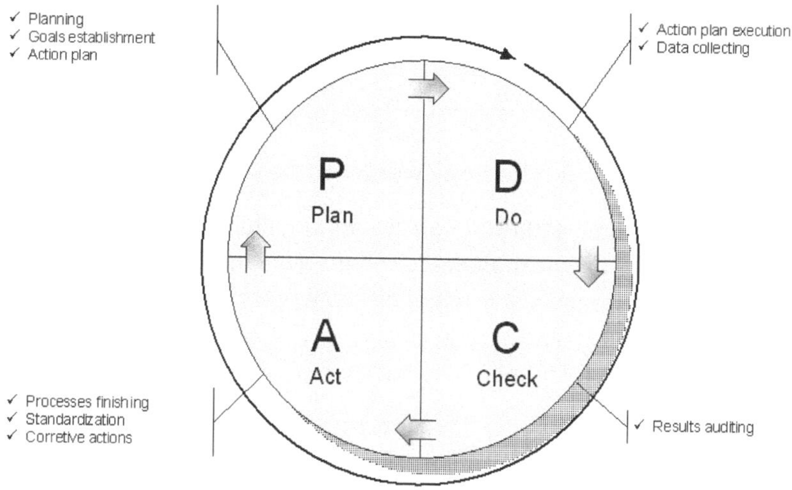

Figure 2. The PDCA cycle [12]

Once all those phases are concluded, PDCA has completed its cycle, which must rotate clockwise, following the sequence of the proposed activities. The most the cycle is completed, the greater the knowledge accumulated by the whole teamwork, in the same way that the rates of quality also will tend to be increasingly towards to excellence.

PDCA cycle is designed to be a dynamic model. The completion of a cycle will flow back in the beginning of the next cycle, and so on. Followed by the spirit of continuous quality improvement, the whole process can always be reviewed and a new process of change can be started [12].

The optimization of processes, reached through the PDCA cycle, promotes a reduction in costs and increasing productivity, what means creating new products, with less use of resources, which yields competitive, once production becomes higher than the competitors'. Likewise, the rate of complexity of information is growing each time the PDCA is fully completed. Each time the process is completed, the involved processes presents indices of improvements, which are represented by the numbers 1, 2, 3 and 4, according to figure 3. The closer to the top of the ramp are the improved processes, the better will be the indexes of maturity of the organization. Andrade (2003) points out that the lessons learned in a PDCA cycle application can be used in a second, third, fourth application (...) that can be more complex and daring, and so on.

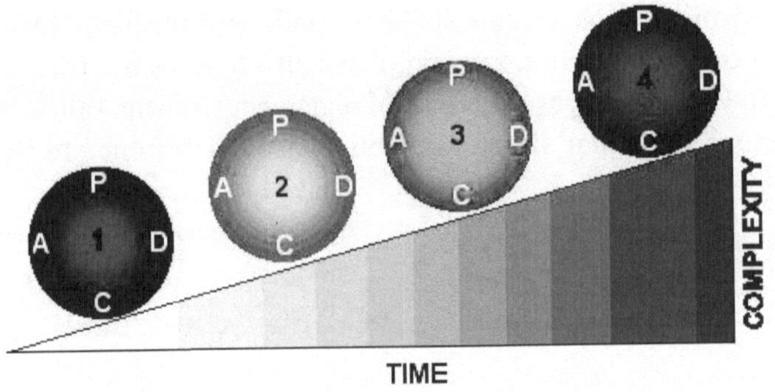

Figure 3. The improvement ramp [17]

The roll of information gathered can be a valuable instrument for the OSH management community, for the moment it is made available for public consultation as a data source of best practices, it may help managers to think globally and act locally. This way, the healthy and safe workplace will not just consolidate into one organization, but in all Brazilian industrial fabric.

There are several instruments for improving the quality of products and processes within an organization. Another instrument that can also be used together with PDCA to optimize the quality of work live is 5S.

3.2. Quality management according to the 5S process

Designed in Japan in 1950 by Kaoru Ishikawa [18], 5S is a method of quality management which aimed to delineate efficient and effective techniques to reach costs reduction, optimization of material, technological and human resources, and also combat waste [19]. It is an educational process, which is often applied as a basis for other management instruments.

The 5S program is a philosophy of work that aims to promote the discipline in the company through awareness and responsibility all to become the workplace a pleasing, safe and productive place.

The Program was named after five Japanese words that synthesize the five steps of the program. They are: Seiri, Seiton, Seisso, Seiketsu, Shitsuke [20].

The goals are to transform the environment of organization and the attitude of people, improving the quality of life of employees, and also reduce costs, waste and raise institutional productivity [18].

The 5S method may imply the improvement of other functions, not directly related in this process, such as production (acquisition of new equipments and adoption of new processes and systems), quality (standards application), human resources (adoption of new organizational policies), as well as hygiene, safety and environment (changes in management process, audits). For this reason, as well as any other innovation within a company, the implementation of 5S requires a detailed analysis on the budget funds to certify that financial aspects will not be an obstacle for the complete implementation of this instrument.

Another point to focus on is the possibility to be applicated simultaneously to other quality management tools, such as the Kaizen method of continuous improvement, safety, environment and quality management tools (NBR ISO 9001, NBR ISO 14001 and OHSAS 18001), Six Sigma, as well as PDCA Cycle itself. It demonstrates that 5S method is an extremely flexible tool that can be combined to other management instruments, what potentizes the results that can be reached by this method.

So, 5S acts in three organizational aspects, which are divided into 5 phases: physical aspects (Seiri, Seiton, Sciso), standardization aspects (Seiketsu) and behavioral aspects (Shitsuke). Each one of the five phases of the method is represented by a sense as follows [21]:

3.2.1. Sense of use and disposal (Seiri)

The sense of use and disposal awakens the workers consciousness to keep in the workplace only what is really necessary for the development of their functions. This provides to the worker a clearer, more accessible and easier to clean workplace, and increase visibility to the material actually used. Furthermore, this practice promotes the evacuation of areas previously occupied by unnecessary materials. Thus there may be an increase in productivity, reduction on bureaucracy processes, and prevention of unnecessary purchases.

The methods applied for materials disposal can be diverse: once separated all the unnecessary material, this can be transferred to other sectors, or returned to the warehouse, sold or donated for recycling entities, or even be disposed as trash, that also should be classified as common or toxic.

3.2.2. Sense of ordering and arrangement (seiton)

For each material, its proper place. The ordering sense recommends that once separated all the useless material, the remaining should be sorted in the workplace, preferably in the closest place to where its use is more frequent, location to which it should be returned as soon as it is not being used anymore. This provides benefits, for example the rationalization of the space designated to material storage, the encouragement of creativity, quickness, and facility in finding documents and objects. This reduces physical efforts and emotional distress in searching for not founded objects, besides preventing occupational accidents that possibly may be caused by a disordered workplace.

3.2.3. Cleaning sense (seisso)

Machine, equipment and any work tool including the workplace itself, must be kept clean and organized. The duty of maintaining hygiene in the workplace is not the sole responsibility of the cleaning crew, but also an obligation of each one of the employees. Cleaning makes the environment becomes more pleasant, and transmits the image of excellence in organization's products. And again, a cleaning workplace is less susceptible to the occurrences of occupational accidents, for example, a fall caused by a slipping way.

3.2.4. Sense of health and hygiene (seiketsu)

Personal hygiene is also essential as well as mental health and life quality in general. The good maintenance of employee's health is like an organizational intangible asset, as a prerequisite for the execution of any task in the company. However, this sense requires an holistic view that permits the manager to extend the meaning of health and hygiene also for the issue of OSS. The ergonomic adequacy of the workplace, the adoption of relaxation techniques, and even the aesthetic and disposal of the workplace are factors that must be considered in this way. As results, it can be achieved higher levels of team motivation, facilities in human relationships, dissemination of the positive image of the sector and also its employees, favorable working conditions to health, and also the occupational accidents and diseases prevention.

3.2.5. Sense of self-discipline (shitsuke)

The last and most individual sense is the sense of self-discipline. This corresponds to the regulator sense, which creates in a company's workers the awareness for the accomplishment of the four previous senses. This provides a natural implementation of all recommendations, the discipline, the moral and ethics, the cultivation of good habits, the promotion of participative management and also the ensuring of the quality of life at work.

The 5S program implementation may provide more efficiency and safety at workplace. The disorganization and lack of asepsis may be considered factors that may culminate in occupational accidents and diseases. The waste generates financial losses. The sense of using and ordering allow a better use of space, and finally, self-discipline, as the name suggests, promotes the discipline among the employees toward a policy of awareness in the use of the company resources. And what is more interesting in this process is that its cost of implementation is considered relatively low.

In fact, 5S may be integrant part of PDCA cycle. Both focus on quality management: 5S, aims to improve the quality of the processes and workplaces, while PDCA focus on strategical and managerial issues, although there is no restrictions for PDCA to cover some areas of 5S.

Both PDCA and 5S are quality management tools which may provide great benefits to an organization. It is also pertinent to point that, the implementation and management of both tools can be much more complex than it seems in this discussion. However, it is also important to consider that discussing these tools in a deeper point of view is not the primary objective of this discussion. These quality management tools were chosen and mentioned because they were considered useful in the sense of promoting the generation of information that may be applied in other workplaces or even other enterprises, through the association with a system of scientific information sharing among all of them. The management of safety and health at work can also be considered a quality management process, once it cares about quality of life at work, quality of safety processes, quality in the workplace, and so on. That does justify the association between quality management and OSH.

4. The fusion between cycle PDCA + 5S = Organizational benefits?

Increasing the quality and productivity while promoting quality of life, simultaneously with competitiveness: that is the great challenge of the manager / entrepreneur of the XXI Century. Whatever are the tools used to improve the quality of the work process within industries, there will be tangible benefits for OSH management.

An OSH quality management process may improve the systems applied in the worker safety and health:

As the organization adopts the audit protocol over time, it has the opportunity to experience the evolution of the performance of the OSH management system as well as the opportunities for improvement.

As it knows the opportunities of improvement, it becomes possible to develop an action plan focused on more expressive and efficient results, continuously improving the performance of the OSH management system [22].

Whatever are the instruments used to improve the quality of the work process within the industries, there will be tangible benefits to the OSH management. Equally, information systems also have their importance in this issue, since they are responsible for storage and make available case studies, which may be useful in identifying diverse points susceptible to improvements in organizational processes as a whole. Case studies, in particular, play a fundamental role in order to disseminate best practices applied by one organization, so other organizations also may be inspired in these cases to promote process improvement and quality of life at the workplace. Nevertheless, they have the potential to get the OSH professionals together, and promote the ongoing discussion about emerging issues in order to promote the prevention of occupational accidents and diseases.

The knowledge is available, the techniques, methods and tools exist and their use has proven results. At the same time, those who study the accident issue in a deeper way probably concluded that the best results only can be obtained from the holistic vision, both of the issue, and specially of the man. Innovating is a need! Dare to actually move toward a more efficient accident prevention which, at last, means to save lives, suffering, reduce costs, increase quality and productivity [23].

Information systems also have their importance in this regard, once several case studies are deposited and published there. These papers are valid to assist in identifying issues for the further improvements in organizational processes as a whole. Information management also is associated to knowledge to management quality [24].

In times when the quality, competitiveness and transparency are considered on determining an organization market share, there is too much to be explored about knowledge management inside the organizations, once explicit knowledge represents a very small portion of what the organization knowledge worth actually. It is also important to invest in capital for the intellectual training of employees in managing occupational safety and health at work. The worker is the most value asset of any organization because it is him who holds the most needed resource to maintain the business "health and force": the human knowledge.

Information, quality, organizational management, OSH, financial returns. Isolated, each one has its particularity, but applied together, these instruments converge to the establishment of organizational improvement programs, which benefit not only the client, but mainly the worker, who is responsible for the final product existence. The value of a worker for an organization is incalculable. Each time a life is lost, it is not only a worker who is not part of the staff of an organization anymore, but also a member of society, whom keeps a range of knowledge, which may still have not been disseminated. In this case, besides the worker, the organization loses a valuable intangible asset: the knowledge.

Effectiveness and efficiency of organizational improvement are linked to innovation to the method applied to available information sources and to the speed and flexibility which they apply, in addition to personal motivation, leadership practiced by the direction of the organization, and the relationship between the organization and stakeholders [25]. That justifies the importance of disseminating knowledge at most. The proposal is that the OSH and knowledge management are convergent in order to ensure that the quality management as a whole, not just the final product, may present strengths to provide better working conditions for workers, being subsidized by the provision of technical-scientific data from both tacit knowledge, and other professionals experience.

The availability and use of scientific information in an organization environment are fundamental to the process natural and safe flowing. And this safety may be built jointly by several professionals, through scientific dialogue, exchange of experiences and joint reflection. Accidents will always happen but if at least one live is saved by something that has been read somewhere, every effort will have been valid [26].

The principle proposed by Maguire seals the proposed link between scientific information and OSH management, and emphasizes the need to reflect on the creation of tools that enable scientific exchange among peers. The Open Archives Initiative

(OAI) philosophy adds up this cause, in making available as much information as possible, specially the gray literature (thesis and dissertations, reports and other documents) which usually hold the results of observations, experiments and empirical reflections that may contribute to the OSH management.

5. The Internet as a source of data in occupational safety and health

Among the many emerging communication forms, the one which excels at terms of enabling the information transferring is the Internet, which is defined as a set of academic, scientific, commercial, and military interconnected networks, which communicate with each other through the use of a common protocol Transmission Control Protocol / Internet Protocol (TCP/IP) [27]. The authors highlight the facilities offered by the Internet to the scientific community like its globalization power, the immediacy of production, release, update and access to published information, the hypermedia capability (hypertext and multimedia), that makes it easy to search and access information as well as the volume of data gathered at relatively low cost, in addition to the possibility of interconnecting users, to eliminate barriers in the geographical, temporal, political, social spheres among other barriers to information transferring.

From the moment information has its value recognized, in the sense of being able to aggregate power to its holders, it became essential to generate resources to facilitate the information flowing among scientific community. The Internet has emerged in this scenery as a tool to add value for produced work, and increase visibility not only for presented results, but also to their respective authors and institutions. The creation of research groups besides the facilities offered by the virtual space, such as discussion groups and even electronic mail, narrowed the relations and facilitated access to the author of recently published works and this generates and enriches the scientific discussion, favoring the maturation of the collective intellectual thought [28].

The advent of Internet assured conditions for significant changes in ways of communicating among people, both in regard to communication media, and also the speed of data transferring.

The virtual ambient offers to the researcher agility in processes of production, evaluation, publication and validation of the essays. In printed media, the communication flows slowly, since the entire process of developing a new publication, which involves reviewing, formatting, graphics constructions and other activities related to the completion of the research are made prior to its publicizing. On the other hand, in digital environment, the processes of review and validation of works, through scientific proof, may be done jointly with the target audience of the work, accelerating the process of information transferring. Thus, the high levels of formality in contact with authors, give way to interactivity since a simply electronic mail replaces the content of an official letter, which could take several days to reach the final destination. The Internet enables immediate and direct contact with members of the community to close relations between scientific researchers.

It is true that the Internet being a network of networks with services of electronic mail and thematic discussions in group facilitates the informal communication between researchers in different areas of Science & Technology. And when one think that informal channels are essential to work at levels of great value added to the information, because they are useful in decision making and allow the creation of strategies from unpublished information, one can have an idea of the value of the communication network [28].

It is worth an interesting comparison between the virtual environment, and the typical scenario of scientific events such as symposiums, conferences, and discussions, exchanges of experiences, and other considerations that both enrich and add value to the collective intellectual thought. The scientific dialogue promoted in digital media, is like an extension of the shared moments of reflection, always present in scientific events, round tables, sessions of questions, debates etc... In this sense, the authors refer to the facilitation of informal communication, once at a conference or other event of this nature, there is the establishment of a direct dialogue between authors and the target audience of your statements. The virtual space in a way, also promotes this feature, since once that there are two computers connected to the Internet, the communication between them can be directly established at any time. A dialogue is formed since there is a transmitter who sends a message to a receiver, who decodes and process the sent message, and generates a response for the transmitter, who assumes the position of receiver at this moment, while the receiver, passes to be the transmitter. It is also true that this not always happen be due to the unavailability of the receiver, be due to human or technological failures in the moment of message sending, but this, not necessarily consists a barrier to scientific dialogue.

Even so, the Internet can be considered a way of direct communication between two or more people which may offer great benefits to the scientific community, when applied properly. By using this computer network, the scientific debates which are usually made in scientific events, are able to extend for an indefinite period, rather than being limited to the 15 or 30 minutes generally reserved for complaints in events. With this, the author-reader interaction enables the clarification of one or another aspect addressed in the work that deserves a greater detailing, creating opportunities for the improvement of qualitative indices of the work presented, since any failure or passages that deserve special attention, will certainly be raised by the community members. Thus, a particular work is no longer individual, but will incorporate the collective thought, expressing the opinion of a community on a given topic.

The consensual opinion of the scientific community, which is one of the most fundamental standards of Science, is that the works should be disseminated by their respective authors [27]. The act of publishing enables that a research result be recognized as scientific truth, through validation of theories presented [29].

Interestingly, the publication and dissemination of information are essential to the development of new studies, since previous publications may be considered as inputs for the development of new studies as a way of providing greater credibility, validate, demonstrate and emphasize the new data submitted. Thus it is formed a scientific dialogue among different authors, which is precisely what makes possible the verification of the premises presented, and encourages the generation of new studies, and instigate researchers to explore emerging topics in their respective areas of expertise.

In this context the Internet is presented as the support which receives such a dialogue, which can also be called scientific debate, giving continuity to the information cycle which is responsible for the generation and improvement of human knowledge. Just as in other areas of knowledge, in occupational safety and health the Internet is a powerful tool while helper and source of research in professional forming, or continuing education in OSH, specially in remote, poor and less assisted geographical areas.

Additionally, the option of publishing in electronic format offers various facilities, both to the author of the study, and the information units, that later will incorporate such publication to their collections, with possibility of eliminating decisive processes of production and dissemination of knowledge that includes the best way of providing, acquiring and disclosing information, how much to spend, and so on. The stages of the publishing process is generally lengthy and costly - factor of impediment to the publication of works for many writers - is broken, since the cost of publication in digital format are lower in relation to the printed publication, not to mention validation activities, which in this case are made collectively by the entire scientific community, speeding up the improvement of publication process, and covering topics since from formatting and standardization until wording and content revision.

From the viewpoint of the information user, the greater and most social visible benefits and probably the most relevant ones provided by digital publishing are:

- ❖ The possibility of simultaneous access by multiple users to the same document;
- ❖ The facilities of handling, transporting and transmitting data;
- ❖ The possibility in converting the digital media in a printed document;
- ❖ The versatility of the media;
- ❖ The possibility of breaking down temporal, geographical and linguistic barriers.

The benefits of electronic communication are numerous and surrender itself a new research. The trend that has been the cause of controversy among scholars is that the volume of information produced in analog format, (paper) decrease significantly. Many publishing companies and other organizations are already producing their journals only in electronic format, which generally corresponds to the CD-ROM or online versions. Besides the benefits discussed, the fact brought the rationalization of physical spaces for storage, cheapening of resources for conservation of material, since the digital media tends to be more resistant to decay in relation to the paper. All this, not considering that traditionally, printed periodical publications only can be accessed inside the information units dependencies, with no possibility of home loan, which can cause potential complications or even impossibility of consulting the material by the researcher. Such restrictions do not occur with electronic publishing, since its contents can be duplicated for another media, and transported for consultation in another location, without making the original document unavailable for immediate consultation in its unity of origin.

From the moment in which the publications are available exclusively in electronic format, there is an increased risk of excluding many people from the scientific discussion, considering all these people that has no access or are not familiar with the digital environment, and thus the whole purpose of facilitating access knowledge, promote scientific integration, and provide visibility to published researches is inversely reached.

A possible solution to this situation may be to provide equipment connected to the Internet in educational institutions, both public and private, with monitors for guidance on access to information. However, the fact is that even though these facilities are available, they are not sufficient to eradicate the digital exclusion. The reasons are diverse, ranging from disinterest of the public to the lack of available equipment. Considering the global trend, driven by globalization, which initiated the computerization era, it becomes necessary to establish a national policy to encourage the generation and use of electronic information, so that the country can keep up with advances worldwide, regarding to the development of information retrieval systems. On the other hand, although the scientific information does not reach 100% of the public involved, managers often have easy access to these kind of data. It is up to them so, to consult such information and transmit them to their whole staff, so that scientific language is properly converted to the worker understanding, which constitutes the focus on OSH management.

However, the fact that the information is available does not mean that it will be accessed and converted into knowledge for further practical application, when applicable, as a way of promoting the advancement of OSH, and consequently the welfare of the worker. Thus, the OSH class institutions, together with the units and information professionals, play a crucial role in order to highlight and promote the use of this vast material, which stores the inputs to the construction of knowledge, and also provides support to the professionals of the area for such actions to achieve remarkable and satisfactory results. This responsibility increases from the moment that is considered the fact that the OSH institutions are the main producers of information on this subject. So, a new necessity arises, besides an encouragement for computerization, creation and scientific knowledge communication. The investigative spirit must be something inherent to the researcher, so that he must feels the need for seeking information involuntarily, and do not expect something to emerge that bring it to him. Hence the need to develop the instinct of independence in these professionals to have greater autonomy to act in terms of information searching and scientific expression.

In this sense, knowledge management plays a key role considering the need to promote the scientific informational capital turnover allowing the maturation processes of knowledge management within organizations. The benefits offered by the Internet allow the process of information transfer to increased speed and efficiency to ensure the safety of workers more effectively, reducing the rates of occupational accidents.

According to tables 1 and 2, official statistical data from the Brazilian Ministry of Social Welfare (Ministério da Previdência Social – MPAS) indicates that from 2008 to 2010, Brazil has registered over 700.000 occupational accidents a year, resulting in more than 8.000 deaths. Important to remember that these numbers do not correspond to reality, once that underreporting is still a problem faced in the workplace. Statistics themselves, better than any other data, justify with great skill the necessity of reflecting on the working conditions not only in Brazil, but all around the world, and, above all, the need for developing effective tools in reducing these rates.

Achieving development is a virtually impossible task if at least the minimum conditions to preserve the physical and mental integrity of the worker are not observed. Therefore, work, health and safety are three components that must walk and advance together. From this perspective, the Internet can be applied as an instrument that enables and facilitate the scientific information flow and also the OSH intellectual capital turning, especially in underdeveloped regions, where the levels of exploitation and misinformation are higher.

Table 1. Number of occupational accidents registered by Brazilian Ministry of Social Welfare, from 2008 to 2010 [30]

Number of occupational accidents registered by Brazilian Ministry of Social Welfare, from 2008 to 2010														
Total number (with and without CAT registered)			With CAT** registered											
	Total with CAT registered			Cause of the accident										
		Typical accident***				Commuting Accidents****			Occupational diseases					
2008	2009	2010	2008	2009	2010	2008	2009	2010	2008	2009	2010	2008	2009	2010
755.980	733.365	701.496	551.023	534.248	525.206	441.925	424.498	414.824	88.742	90.180	94.789	20.356	19.570	15.593

Table 2. Number of occupational accidents in Brazil, by consequence, from 2008 to 2010 [30]

Number of occupational accidents in Brazil, by consequence, from 2008 to 2010							
Years	Total	Consequence					
		Medical assistance	Temporary disability			Permanent disability	Deaths
			Total	Less than 15 days	More than 15 days		
2008	774.473	105.249	653.311	317.702	335.609	13.096	2.817
2009	752.121	103.029	631.927	306.900	325.027	14.605	2.560
2010	720.128	97.069	606.250	299.928	306.322	14.097	2.712

So, the Internet can be considered an efficient tool for communication and dissemination of organizational knowledge, which assists in the process of homogenization of knowledge among employees of an organization. Information is essential to any organization, including all its workers, be from top management, be from shop floor, and emphasize the need for rapid access to data to reach the open market, competitiveness and also quality [31]. For this it is necessary to reveal the way in which information technology interact with organizations, since according to the literature, it is by this process that occurs cultural changes what includes the interlocutors behavior, which yet can involves business results, motivation and well-being of employees.

Information technology is a powerful tool to maximize the number of people reached, especially in shop floor, where can be found workers that have all the conditions to opine on a subject, resulting in new ideas, aiming to optimize ways of working and even getting better relationships among the staff. It is up to OSH professionals, to recognize the resources applied by information technology as a tool for OSH management in Brazil, so there will be a really valid and visible contribution the worker point of view.

1. It should be generated by means of empirical thinking;
2. It must be liable of understanding and proof;
3. It must be formerly communicated to scientific community.

The most common form of communicating scientific information is by written communication (publication of books, articles, essays and so on) or by oral communication (informal conversation or oral presentations in events like congresses, seminars, symposiums and so on). Which one is better in OSH field is a hard question, because it depends on several factors, including the nature and complexity of information to be transmitted. However, written media is still the easier way to access scientific information, once participation in scientific events may become unviable due to geographic limitations or even budgetary restraints. So, an alternative to overcome these barriers to information access is to create information systems designed to store, preserve, disseminate, and promote access to intellectual output of the scientific communities around the world, preferably in an open access ambient, to overcome another barrier to scientific information: its costs.

Considering the opportunity of promoting collective thinking on OSH and the conveniences offered by digital age, some reflections emerge in order to build and encourage the use of instruments that allow scientific communication flowing. Therein, the Open Archives Initiative (OAI), coupled with the proposed mechanisms for quality management, emerges as an instrument with high potential for promoting a healthy and safe work environment for workers, overcoming geographic, temporal, and possibly linguistics barriers, facilitating access and dissemination of information.

So, scientific information reliability and validation is based on three points:

6. Digital information access resources under the open archives Philosophy

Success and professional recognition are the targets of any professional, but only can be achieved against great deeds. For that, the main additive is a good quality information made timely accessible. Precisely for this reason, information has been increasingly valued and pursued by all professionals.

The society has watched to an informational explosion along the last few decades, both in qualitative and quantitative aspects. Many facts should have collaborated to it, for example, the need of publishing, to get or to maintain a job [2]. This is a common practice in the academic world. However, regardless of the reason, information is there, everywhere, to whom it may concern. Once available, information has to be communicated, for it to have its worth, and to achieve its objectives. Internet has contributed significantly for that, as an effort to facilitate the access for information, and communication processes among scientists and researchers, that is, the scientific society, defined as a social group formed by of individuals whose profession is scientific and technological research [2].

But how do these scientific communities work? By the same way that did the primitive societies: by donation system. Scientists transfer their knowledge spontaneously and gratuitously to the scientific community, without expectation of receiving any economic compensation in return [2].

However, the existence of use and production of information become useless, if whomever may need it do not know about its existence or can not retrieve it, whatever the reason. Just for this not to happen, any effort towards making information available to as many people as possible is valid. Thus, facing the scenery described, arises the incentive to free access to scientific information that

as already extensively reported in the literature, is both the result: (a) of a reaction of researchers to the business model of scientific journals commercial publishers (and their increasingly high prices for [...] signing and (b) increasing awareness of the impact caused by availability of scientific papers free from access barriers. The motto of the worldwide movement in favor of the Open Access search results, therefore, is the wide and unrestricted dissemination of the researches supported by public resources [32].

Considering that information can play the noble task of generating and disseminating knowledge, which through an empirical combination (scientific communication), may culminate in the development of increasingly effective tools for building and maintenance of healthy and safety work environments, any expenses for this purpose, either with job training (input) or with the access and dissemination of information (output) should be considered as an investment.

Facing this, the Open Archives Initiative may be applied as a tool for adding value to quality management in OSH, while providing information, including successful reports, which may be adapted and possibly applied in the workplace, aiming to promote better working conditions for workers.

Open archives (OA), which can also be considered an information quality management tool, are seen as an innovative concept designed to provide the text as quickly as possible, promoting and popularizing free access to electronic publications, so that there is a weakening of the monopoly detained by the publishers upon any scientific publication until then [33]. The OA established a set of standards that enable interoperability between different digital repositories, which [34], are defined as a form of storage for digital objects that have the ability to maintain and manage material for long periods of time and provide proper access. Becoming the main forum for dissemination of scientific results and discussions [35].

Digital repositories of information are information systems which are intended to storage, preserve, disseminate and promote access to the intellectual output of the various scientific communities around the world. The main features are the extensiveness of public access, the variety of documents regarding to form and content, including aspects of interdisciplinary and preservation of digital data stored in there, and also the storage of this kind of informational support, that dispense the formation of the traditional printed collections, that demands above all, physical spaces for the accommodation of the archived material. However, the differentiating feature of open archives is the ability to provide simultaneously, freely accessible documents, using the philosophy of self-archiving, and also more restricted documents, made available only to a select audience, such as researchers engaged in a specific educational institution, for example. This way, digital repositories may present both interfaces: one widely and opened accessible, and another one, more restrict, gathering both the philosophy of free access, and the digital repositories philosophy itself. The policy used in the process of adding new documents collection ever assembled determines the repository as open archive, or digital repository. In the first case, the name refers that the information access is completely open, allowing the authors to remove or alter documents already posted whenever they found it convenient. In digital repositories, instead, access is restricted to a specific community, such as university professors or researchers from an institution.

Since digital repositories are constructed within the philosophy of open access, the final user of this kind of information source is able to search, copy, print, and also send a document to another researcher besides the possibility of using the full text documents in other discussions, since the source and authorship is indicated. The data reliability is ensured through quality policies that establish minimum standards of quality of the archived information. A way of applying this sense is to have a specific judging commission, which should be responsible for examining the papers submitted for publication before making them available in the network.

The popularization of the Internet contributed to the depreciation of the degree of reliability of the data available in there, since anyone that dominates some basic features of information technology are able to add and/or change data on the network, inserting new documents and information that are not always reliable or truthful. The thematic digital repository under the philosophy of open archives, however, can mitigate such effects, recovering the initial principle of scientific communication in digital format, which is to facilitate the scientific information exchange between professionals.

Usually maintained by renowned institutions, thematic repositories adopting the philosophy of open archives gather in one place plenty of documents on a specific issue. That saves the researcher free time, and eliminates the need of displacement until an information unit or a library to have access to such data. Therefore, rather than the researcher looks for information, the opposite is what happens: the information comes to the table of the researcher, through a computer, bringing with it the backing of the institution responsible for its authorship.

The establishment of the open archives initiative in Brazil, may contribute to the breakdown of the barriers to access to OSH information, and also promote joint reflection between the members of the scientific community. The self-archiving, defined mechanism that allows the authors to submit or deposit their papers in a digital repository [36] besides being a way of democratizing forms of publication and access to scientific information, is also a way of promoting the continuing debate about a subject among scholars. The contribution to the construction and sharing of knowledge in this scenario becomes an automatic process, in order to contribute significantly to the evolution of science and improvement of various forms of labor relations in the OSH context.

7. OSH information management in São Paulo city, Brazil, under the professionals perception

During the year of 2008, Santos realized a study (presented as a master dissertation) in the south area of São Paulo City, in Brazil. The objective of this study was to detect the perception of the OSH professionals regarding to the availability of scientific information in occupational safety and health, and ways of accessing it, in their perception. It was also investigated the perception of the OSH professionals about how OSH scientific information and the maintenance of a healthy working environment could contribute to increase quality indicators, and consequently the organizational profits, that corresponds to the managers and stakeholders main interest.

It was randomly selected 23 organizations of the south area of São Paulo city, which OSH professionals were interviewed. 52% of the studied population consisted of safety technicists, while safety engineers consisted 21% of the total sample. The profile of the selected community was characterized predominantly male (82.6%), ranging in age from 36 to 50 years and over 10 years experience in the area.

The analysis of the answers to the 17 questions questionnaire that was applied to these professionals indicated that there is still a great job to be done to improve OSH information access and application in companies in São Paulo. It was found that the virtual environment for research is not yet fully exploited by the respondents, which is a fact that arises some concern to OSH managers, considering all the benefits of this knowledge transfer resource.

With regard to the worker welfare, the results clearly show that access to information is essential for safety processes improvement. Workers physical and mental health protection depends largely on access to the occupants of senior professionals have to the scientific information in general. According to respondents, cross-check between one and another company is very important in order to improve techniques and processes to assist in decision making or problem solving.

The study results also showed the need to promote the practice of OSH management as a means of protecting workers, and not as an inspection agent with punitive power to those who fail to comply with its recommendations. The worker must have a different view of safety at work, as well as professionals working in the area. Therein, integration and awareness of employees through information once again revealed itself effective. Information by itself is not the key to all problems solving, but it may be a major step in building and maintaining the ideal working environment. Data exchange among companies provides the sharing of experiences, improvement of processes and support in the prediction of harmful events, the accident itself.

The initial premises, which were confirmed at the end of the study, indicate that the philosophy of dissemination of scientific information used by open archives is considered by the interviewed professionals as a support tool for OSH management. Regarding the attitude of these professionals, the acceptance of this new form of dissemination of scientific information is plausible, although the reliability and familiarity of the public with digital instruments of research is relatively low, what creates the need of promoting the scientific culture, which stirs up both production and sharing of information, and above all, the scientific papers publication.

The discussion emphasized that scientific information can be viewed and used as a support tool to OSH management in general, which is considered a pillar to occupational safety and health management. An information resource would help professionals in the area, be in decision making processes, be in technical aspects. Allied to the mechanisms for quality management, benefits are enhanced with the possibility of achieving significant results to organizations.

8. Final considerations

It is important to establish the alignment of the alliance among scientific information, quality management and occupational safety and health. This combination, stands for the interests of OSH professionals, workers (including shop floor), and managers. OSH professionals, through the access of information are able to improve their professionals skills and apply methods of accident preventions never experienced before, or experienced in other organization, while quality management tools, like PDCA and 5S may help them to evaluate the adopted measures, making easier the identification and correction of any fail, and also to provide a clear, organized and healthy work environment, through the 5S recommendations. Workers, may enjoy a better (friendly) work environment, ergonomically correct, cleaned and organized job stations, not to mention the perception of the importance of him and his job to the organization, and also the improvement of occupational hazards management, once a suitable working environment reduces the chances of the occurrence of occupational accidents. For the organizational manager, all of this represents profits increasing. Once the OSH professional has conditions to offer a proper workplace in the organization, the worker is able to produce more, and more efficiently (profit). The worker, less susceptible to occupational accidents, requires less spending with medical assistance and rehabilitation costs (profit). Conscientious use and disposal of organizational raw material, also represents saving money (profit). PDCA also has its contribution in this sense through the provision of the constantly evaluation of the production process, what allows the organization to keep a continuous evaluation and improvement process, towards excellence, which may represent the conquest of market share (profit).

Increasing production and quality indicators, decreasing occupational accidents, using raw material properly (including human resources), and all of this supported by scientific information, meet the interests of any organization manager which is profit. In this scenery, the effective use and application of scientific information and communication may be considered as one of the major challenges, to be reached by occupational safety and health professionals. Since OSH is a branch of knowledge that deals with human life, comfort and safety become priority issues in order to provide the quality of work life, especially in activities that exposes the worker to occupational hazards, which risks both safety and even the workers lives.

However most needed material goods, we must not forget the following: ALL OF THIS EXISTS FOR THE MAN. And he can not have his health unprotected. The workplace control is obligatory. This is a matter of morality. If the work does not contribute to your happiness, it will be a huge and lamentable loss [37].

In work accidents field any kind of adaptation must be unacceptable. After all, a worker who is mutilated by an occupational accident, hardly will to adapt to society and the workplace in order to carry out their routine activities. The accident scene will never be erased from his memory. Neither compensation, (no matter the value) nor any other social benefit will aid the psychological trauma of an injured worker, or replace the activities of which he was deprived as a result of an accident. Therefore, it is inadmissible to use the term adaptation when it is related to safety and health at work, since the health and integrity of the employee is an acquired right that must be respected by the employer.

Knowledge attracts money, while the converse is not true. Whereas the non-conformities in the workplace may culminate in damages to the organization, investments in the development of the intellectual capital and management knowledge should be priorities to the organizations and be considered as support tools for managing health and safety at work.

Quality management should promote effective changes in the workplace, and these changes should reach and invigorate in all working environments, so that the workers lives, which represents the strength of a nation, be respected and preserved.

References

1. BarreirosD.Gestão da segurança e saúde no trabalho: estudo de um modelo sistêmico para as organizações do setor mineral. [thesis]. [São Paulo]. Universidade de São Paulo, 2002
2. Le CoadicF. Y. A.Ciênciada.Informação2nd.edRio.de JaneiroBriquet.de Lemos; 2004
3. Silva JPC, Machado, FO. A qualidade como estratégia empresarial: um estudo conceitual. Ingepro. 2011, 3:35-46. [cited 2012 Feb 10]. Available from https://ojs.ingepro.com.br/index.php/ingepro/article/view/473/388.
4. MagrettaJ. O.queé.gerenciare.administrarRio de Janeiro: Elsevier; 2002
5. BragaM.Prevenção deacidentes.investimentoou.custo?Cipa.2007Oct; 28(355)92.
6. Silveira AA. A gestão do conhecimento como ênfase na aprendizagem organizacional: um estudo de multicaso no contexto bancário. [dissertation]. [Itajubá]. Universidade Federal de Itajubá, 2004. [cited 2012 Feb. 23] Available from 8. http://adm-net-a.unifei.edu.br/phl/pdf/0031263.pdf
7. AlmeidaD. A.LealF.PinhoA. F.et al.Gestão do conhecimento na análise de falhas através de sistema de informação. Produção. 2006Jan./Apr. 1617188
8. Pereira FCM. O Processo de conversão do conhecimento em uma escola de atendimento especializado. Enc. BIBLI: R. eletrônica de Bibl. Ci. Inform. 2005 20. [cited 2011 Nov. 9] Available from 11. http://www.encontrosbibli.ufsc.br/Edicao_20/3_mafra_pereira.pdf.
9. de Referência em Inteligência Empresarial.Gestão do Conhecimento. [cited 2012Jan. 4] Available from: 13. http://portal.crie.coppe.ufrj.br/portal/main.asp?ViewID=%7B32E72BC9-F838-4577-AF25-A5803179DF22%7D.
10. NonakaI.TakeuchiH.Criação de conhecimento na empresa: como as empresas japonesas geram a dinâmica da inovação. Rio de Janeiro: Campus; 1997
11. Campos VF.Gerenciamento da rotina de trabalho no dia-a-dia. 7th. ed. Belo Horizonte: Desenvolvimento Gerencial; 2001

12. Andrade FF. O método de melhorias PDCA. [thesis]. [São Paulo]. Escola Politécnica da Universidade de São Paulo, 2003. [cited 2012 Feb. 14] Available from 17. http://www.teses.usp.br/teses/disponiveis/3/3146/tde-04092003-150859/.
13. SouzaR.MekbekianG.Metodologia de gestão da qualidade em empresas construtoras. EdUSP. Anais do ENTAC93: encontro nacional de tecnologia do ambiente construído: avanços em tecnologia e gestão da produção de edificações. 1993127131
14. Centro de Tecnologia de Edificações.Sistema de gestão da qualidade para empresas construtoras. São Paulo: SINDUSCON; 1994
15. Parceria do Distrito Industrial em Ciências Ambientais.Ciclo PDCA. [cited 2010May 22] Available from: http://www.pdca.org.br/.
16. Mattos RP Eletricidade: perigo de morte. Cipa.2007283339091
17. Unlimited Learning Resources.The Deming Cycle (PDCA). [cited 2012Feb 13]. Available from: http://www.4ulr.com/products/productquality/deming_cycle.html. programaI. P. E. M. O.esses.[citedfrom.2012Mar. 20] Available from: 26. http://www.ipem.sp.gov.br/3emp/5esses.asp?vpro=abe.
18. GomieroF.Os cinco sensos do combate ao desperdício. [cited 2011Sep 20]. Available from: http://www.janelanaweb.com/digitais/gomiero24.html.
19. Programa 5S. [cited2011Nov 5]. Available from: http://www.pucrs.br/feng/5s/capa.php.
20. ANVISA. O Método 5S. Brasília: ANVISA; 2005. [cited 2010 July 15]. Available from:
21. http://www.anvisa.gov.br/reblas/procedimentos/metodo_5S.pdf.
22. CEGarciaAmaral. M.Auditoria como instrumento de avaliação do desempenho do sistema de gestão da saúde e segurança no trabalho. Anais do SIMPEP. 2005
23. CoutoH.Rompendolimites. a.neurolinguísticaé.umatécnica.comaplicação. naprevenção.de acidentesProteção. 199811793839
24. Hékis HR et al.Estratégia para crescimento através da gestão da qualidade alinhada com a gestão da informação e do conhecimento: um estudo de caso em uma empresa do ramo da construção civil em Natal/ RN. Ingepro.3.111Cited 2010Aug. 10]. Available from: https://http://ojs.ingepro.com.br/index.php/ingepro/article/view/427.
25. 3038Penabad AA et. Al. Consideraciones para la mejora de los sistemas de gestión de la calidad en la Industria Biofarmacéutica. Vaccimonitor. 2010. 19.30-38. [cited 2011 May 5] Available from http://redalyc.uaemex.mx/src/inicio/ArtPdfRed.jsp?iCve=203415359006&iCveNum=15359.
26. MaguireR.Safety cases and safety reports: meaning, motivation and management. Hampshire: Ashgate; c2006
27. SilvaE. L.MenezesE. M.BissaniM. A.Internetcomo.canal decomunicação. científicaInf. & Soc.: Est. 2002121290320
28. AraújoV. R. H.FreireI. M. A.redeInternet.comocanal.de comunicaçãona.

perspectivada.Ciênciada.InformaçãoTransinformação. 1996824555

29. VelhoL. A. A.Ciênciae.seupúblico.Transinformação. 1997931632
30. Brasil, Ministério da Previdência Social. Anuário estatístico da Previdência Social. Brasília; 2010. [cited 2012 Fev 22] available from http://www.mpas.gov.br/arquivos/office/3_111202-105619-646.pdf.
31. RosiniA. M.PalmisanoA.Administração de sistemas de informação e a gestão do conhecimento. São Paulo: Thomson; 2003
32. Baptista AA, Costa SMS, Kuramoto H et al. Comunicação científica: o papel da Open Archives Iniciative no contexto do acesso livre. Enc. Biblio: R. Eletr. Bibliotecon. Ci. Inf. special edition, 2007. [cited 2010, Aug 26] Available from http://journal.ufsc.br/index.php/eb/article/viewFile/1518-2924.2007v12nesp1p1/435.
33. TriskaR.CaféL.Arquivosabertos.subprojetoda.BibliotecaDigital.BrasileiraCiência da Informação. 2001
34. VianaC. L. M.MárderoArellano.MAShintakuM.Repositórios institucionais em ciência e tecnologia: uma experiência de customização do Dspace. [cited 2010Mar. 22] Available from: http://eprints.rclis.org/archive/00005563/01/viana358.pdf.
35. Machado M. Open archives: panorama dos repositórios. [dissertation]. [Santa Catarina]: Universidade Federal de Santa Catarina, 2006. [cited from 2012 Jan 10] Available from http://www.tede.ufsc.br/teses/PCIN0015.pdf.
36. KuramotoH.Informaçãocientífica.proposta deum.novomodelo.parao.BrasilCiência da Informação. 200635291102
37. SaadI. F. S. O.adicional deinsalubridade.apósa.novaN.R-Revista ABHO de higiene ocupacional. 200651311

5

Reducing Mirror Slippage of Nightstand with Plackett-Burman Doe and Ann Techniques

1. Introduction

Understanding the behaviors of the systems is possible by conducting some experiments. But, experiments have to give all needed information in a reasonable manner; especially time and budget for the experiments are main constraints in a system observation. Identifying the key factors of experiments which have impact on the process and conducting all trials relevant with key factors are very difficult. It is difficult to make analytical and numerical modeling in complex processes and the real life problems which have many parameters that affect the systems. To prevent all these negativities, design of experiment (DOE) is an indispensable method which observes systems behavior. It is a systematic approach to find how inputs affect the system outputs and also provides a continuous improvement for processes and explores relationship between key input variables and output performance characteristics by existing and performing a designed set of experiments. Key input variables can be defined as factors (independent variables) and outputs can be defined as response (dependent variables)[1]. A well organized experimental design needs a design matrix (Orthogonal) prior to experiment. The design matrix contains all the settings of factors at different levels and also gives the order of experiments. The number of factors and interactions, levels of factors, budget and resources define the size of experiment[2]. Design of experiment was firstly used for agricultural applications in 1920s and became a popular tool observing systems and processes. Many books and journals have been published about the applications of Chemistry[3], Biology[4], Statistics[5] and Engineering[2] for years. Optimizing the levels of factors, finding interaction relationship between factors and response, making a fast decision, improving the system performance are some advantages of DOE[6].

Some design of experiment methods used in the literature are Full Factorial Design, Fractional Factorial Design, Plackett-Burman Design, Central Composite Design, Box- Behnken Design, Robust Parameter Design, Computer Aided Design and Taguchi. Taguchi method has a wide usage in the engineering applications but criticized in the literature about not concerning with the interactions between factors while focusing on evaluation of main effects totally[7]. It is one of the most important tools that can be used for improving steps of Six Sigma loop[8].

This application study which was done in furniture manufacturing firm in TURKEY to prevent the mirror slippage of nightstand is analyzed from the point of Six Sigma philosophy view. The structure of the rest of the paper is given as follows: In Section 2, it is described in detailed about the concept of six sigma. In Section 3, the application of the artificial neural network is presented. In Section 4, conclusion is given as evaluation of the results.

2. Application of Six Sigma

Manufacturing sector has to provide perfect products and produce fast, quality and economical products to their consumers because of severe competition in the market. As all businesses reach this purpose, their processes must be stabilized or kept under control with continuous improvement. Six Sigma that provides continuous improvement is a key to being successful in business. This management philosophy which uses some statistical and systematic approaches will prevent defects. Providing continuous improvement in a system is a result of applying the loop of Six Sigma in correct order and identifying relations between inputs and outputs. This loop is made up of the steps of Define, Measure, Analyze, Improve and Control. Some quality and statistical methods are being applied in all these steps[9]. The most used Six Sigma tools are Design of Experiments, Response Surface Method, Robust Design, Statistical Process Control, Quality Function Deployment, Failure Mode and Effect Analysis, Capability Analysis, Hypothesis Testing, Analysis of Variance, Regression Analysis[10]. Sigma (σ) is a letter which represents the variability and standard deviation in the processes. On the other hand, Six Sigma identifies how a deviation is shown from the perfect conditions[11]. The significant factor in six sigma process is to share responsibility in an organization. For this reason, everyone has different responsibilities depending on their education[10]. Organization structure can be addressed as Leadership, Champions and Sponsors, Black Belt, Green Belt, and Master Black Belt[12]. Applying Six Sigma tools in manufacturing processes provides reduction of defect rate, cycle time, manufacturing costs and improvement in quality and productivity[13].

A Six Sigma application study was done in a furniture manufacturing firm in TURKEY. Steps following for the application of Six Sigma by considering DMAIC loop is given below.

This six sigma application study was simulated with a mechanism which rotates the mirror nightstand (Figure 2) having constant speed. Thus, slippage resistant of mirror was measured. While running the system, if it stands out against slippage more than 72 hours (based on expert opinion), this situation is called as "no slippage".

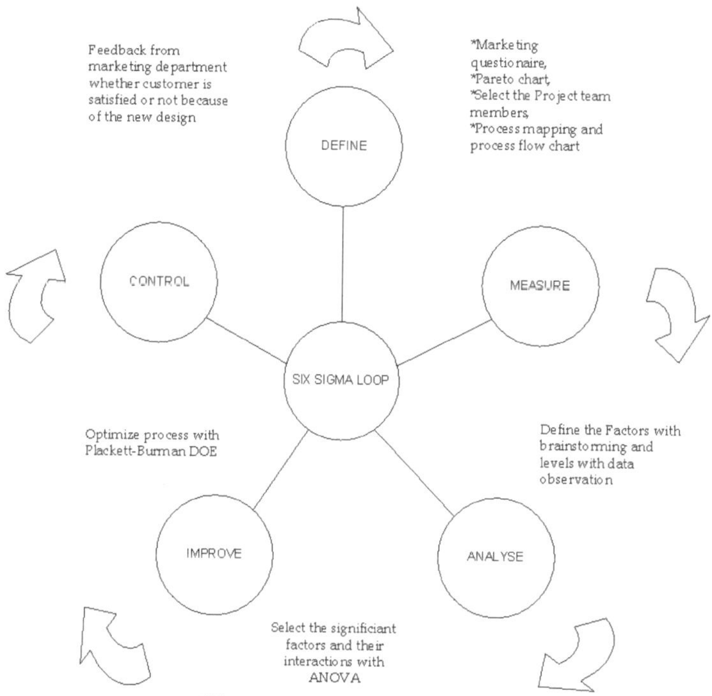

Figure 1. Six sigma loop

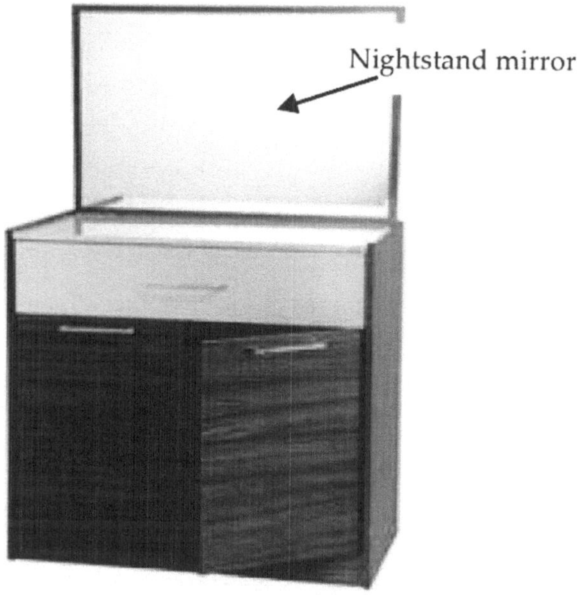

Figure 2. A Nightstand Sample

2.1. Define

In this stage, following steps should be applied;

- Required studies are done to define systems' problem,
- A team is constructed for solution,
- Follow-up method is selected
- Goal is defined by analyzing process carefully.

In the application, a marketing questionnaire was applied to customers of the firm to find the reason of the problem. This project's objective is reducing the customer complaints of the manufacturing firm about nightstand mirror slippage. As a result of questionnaire, pareto chart shown in Figure 3 was constructed by using data about product returns. It is seen that one of the significant complaints (%7) are dealing with slippage of glass on the nightstand and we selected this problem for obtaining a fast improvement. A quality engineer, a foreman from the plant and two persons from university were formed a team for this six sigma application.

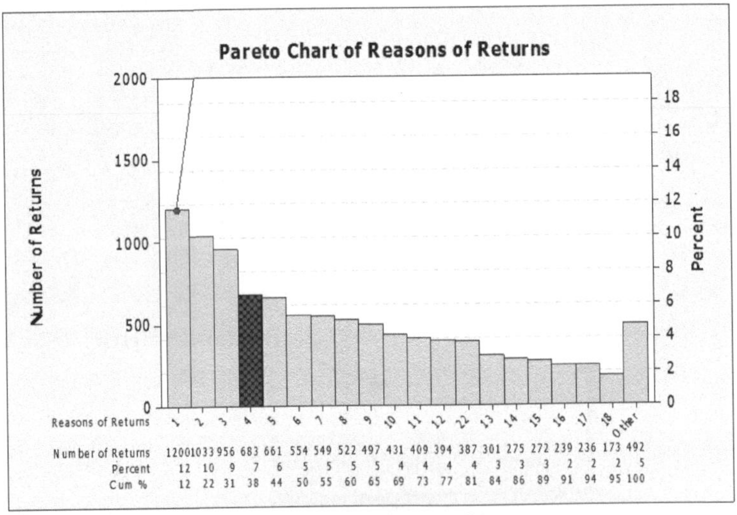

Figure 3. Pareto chart of Returns

Slippage of mirror nightstand is a reason of customer dissatisfaction. The aim is to increase customer satisfaction by reducing variance and maximizing mirror slippage time. After problem definition and forming a team, the next step is to perform a detailed process analysis. A process chart was constructed by the team and is given in Figure 4.

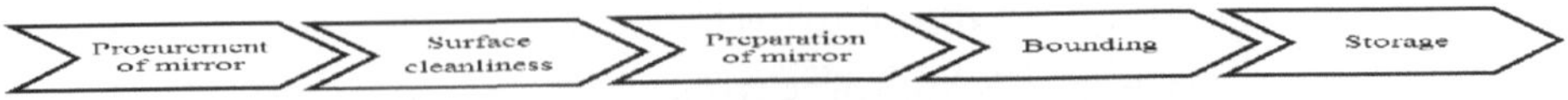

Figure 4. Process chart

2.2. Measure

Key input variables (factors) and their levels are defined to determine impacts of variables on process performance. The steps which will be followed in this phase are:

- Identifying key quality characteristics (dependent and independent variables),
- Measuring system capability,
- Finding a solution to data collection.

In measure phase, most important factors and levels were defined clearly for the slippage problem. The number of independent variables (or factors) were found as 13 and given in Table 1. 13 factors which are affected glass-mirror slippage were decided by brainstorming of team members and it was decided to evaluate all factors as two levels.

Table 1. List of experimental control factor

Control Factors	Factor Effects	Min. Level (-1)	Max. Level (+1)
Environment temperature (C°)	A	- 20	60
Surface Cleanliness	B	Dirty	Clean
Temperature Difference in working place (C°)	C	14	28
Storage Position	D	Bottom	Top
Mirror Cleanliness	E	Dirty	Clean
Waiting time (hours)	F	0	2
Anti slip tape quality	G	ELSEM*	TESA*
Styrofoam	H	Nonexistence	Existence
Cleaning liquid	J	Existence	Nonexistence
Dust in the working place	K	Dirty	Clean
Double sided anti slip tape quality	L	Existence	Nonexistence
Anti slip tape usage	M	Undismantle	Dismantle
Anti slip tape usage rate (centimeter)	N	6	12
*ELSEM and TESA are the names of anti-slip tapes			

2.3. Analyse

Which tools will be used is defined and factor analyzes are done by using some statistical methods. The steps which will be followed in this phase are:

- Identifying causes of defects and variation,
- Conducting multi-variable analysis,
- Collecting process data about dependent and independent variables,
- Using regression analysis to define relationship between dependent and independent variables.

Table 2. Orthogonal Matrix of PB design

Exp. Number	A (Co)	B	C (Co)	D	E	F (hours)	G	H	J	K	L	M	N (cm)	Slippage (hours)
1	60 (1)	Clean (1)	14 (-1)	Bottom (-1)	Clean (1)	2 (1)	Tesa (1)	Existence (1)	Existence (-1)	Dirty (-1)	Existence (-1)	Undis-mantle (-1)	12 (1)	No slippage
2	-20 (-1)	Clean (1)	28 (1)	Top (1)	Dirty (-1)	2 (1)	Elsem (-1)	Existence (1)	Nonexis-tence (1)	Dirty (-1)	Existence (-1)	Undis-mantle (-1)	6 (-1)	No slippage
3	60 (1)	Dirty (-1)	28 (1)	Bottom (-1)	Clean (1)	2 (1)	Elsem (-1)	Existence (1)	Nonexis-tence (1)	Dirty (-1)	Nonexis-tence (1)	Dismantle (1)	12 (1)	16
4	-20 (-1)	Dirty (-1)	28 (1)	Top (1)	Dirty (-1)	0 (-1)	Tesa (1)	Nonexis-tence (-1)	Existence (-1)	Dirty (-1)	Nonexis-tence (1)	Dismantle (1)	12 (1)	No slippage
5	-20 (-1)	Clean (1)	14 (-1)	Bottom (-1)	Dirty (-1)	2 (1)	Elsem (-1)	Nonexis-tence (-1)	Nonexis-tence (1)	Clean (1)	Nonexis-tence (1)	Undis-mantle (-1)	12 (1)	No slippage
6	-20 (-1)	Clean (1)	28 (1)	Top (1)	Clean (1)	0 (-1)	Tesa (1)	Existence (1)	Existence (-1)	Clean (1)	Nonexis-tence (1)	Undis-mantle (-1)	12 (1)	No slippage
7	-20 (-1)	Dirty (-1)	14 (-1)	Top (1)	Dirty (-1)	2 (1)	Tesa (1)	Existence (1)	Nonexis-tence (1)	Dirty (-1)	Existence (-1)	Dismantle (1)	6 (-1)	No slippage
8	-20 (-1)	Clean (1)	28 (1)	Bottom (-1)	Clean (1)	2 (1)	Tesa (1)	Nonexis-tence (-1)	Nonexis-tence (1)	Clean (1)	Nonexis-tence (1)	Dismantle (1)	6 (-1)	No slippage
9	60 (1)	Clean (1)	14 (-1)	Top (1)	Dirty (-1)	2 (1)	Tesa (1)	Nonexis-tence (-1)	Nonexis-tence (1)	Clean (1)	Existence (-1)	Dismantle (1)	12 (1)	No slippage
10	-20 (-1)	Clean (1)	14 (-1)	Bottom (-1)	Clean (1)	0 (-1)	Tesa (1)	Nonexis-tence (-1)	Existence (-1)	Dirty (-1)	Nonexis-tence (1)	Dismantle (1)	6 (-1)	No slippage
11	-20 (-1)	Clean (1)	28 (1)	Top (1)	Clean (1)	2 (1)	Elsem (-1)	Nonexis-tence (-1)	Existence (-1)	Clean (1)	Existence (-1)	Undis-mantle (-1)	6 (-1)	No slippage
12	-20 (-1)	Dirty (-1)	28 (1)	Bottom (-1)	Dirty (-1)	2 (1)	Elsem (-1)	Existence (1)	Existence (-1)	Clean (1)	Existence (-1)	Dismantle (1)	12 (1)	No slippage
13	60 (1)	Dirty (-1)	28 (1)	Top (1)	Clean (1)	0 (-1)	Tesa (1)	Nonexis-tence (-1)	Nonexis-tence (1)	Dirty (-1)	Existence (-1)	Undis-mantle (-1)	12 (1)	No slippage
14	-20 (-1)	Dirty (-1)	14 (-1)	Top (1)	Clean (1)	0 (-1)	Tesa (1)	Existence (1)	Nonexis-tence (1)	Clean (1)	Existence (-1)	Undis-mantle (-1)	6 (-1)	No slippage

Exp. Number	A (Co)	B	C (Co)	D	E	F (hours)	G	H	J	K	L	M	N (cm)	Slippage (hours)
15	60 (1)	Dirty (-1)	14 (-1)	Bottom (-1)	Clean (1)	0 (-1)	Elsem (-1)	Existence (1)	Existence (-1)	Clean (1)	Existence (-1)	Dismantle (1)	6 (-1)	2-3
16	60 (1)	Dirty (-1)	28 (1)	Top (1)	Dirty (-1)	2 (1)	Tesa (1)	Existence (1)	Existence (-1)	Clean (1)	Nonexistence (1)	Dismantle (1)	6 (-1)	No slippage
17	60 (1)	Clean (1)	28 (1)	Bottom (-1)	Dirty (-1)	0 (-1)	Tesa (1)	Nonexistence (-1)	Nonexistence (1)	Dirty (-1)	Existence (-1)	Dismantle (1)	6 (-1)	72
18	60 (1)	Clean (1)	14 (-1)	Top (1)	Clean (1)	2 (1)	Elsem (-1)	Nonexistence (-1)	Existence (-1)	Dirty (-1)	Existence (-1)	Dismantle (1)	12 (1)	16
19	60 (1)	Dirty (-1)	14 (-1)	Bottom (-1)	Dirty (-1)	2 (1)	Tesa (1)	Nonexistence (-1)	Existence (-1)	Clean (1)	Nonexistence (1)	Undis-mantle (-1)	6 (-1)	72
20	60 (1)	Dirty (-1)	14 (-1)	Top (1)	Dirty (-1)	0 (-1)	Elsem (-1)	Nonexistence (-1)	Nonexistence (1)	Clean (1)	Nonexistence (1)	Undis-mantle (-1)	12 (1)	2-3
21	-20 (-1)	Dirty (-1)	28 (1)	Bottom (-1)	Clean (1)	0 (-1)	Elsem (-1)	Nonexistence (-1)	Nonexistence (1)	Clean (1)	Existence (-1)	Dismantle (1)	12 (1)	No slippage
22	60 (1)	Clean (1)	28 (1)	Bottom (-1)	Dirty (-1)	0 (-1)	Elsem (-1)	Existence (1)	Nonexistence (1)	Dirty (-1)	Nonexistence (1)	Undis-mantle (-1)	6 (-1)	2-3
23	60 (1)	Clean (1)	28 (1)	Bottom (-1)	Dirty (-1)	0 (-1)	Tesa (1)	Existence (1)	Existence (-1)	Clean (1)	Existence (-1)	Undis-mantle (-1)	12 (1)	No slippage
24	-20 (-1)	Dirty (-1)	14 (-1)	Bottom (-1)	Clean (1)	2 (1)	Tesa (1)	Existence (1)	Nonexistence (1)	Dirty (-1)	Nonexistence (1)	Undis-mantle (-1)	12 (1)	No slippage
25	60 (1)	Clean (1)	14 (-1)	Top (1)	Clean (1)	0 (-1)	Elsem (-1)	Existence (1)	Nonexistence (1)	Clean (1)	Nonexistence (1)	Dismantle (1)	6 (-1)	2-3
26	-20 (-1)	Dirty (-1)	14 (-1)	Bottom (-1)	Dirty (-1)	0 (-1)	Elsem (-1)	Nonexistence (-1)	Existence (-1)	Dirty (-1)	Existence (-1)	Undis-mantle (-1)	6 (-1)	No slippage
27	60 (1)	Dirty (-1)	28 (1)	Top (1)	Clean (1)	2 (1)	Elsem (-1)	Nonexistence (-1)	Existence (-1)	Dirty (-1)	Nonexistence (1)	Undis-mantle (-1)	6 (-1)	16
28	-20 (-1)	Clean (1)	14 (-1)	Top (1)	Dirty (-1)	0 (-1)	Elsem (-1)	Existence (1)	Existence (-1)	Dirty (-1)	Nonexistence (1)	Dismantle (1)	12 (1)	No slippage

In this phase, we need to collect data about process variables to find the resource of variation in the system. By this way, an experimental design will be performed. If a full factorial experimental design is run, we need 2^{13} (8192) experiment for trying all possibilities. This condition takes too much time and effort. On the other hand, a Plackett Burman fractional factorial design can give faster results in a short time. For this reason, application of fractional factorial experimental design was decided and by considering all the advantages above, Placket-Burman L28 experimental design was used in this study. Orthogonal (Design) matrix is given in Table 2.

Table 3. ANOVA for factors

Source	DF	Seq SS	Adj SS	Adj MS	F	P
A	1	11100,2	11100,2	11100,2	35,12	0,000
B	1	110,0	110,0	110,0	0,35	0,565
C	1	279,7	279,7	279,7	0,89	0,363
D	1	118,1	118,1	118,1	0,37	0,551
E	1	885,9	885,9	885,9	2,80	0,116
F	1	390,0	390,0	390,0	1,23	0,285
G	1	8349,0	8349,0	8349,0	26,42	0,000
H	1	325,7	325,7	325,7	1,03	0,327
J	1	299,0	299,0	299,0	0,95	0,347
K	1	18,1	18,1	18,1	0,06	0,814
L	1	1627,9	1627,9	1627,9	5,15	0,040
M	1	157,9	157,9	157,9	0,50	0,491
N	1	192,9	192,9	192,9	0,61	0,448
Error	14	4424,6	4424,6	316,0		
Total	27	28279,2				

ANOVA was performed to verify whether the main effects exist or not between variables and given in Table 3. Confidence level of the model was selected as 99 %. Main effects plotted for mean response is as shown in Figure 5.

2.4. Plackett-burman in screening design

This design is used for screening a large number of process factors to identify bthe most important parameters that have a significant impact on the process performance. It provides a reduction in the number of factors that needed to be observed. PB design includes in a family of screening design. The natures of interactions among the factors are not interested in screening designs. Geometric PB designs contain M number of experiment which is a power of two (4, 8, 16...etc.). On the other hand, non-geometric designs contain M number of experiment which are multiples of four but not power of two (12, 20, 24, 28...etc.)[2]. All main effects have same precision in PB designs.

After defining the key parameters, if necessary, full factorial, fractional factorial designs and response surface method, subsequent experiments can be run. Full factorial design needs all possible combinations of levels for all factors. PB designs which focus on the main effects by considering the interactions among the factors are negligible[14]. It can be applied effectively to industrial studies which need to analyze many factors in the same manner[15]. The main aim is not to see interactions among the factors, but this doesn't mean that we couldn't have any idea about the interactions among factors. It is proved that, PB can be used even two sided interactions and main effects are confounded[16]. In PB designs, M+1 experiment are conducting to analyze M factors and generally it is advised to use for main effects.

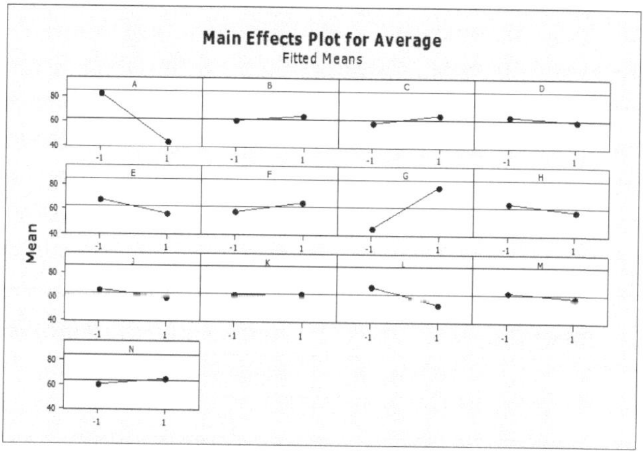

Figure 5. Main effects plot for average

The 12, 20, 24, 28- run PB designs have a special importance, because they fill the gap in the standard design. On the other hand, these designs have a high alias structure. If it is assumed that there are no interactions between the factors, this alias structure can be prevented. But this assumption is not possible in real life systems. To get away with alias structures, it is provided to make a second run, but not an exact replication[17]. In a research, it is proved that using 28-run PB gives best results for a system between 32-run design fractional factorial, 28 experiment design PB and 28 experiment design a random balance[18].

2.5. Improve

Improvement points which yield an optimum process are selected and system model is constructed by considering the optimum condition. The steps which will be followed in this phase are:

- Conducting design of experiments.
- Trying to improve process by eliminating variation.
- Optimizing process.

Slippage time of mirror was analyzed by using Plackett-Burman experimental design matrix. After the experiments were conducted, some statistical analyses were done to define parameters which have a great variance. Also, two way interactions were taken into consideration. An optimum setting which contains different factor levels was found. This provides maximum slippage time and extracted data was used in the real manufacturing of mirror nightstand.

Our main goal in this study is decreasing process variability by defining most effective factors and levels on quality characteristics. PB experimental design module of MINITAB 15.0 software was used. As a result, optimal factor setting obtained is -1,1,1,1,-1,1,1,1,-1,1,-1,-1,1 (Environmental Temperature= -20 C°; surface cleanliness = clean, temperature difference (C°) in working place = 28 C°; storage position= top; mirror cleanliness= dirty; waiting time= 2 hours; anti slip tape quality= Tesa; styrofoam= existence; cleaning liquid= existence; dust in the working place = clean; double-sided tape usage = existence; anti slip tape usage= undismantle; anti slip tape usage rate= 12cm). If these conditions are applied to process/product, a quality product having a new design reducing complaint of customers comes out and mirror nightstand slippage will not occur.

Table 4. Factors and interaction table

Factors	Effect Values	Coef.	Interaction	Effect Values	Coef.
Fixed values		**56,82**	**Fixed Values**		**56,82**
A	-32,36	-16,18	A*B	1,48	0.74
B	1,48	0,74	A*C	3,48	1,74
C	3,48	1,74	A*D	-1,26	-0,63
D	-1,26	-0,63	A*E	2,40	1,20
E	2,40	1,20	A*F	2,07	1,04
F	2,07	1,04	A*G	31,62	15,81
G	31,62	15,81	A*H	-3,05	-1,52
H	-3,05	-1,52	A*J	-1,71	-0,86
J	-1,71	-0,86	A*K	1,00	0,50
K	1,00	0,50	A*L	1,50	0,75
L	1,50	0,75	A*M	1,33	0,67
M	1,33	0,67	A*N	2,17	1,08
N	2,17	1,08	B*C	-0,00	-0,00
			B*D	-0,00	-0,00

When Table 4 is analyzed, A, G factors and their interaction values have a significant difference from others. In order to decide which effects are statistically meaningful, normal plot of standardized effects was used. While interpreting the normal plot of standardized effects, important factors can be shown under the plot and left of the line, or above the plot and right of the line. As shown in the Figure

6, A (Environment temperature), G (Anti slip tape quality) and A-G interaction are significant factors for this study. While optimizing dependent variable, these factors must be considered together.

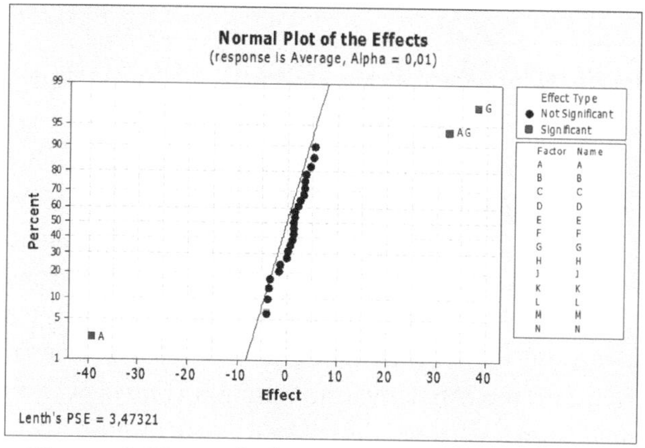

Figure 6. Normal plot of standardized effects

We can also see the importance of A and G factors from pareto chart of standardized effects as shown in Figure 7.

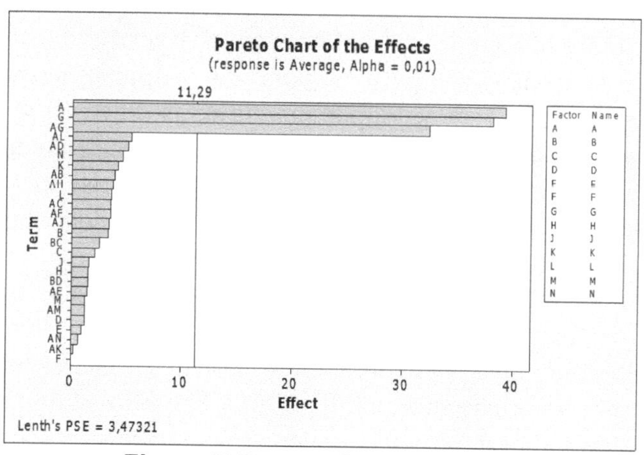

Figure 7. Pareto chart of effects

2.6. Control

New responsibility distribution is done as a result of optimization, and data is collected to achieve continuous improvement[2]. The steps which will be followed in this phase are:

- Checking new system performance,
- Monitoring improvements,
- Implementing control plans and trying to construct continuous improvements in the system.

In order to test the model verification regarding whether this design gives optimal result or not, optimal factor setting was applied in the process and process data of 6 months were collected. We realized that Product-complaints were reduced by 60 %.

3. Application of ANN (Artificial Neural Network)

ANN can perform modeling in nonlinear systems as well as linear systems. Nonlinear relations between data are extracted by ANN which is inspired with human nervous system. In a neural network, three have been three types of units: input units which receive data from outside, output units which send data out of the neural network, and hidden units where the signals are transferred inside of the neuron. Weights can be represented as the strength of the connection between an input and a neuron. The neuron can be defined as processing units, the place where actual activity occurs. An activation function controls the value of the input while transferring it from the layers, until it reaches to output. Each neuron receives input from neighbours and use these inputs while computing an output signal[20]. When this data flow occurs between the units, weights are adjusted. There can be different types of ANN. Multilayer Perceptron (MLP) and Radial Basis Function are the most known types of ANN. MLP networks consist of typically an input layer, single or more hidden layers, and one output layer. Hidden layers have one or more hidden neurons which perform nonlinear mapping between inputs and outputs[21]. Relationship between input and output is constructed by using some methods that is needed for adjusting the weights during the training session. This can be represent as learning algorithm. Choosing the proper learning algorithm is also very important while training the networks. The most common learning algorithm is called Back Propagation (BP). There are several types of optimization techniques for ANN using the backpropagation algorithm such as gradient descent method, gradient descent with momentum, conjugate gradient method (Fletcher-Reeves, Polak-Ribiere), quasi-newton method (Broyden-Fletcher-Goldfarb-Shanno - BFGS). Generally, levenberg-marquardt method is used for the networks which contain up to a few hundred weights, converges faster and uses second derivatives for solutions. For the moderate size networks, quasi-newton method are often the next fastest algorithms[22]. BFGS is the most successful quasi-newton method employed in our study[23].

The optimal number of nodes in the hidden layer is generally computed by a trial-and-error approach[24]. To find the best neural network, networks which contain different number of hidden neurons are compared with each other[25]. Mean square error (MSE) is the basic criteria while judging the capability of networks. If we use too much hidden neurons, this leads us too much flexibility and over fitting. On the other hand, if we use too few hidden neurons, this prevents the learning capability[26]. Therefore, one of the biggest problems is to find the proper number of hidden neurons. Furthermore, the simplest architecture is better than others[27]. Thus, single hidden layer can be chosen and it is sufficient for many continuous nonlinear mapping.

In this study, after the design of experiment study was conducted, ANN application was done for modeling the system. PB experiment data were used as input to ANN. To prevent alias structure of PB, a second run was used as input; but not an exact replication. Data were divided into two parts as training (80%), and testing (20%), respectively. The Neural Network module of STATISTICA 9.0 software was employed in modeling the ANN. In the network, there were 13 inputs and one output as defined before. A trial and error method was used in deciding the best model for the system. 500 different combinations of activation functions and neuron numbers were tried by considering the fitted model MSE. The performance of best five models was evaluated and MLP 13-5-1 model was chosen as shown in Table 5. The MAE and MSE of the selected topology are 2.957247 and 15.54706, respectively. Generally, using the sigmoid (logistic) function in ANN topology provides a good nonlinear input–output mapping capability[28]. A weight decay method was used to reduce the overfitting problem. This option encourages the development of smaller weights, so it potentially improves performance of the network and modifies the network's error function to penalize large weights[29].

Table 5. Best Network Architecture

Network Architecture	Training perf.	Test perf.	Training error	Test error	Training algorithm	Hidden activation function	Output activation function
MLP 13-7-1	0.986217	0.988856	21.77105	8.85785	BFGS	Exponential	Logistic
MLP 13-11-1	0.986836	0.989038	20.68673	10.50230	BFGS	Exponential	Logistic
MLP 13-9-1	0.980550	0.987936	23.20724	11.52654	BFGS	Exponential	Logistic
MLP 13-10-1	0.984314	0.986702	18.20639	12.64308	BFGS	Exponential	Logistic
MLP 13-5-1	0.986941	0.990763	23.59058	7.77353	BFGS	Exponential	Logistic

ANN model was controlled whether the slippage occurs or not for PB's optimal set. We showed that slippage time was predicted as 89 hours in the ANN. We wanted to see which parameters are significant for ANN as in PB. On account of this, sensitivity analysis was performed in ANN. There have been some researches about determining the effects of the input parameters on the response variable[30] in terms of sensitivity. Sensitivity analyzes can be applied to a network after the training session is completed. It tries to see what the effect would be of changing the inputs of the model on the overall model fit. While an input of the artificial neural network model is eliminated from the model, sums of squares residuals for the model are analyzed and the inputs can be sorted by their importance[29]. The larger the variance in the error after the parameter is omitted, the more important the variable is[31]. As a result of the sensitivity analysis, it proves that A and G are really the most important parameters for the ANN model. This sensitivity analysis values were normalized as given in Figure 8.

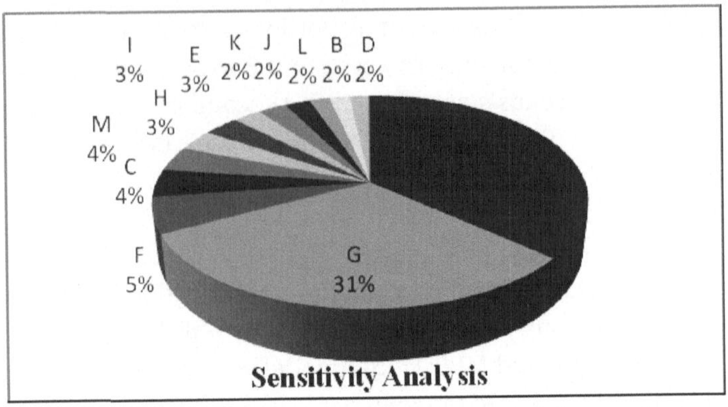

Figure 8.Pie chart of normalized sensitivity analysis

4. Conclusion

In this study, PB experimental design method was used from the point of view of Six Sigma philosophy. Besides, ANN was employed for the solution of same problem and results were compared whether nightstand mirror slippage occurs or not. If the mirror of nightstand can stand out against rotate without slipping as long as 72 hours, it is called as 'no slippage'. In this context, the modeling of PB and ANN came to conclusion as 'no slippage'. This is the first time we used PB and ANN in the literature for a furniture manufacturing firm.

We defined significant factors with ANOVA used for PB and sensitivity analysis of ANN was performed for verification. Environment temperature, anti slip tape quality and their interactions were the significant factors for both modeling. Customer complaints were decreased by 60 % in products / processes with real production of optimal factor settings. Thus, we found applicability of PB and ANN for nightstand mirror slippage problem. In the future, optimum solutions can be compared by using different meta- heuristics such as Genetic algorithm, Particle swarm optimization and Bees Algorithm.

References

1. PronzatoL.Optimal experimental design and some related control problems. Automatica 200744303325DOI:j.automatica.2007.05.016
2. AntonyJ.Design of Experiments for Engineers and Scientists. Elsevier Science & Technology Books: 20034448
3. Lazic ZR.Design Of Experiments In Chemical Engineering: A Practical Guide, Wiley: Mörlenbach; 2004
4. SwinnenI. A.BernaertsK.GysemansK.Van ImpeJ. F.Quantifying microbial lag phenomena due to a sudden rise in temperature: a systematic macroscopic study. International Journal of Food Microbiology 20051008596

5. Box GEP, Hunter JS, Hunter WG. Statistics for Experimenters: An Introduction to Design, Data Analysis and Model Building, Wiley Interscience: NewYork; 1978
6. TayK. M.ButlerC.Methodologies for experimental design: A survey, comparison, and future predictions. Quality Engineering 199911343356
7. ZeydanM.Modelling the woven fabric strength using artificial neural network and Taguchi methodologies. International Journal of Clothing Science and Technology 200720104118DOI:
8. PeterkaP. '.Design of Experiments for Six Sigma', (http://ezinearticles.com/?Design-of-Experiments-for-Six-Sigma&id=101212)
9. LindermanK.SchroederR. G.ZaheerS.ASChooSix.SigmaA.Goal-TheoreticPerspective.Journal of Operations Management 200321193203DOI :S0272-6963(02)00087-6
10. Hoerl RW.Six Sigma Black Belts: What do they need to know? Journal of Quality Technology, 200133391417
11. CaulcuttR.Why is Six Sigma so successful? Journal of Applied Statistics 200128301306
12. PyzdekT.SixSigma.HandbookCompleteA.Guidefor.GreenBelts.AckBelts. Managersat.alllevels.Mc Graw Hill: New York; 2003
13. KumarM.AntonyJ.AntonyF. J.MaduC. N.Winning customer loyalty in an automotive company through six sigma: a case study. Quality and Reliability Engineering International, 200723849866DOI:qre.840
14. MillerA.SitterR. R.Choosing columns from the 12-run Plackett-Burman design. Statistics & Probability Letters 2004; 67193201DOI:j.spl.2004
15. MeyerR. D.SteinbergD. M.BoxG.Follow-up designs to resolve confounding in multifactor experiments. Technometrics 199638303313
16. DengL. Y.TangB.Generalized resolution and minimum aberration criteria for Plackett-Burman and other nonregular factorial designs. Statistica Sinica 1999910711082
17. Anderson MJ and Whitcomb PJ. DOE Simplified- Practical Tools for Effective Experimentation.Productivity Inc.: Portland; 2000
18. Williams KR. Comparing screening designs.Industrial Engineering Chemistry 1963DOI:ie50642a004
19. TancoM.VilesE.IlzarbeL.MJAlvarezImplementation of design of experiments projects in industry. Applied Stochastic Models in Business and Industry 200925478505DOI:asmb.779
20. http://www.learnartificialneuralnetworks.com\,"Neural Networks- Training Models and algorithms".
21. Lin HL, Chou CP.Modeling and optimization of Nd:YAG laser micro-weld process using Taguchi Method and a neural network. International Journal of Advanced Manufacturing Technology 200837513522DOI:s00170-007-0982-3

22. SantiagoD. F. A.PederivaR.Comparison of optimization techniques of neural networks training for faults diagnostic of rotating machinery. Mercanica Computacional 20022119121921
23. CınarM.EnginM.EnginE. Z.AtesciY. Z.Early prostate cancer diagnosis by using artificial neural networks and support vector machines. Expert Systems with Applications 20093663576361DOI:j.eswa.2008.08.010
24. TuJ. V.Advantages and disadvantages of using artificial neural networks versus logistic regression for predicting medical outcomes. Journal of Clinical Epidemiology 19964912251231DOI:10.1016/S0895-4356(96)00002-9
25. Wu SJ, Shiah SW and Yu WL.Parametric analysis of proton exchange membrane fuel cell performance by using the Taguchi method and a neural network. Renewable Energy 200934135144DOI:10.1016/j.renene.2008.03.006
26. HaykinS.Neural-ANetworks.ComprehensiveFoundation.MacmillanCollege. PublishingNewYork: 1994
27. DidierC.FornoG.EtcheverrigarayM.KratjeR.GoicoecheaH.Novel chemometric strategy based on the application of artificial neural networks to crossed mixture design for the improvement of recombinant protein production in continuous culture. Analytica Chimica Acta 200965016774DOI:10.1016/j.aca.2009.07.051
28. Lo YL and Tsao CC.Integrated taguchi method and neural network analysis of physical profiling in the wire bonding process. IEEE Transactions On Components And Packaging Technologies 200225270277DOI:TCAPT.2002.1010017
29. The Statsoft Website. http://www.statsoft.com/products/statistica-Automated-Neural-Networks/
30. Olden JD, Joy MK, Death RG.An accurate comparison of methods for quantifying variable importance in artificial neural networks using simulated data. Ecological Modelling 2004178389397DOI:j.ecolmodel.2004.03.013
31. DjavanB.RemziM.ZlottaA.SeitzC.SnowP.MarbergerM.Novel artificial neural network for early detection of prostate cancer. Journal of Clinical Oncology 200220921929

6

Redesigning the Service Process for Total Quality in Government Hospitals: Evidence From Kwara State

1. Introduction

Total Quality Management (TQM) application in Public Health sector is recent and it has begun to redefine the administrative infrastructure and mindset of managers in different fronts. The old management models in public sector that hardly differentiate between management and leaderships are now beginning to change through identification and focus on meeting and exceeding the needs of customers (citizens). The application of TQM tools and techniques has begun to re-orientate managers and policy makers' arbitrary command and control methodology with strong emphasis being placed on leadership, teamwork and continuous learning. The old culture of top-down management where the subordinates must wait endlessly for their superiors before they can perform has impinged the implementation of TQM in government hospitals.

Research has found that 4 percent of hospital patients suffer an avoidable injury, 7 percent experience medication error, and 45 percent experience some medical mismanagement (Andrew et al 1977). In a recent study conducted by Lagasse et al (1995), 8 percent anesthetics error were found to be due to human error and 92 percent due to system error.

Health care organizations therefore can use service process redesign to minimize the system error. The quality of service delivered is contingent on the service process that is put in place in organizations. Service process has been therefore identified as the core of service delivery, not only frequently visible to customers (patients), but often constituting the very service itself (Johnston and

Clerk, 2000). Unfortunately, many private organizations and public outfits devote little attention to understanding and designing this very aspect of their businesses.

Once the service process is erroneously designed, it has a way of affecting the quality of service delivered to the patients of government hospitals.

Health care redesign can be broadly defined as thinking through from scratch, the best process to achieve speedy and effective care from a patient perspective, identifying where delays, unnecessary steps or potential error are built in the process and then redesign the process to remove them and dramatically improve the quality of care (Locock, 2003). Service process redesign therefore is used to tackle variation in the quality of care and improve patients' satisfaction. Service process redesign is therefore a radical challenge to traditional assumptions and practices which involves thinking through the best process to achieve speedy and effective patients' care.

Redesigning the service process in an organization therefore involves assessing how the current functions, structures and responsibilities are operating and identifying what would be done to improve the efficiency and effectiveness of operation within the unit.

Ching chow (2003) conducted a survey for a hospital to identify the major obstacles to TQM implementation in health care industry as;

- Organizational structure: Traditionally, the health care organizations use functional-hierarchical structure as the base which may cause poor communication between sections.
- Leadership style: Most leaders of health care organizations are specialists in their profession with authority. Hence, the unchallengeable leadership style does not allow health care professionals to accept the opinions of their subordinates.
- Organizational culture: The health care organizational structure and leadership style create a highly hierarchical, bureaucratic and authoritarian culture which conflict with the idea of empowerment of subordinates.
- Professional autonomy: The physicians, medical technicians, and clinical professionals work independently in their fields which makes the issue of teamwork (which is the basis of TQM), impossible. Different departments might have different views on TQM.
- Lack of Consensus: Misunderstanding might ensue between the physicians with respect to application of TQM. While TQM practices may be considered as significant for administrative efficiency and service quality, it cannot be applied for medical treatment. This explains why health care professionals do not have strong enthusiasm for its adoption.

- Internal requirement domination: The health care organizations tend to focus on their internal requirement for medical treatment rather than the patients' needs. The ethics of the profession are strictly followed. The emergency situation of a patient into the hospital will not make a physician to violate the process of diagnosis.

- Efficiency-oriented: Efficiency is the hallmark of medical profession and not how much income they would make from surgeries and chemical treatments. The objective is qualitative performance. Hence, such bias tends to affect their understanding of patients' conditions and improper judgment resulting in poor quality.

- Manpower shortfall: Insufficient manpower in hospitals means that professional specialists are overloaded in their work. Since the physicians are much more involved in their routine work, participating in TQM implementation is considered as an added responsibility that they may not have time for.

The objective of this paper therefore is to investigate how redesigning of the service process can engender total quality in Kwara state government hospitals as well as removing barriers to TQM implementation.

2. Problem statement

The implementation of Total Quality Management in Kwara State Hospitals is fraught with some difficulties. These difficulties have made the attainment of quality goals by the government elusive. TQM has been described as a co-operative form of doing business which relies on the talent and capabilities of both labour and management to improve quality and productivity continually using teams (Plek 1995). This definition has three ingredients necessary for TQM to flourish in any service organization. They are:

1. Participative Management
2. Continuous Process improvement and
3. The use of Teams.

Redesigning the hospital process falls under the second ingredient. Hence, successful TQM implementation is dependent on employee participation, employee training, technology, employee-patients' relationship, drug administration, employee motivation and care takers.

The findings and pricing, supervision conformity to standard and feedback from patients are other factors that infringed the hospital process. This study sees redesigning of the hospital process as a leeway to checkmate the poor quality service of the hospitals with the intent of satisfying the patients of the hospitals.

3. Methodology

There are sixteen government hospitals in Kwara state which constitute the population. But for the sake of the study, 4 specialist hospitals in Ifelodun, Ilorin East, Moro and Offa are the sampled hospitals for this study. This is because these four specialist hospitals constitute the referral centres for General Hospitals and Community health centres that exist in the state.

50 copies of questionnaire were served to 50 patients per sampled hospitals. A Likert scale of 5 points was used to measure the level of agreement and disagreement by the respondents. The response format is as follows:

SA – Strongly Agree

A – Agree

N – Not sure

D – Disagree

SD – Strongly Disagree

Frequency Distribution was used to analyze the data collected and to examine the pattern of response to each variable under investigation.

4. Data analysis

Since the study seeks to investigate how redesigning the service process can engender quality service in Government hospitals which will consequently affect the patients' satisfaction, the frequency counts were used to capture the responses of the respondents. From the gender distribution of respondents below

Table 1. Gender distribution of respondents

Specialists Hospitals	Male	Female	Total
Ifelodun	18	32	50
Ilorin East	21	29	50
Moro	22	28	50
Offa	26	24	50
Total	87	113	200

From the table above, 43.5% of the respondents were males while 56.5% were females. In other words, across the four local governments where the specialist hospitals were, there were more female patients than males. This is actually evident from the table as it was only in Offa specialist hospital that we had more male patients than females. This trend of female patients' dominance except in Offa LGA may not be unconnected with gynecological related sicknesses. The eagerness of the patients to fill the questionnaire was a sign of service dissatisfaction which in turn suggests the need for redesign of the service process.

Table 2. Respondents' age-range

Age classification	Frequency (n)	Percentage (%)
21-30 years	40	20.0
31-40 years	30	15.0
41-50 years	92	46.0
51-60 years	15	7.5
61 and above	23	11.5
Total	200	100.0

From the respondents' age classification above, 46.0% of the respondents were within the age of 41-50 years while 20% of the respondents fall within the age range of 21-30 years. The reason for the high rate of patients within the 41-50 years category may not be unconnected with the Menopausal issues since the respondents are female-dominated. 81% of the respondents were between the age of 21 and 50 years which is an indication of level of literacy of the respondents. In other words, they seem to understand the process of service quality and could say categorically whether they are satisfied or not.

Table 3. Respondents' marital status

Marital status	Frequency	Percentage (%)
Single	50	25.0
Married	100	50.0
Widowed	20	10.0
Divorce	30	15.0
Total	200	100.0

With reference to the above table, 25% were singles, 50% married, 15% divorcee & 10% widows and widowers.

Table 4. Distribution of Respondents on service process infringements

Process infringements	Frequency	Percentage (%)
Employees participation	18	9.0
Employees Training	20	10.0
Technology	14	7.0
Employee-patients' relationship	25	12.5
Drug administration	16	8.0
Employees motivation	20	10.0
Care-takers	26	13.0
Language barriers	12	6.0
Funding and pricing	14	7.0
Supervision	10	5.0
Conformity to standards	15	7.5
Feedback from patients	10	5.0
Total	200	100.0

Four process infringement factors featured prominently in the respondents' responses to service process infringements. They are caretakers, employee-patients' relationship, employees' training and employees' motivation. The care-takers i.e., those who stay with the patients in the hospital constitute 13% of the process infringements which in-turn, impact negatively on the quality of services in the specialist hospitals. This is followed by employee-patients' relationship. This may be referred to as the interpersonal factors that boarder on Professional-patients' relationships. Employee training and employee motivation can also affect the quality of service where they are not available.

Table 5. Ranking of process infringement factors

Process infringement factors	Percentage (%)	Ranking
Care-takers	13.0	1
Employee-patients' relationship	12.5	2
Employees' training	10.0	3
Employees' motivation	10.0	3
Employees' participation	9.0	5
Drug administration	8.0	6
Conformity to standards	7.5	7
Technology	7.0	8
Funding & pricing	7.0	8
Language barriers	6.0	10
Supervision	5.0	11
Feedback from patients	5.0	11
Total	100	

Table 6. Distribution of respondents on methods of overcoming the process infringement factors

Methods	Frequency	Percentage (%)
Detection	20	10
Process mapping	50	25
Statistical process control	60	30
Failsafing	70	35
Total	200	100

From the data on table 6, 70 respondents favoured 'failsafing' which is a simple but effective technique to reduce the likelihood of process failure and ensure that both employees and patients do the right thing. The 70 respondents accounted for 35% solution to process infringements. 30% of the process infringements can be overcome with statistical control.

5. Hypothesis testing

The only hypothesis tested in this study is to find out whether redesigning the service process has a way of enhancing total quality in Kwara state specialist hospitals.

Ho: Redesigning the service process does not affect quality of services in government hospitals.

H1: Redesigning the service process significantly affect the quality of service.

Table 7. Distribution of respondents on redesign of the service process for quality service

Options	SA	A	N	D	SD	Total
Yes	110	12	3	10	6	141
No	20	15	5	11	8	59
Total	130	27	8	21	14	200

Table 7 is the observed value of respondents of service process redesign and quality service. To obtain the expected value, we simply use the formulae

Fe=Row total x Column total/Grand total

Table 8. Frequency of service process redesign on quality

fo	fe	(fo-fe)	(fo-fe)2/fe
110	91.65	336.72	3.67
12	19.04	49.56	2.60
3	5.64	6.96	1.23
10	14.8	21.12	1.42
6	9.87	14.98	1.51
20	13.4	338.56	8.82
15	7.96	49.56	6.22
5	2.36	6.96	2.95
11	6.19	23.14	3.74
8	4.13	14.97	3.62
		Total X2 = 35.8	

From the table 8 above, the X2 calculated is 35.78 while the X2-tabulated at 5% significance level and at (2-1) (5-1) = 4 degree of freedom i.e. 14.9. Since X2 calculated is greater than X2 tabulated, we reject the null hypothesis that says redesigning the service process does not affect the quality of service and accept the alternative hypothesis that says redesigning the service process significantly affect the quality of service rendered by hospitals.

From the hypothesis tested, redesigning the service process has a way of affecting the quality of service as well as patients' satisfaction.

6. Recommendations

As we have rightly observed in the literature review, all stakeholders in health care have a role to play in ensuring that the process infringements are checkmated and minimized.

The following recommendations are important for the improvement of the service process:

- ❖ The dichotomy between doctors and nurses in terms of professionalism which make certain area of Medicare a 'no go area' for nurses must be checked, so that when doctors are not available, the nurses can to a certain extent come in to save the life of the patients.
- ❖ Employees' training and refresher courses are very important to sensitize the workers on latest technological evolution.
- ❖ Employees-patients relationship must be cordial because whatever injury suffered by the patient has a way of affecting the hospitals.
- ❖ Care-takers that will stay with the patients must be educated such that a wrong administration of drug is not given. Alternatively, the hospital employees particularly the nurses should be charged with the responsibility of administering the drugs.
- ❖ Where language barrier is to infringe the service process, an interpreter may be immediately hired to serve as a stop-gap for wrong diagnosis.
- ❖ Government should appoint supervisors or agency who will not compromise quality for material gain. Those supervisors and agencies can interact with the patients to know what their misgivings are.
- ❖ The agency in (IV) above should create a way through which those whom have been discharged from the hospital will give a feedback to the hospitals.
- ❖ Government as a matter of policy should develop a motivational formulae that will serve as an impetus for greater and efficient service.
- ❖ Government should release funds for the management of hospitals

7. Conclusion

This paper is of the opinion that what we refer to as infringement factors to the service process in developing nations like Nigeria may not actually be the infringement factors in developed nations. Hence, this study would provoke further investigations into the service process infringements in developed world. If patients' satisfaction is the basis for establishing specialist hospitals, anything that will make that objective unrealizable must be removed. This exactly is what service process redesign intends to achieve.

References

1. AlexanderC. 1971 Notes on the synthesis of form. Cambridge, MA: Harvard University Press.
2. AndrewsL.StockingC.KrizekT.GorHieb. L.KrizekC.VargishT.SieglerM. 1997 "An alternative strategy for studying adverse events in medical care". The Lancet. 349 309313 .
3. BuurJ.BinderT.et al. 2000 "Taking Video beyond "Hard Data" in User Centred Design." Design. Participatory Design Conference (PDC 2000).
4. BuurJ.SoendergaardA. 2000 "Video Card Game: An augmented environment for User Centred Design discussions." Designing Augmented Reality Environments (DARE 2000), Helsingor.
5. Yang, Ching-Chow 2003 "Six sigma and Total Quality Management"
6. CollierD. A. 1994 Research opportunities in service process design "
7. GaverB.DunneT.PacentiE. 1999 "Design: Cultural Probes." Interaction 6 1 2129 .
8. LindsayC.RocchiS. 2003 "'Highly Customerised Solutions'- The Context of Use Co-Research Methodology". Innovating for Sustainability. 11th International Conference of Greening of Industry Network, San Francisco.
9. Johnston and Clark 2000 "The service concept: the missing link in service design research"
10. KumarV. 2004 User Insights Tool: a sharable database for user research. Chicago, Design Institute at IIT.
11. LagasseR.SteinbergE.KatzR.SaubermannA. 1995 "Defining Quality of Perioperative Care by Statistical Process Control of Adverse Outcomes", Anesthesiology, 82 5 11851188 .
12. LocockL. 2003 "Healthcare redesign: meaning, origins and application." Department of Social Policy and Social Work, University of Oxford.
13. MorelliN. 2006 "Designing product/service systems: A methodological exploration. Design (3 EOF17 EOF), 3-1
14. PlekP. E.Techniques".formanaging.quality" Hospital and Health Services Administration, 40 40 1 spring 1995.
15. ShostackG. L. 1982 How to design a service." European Journal of Marketing, 16 1 4963 .
16. ShostackG. L. 1984 "Designing services that deliver." Harvard Business Review, 62(1), 133-139
17. SimonH. A. 1969 The sciences of the artificial (1st Ed.) Cambridge, MA: MIT Press.

18. ShortP. J.adRahim. M. A. 1995 "Total Quality Management in Hospitals", ShortP. J.adRahim. M. A. (1995), "Total Quality Management in Hospitals", Total Quality Management, Vol.6 No.3, Pp.255-63., 6 3 255263 .
19. Sureshchandar, G. S. (2001), "A Conceptual Model for total quality management in service organization", Total Quality Management, Vol. 12 No.3, Pp.341-63.
20. TerziovskjM.SohalA.SamsonD. 1996 "Best Practice Implementation of Total Quality Management: Multiple Cross Case Analysis of Manufacturing and Services Organizations", Total Quality Management 7(5) 459481
21. ThompsonK. R. 1998 "Confronting the paradoxes in a Total Quality Environment", Organizational Dynamics, winter, 6274 .
22. WaltonM. 1986 "The Deming Management Method" (Perigee Books).
23. WilkinsonA.Willmolt-II. 1996 "Quality Management Problems and Pitfalls a critical perspective", International Journal of Quality & Reliability Management 13(2), 5565 .
24. YangC. C. 1997 "Quality is the Best Strategy in Competition." APEX International Management Consulting Co., Taiwan.
25. ZairiM. 1999 "Managing Excellence: Policy and Strategy", the TQM Magazine, 11(2), 7479 .

7

Some Applicable Methods to Analyze and Optimize System Processes in Quality Management

1. Introduction

The complexity of present man-made systems has reached an unprecedented level. In fact any system is grounded on computer technologies in the sense that it contains computer elements or is modeled or supported with the help of computer. This trend resulted in new opportunities and at the same time caused additional difficulties. Shortcomings in integration of scientific, engineering, management, and financial areas, which are used to ensure an effective system development and employment, now become more obvious. Today processes of system life cycle in different conditions and threats are the main objects for forecasting, analysis and optimization. Indeed these objective changes become the main reason for establishing the first system engineering ISO/IEC standard ISO/IEC 15288 "System Engineering - System Life Cycle Processes" (since 2002). Covering systems in industrial, energetic, transport, aerospace, military and other fields, this standard recommends to perform only the actions that were substantiated and not to act in the directions, which were not estimated and justified. It means that feature of our time is the turn to system engineering – see Figure 1.

Up-to-date approach to system maturity refers us to international standards ISO 9001 and 9004, ISO/IEC 15026, 15504, CMMI etc. It is clear without "system analysts" there is not achievable the highest level "Optimizing", but also a previous "Predictable" level. However many customers and Chief Designers often fail to take quantitative system requirements into consideration, they do so wittingly or through an oversight. From now on these omissions do not conform to the requirements of the international standards. It is only the beginning. What will be the continuation?

Nowadays if comprehensive quantitative system requirements were not established in quality management, the system efficiency and customer satisfaction can not be controlled and confirmed. To great regret in many application areas the system requirements do not allow to understand the true reasons of failures. However the degree of processes influences on a final result should be estimated and often may be managed! Let's consider information system. Standards recommend to propose requirements for system reliability during given period, for the information timeliness, completeness, actuality, faultlessness after checking, correctness of processing, protection against dangerous influences and unauthorized accesses, and if needed information confidentiality. It means that those system requirements should be set that are focused on customer satisfaction according to used information.

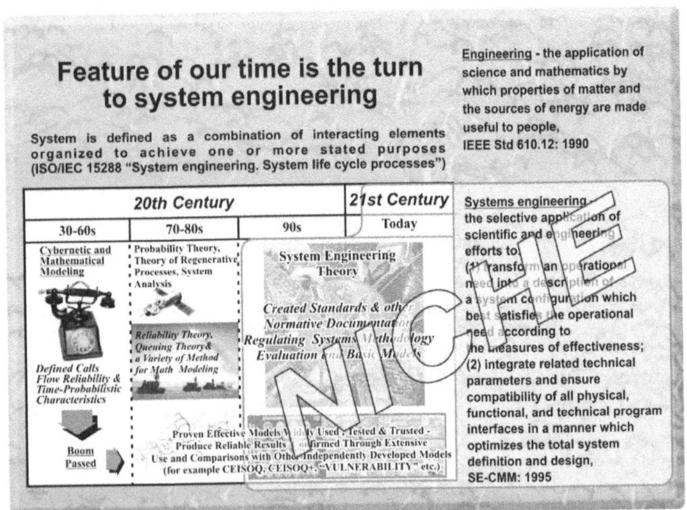

Figure 1. Now applicable models, methods, software tools and technologies support standard system requirements

Unfortunately many graduates from technical colleges and universities do not use the foundations of "system analysis", "operations research", "mathematical modelling" in quality management, they missed the particulars of existing methods and models. And without use of mathematical models they can only dream about deep logical investigation to predict forthcoming effects. That pawns the doubtful base of high risks to the future systems. There is more deeper increasing break between needs for competitive system quality and methodical opportunities of the experts, called these needs to estimate, substantiate and satisfy without wasted expenses. Time has come to make again more popular mathematical models for rational solving the problems of quality management.

The goal of this work is to propose models, methods, and software tools well-tested in practice, for forecasting quality and risks as applied to newly developed and currently operated manufacture, power generation, transport, engineering, information, control and measurement, food storage, quality assurance and security systems. Presented work is devoted to the researches of standard processes for providing effective quality management in systems life cycle. It covers logically closed contour: « system requirements of standards – supporting mathematical models to estimate probabilities of success, risks, profits and damages – ways of rational management». Thereby the reader can substantiate answers on system engineering questions: «How to reach in quality management the level of international standards?», «Is expected quality achievable?», «Can be the system requirements met?», «How much safe are those or others scenarios?», «What about the real risks, profits and possible damages?», «What choice is rational?», «What measures are more effective?», «What rational measures should lead to estimated effect without waste expenses, when, by which controllable and uncontrollable conditions and costs?» and others. The answers may be received before critical events (not only after these events).

While reading this work, the reader will be shipped in logic of standard processes, which can spend resources and be compared on a timeline in different conditions. This implies an understanding of system requirements, strength of mathematical models and optimization methods, i. e. everything that is vitally important but has never been in the focus of attention of either technical specialists or students. Certainly, in the justification of such indifference it is possible to refer to doubts of the famous physicist Albert Einstein. He has spoken, as far as the laws of mathematics refer to reality, they are not certain, and as far as they are certain, they do not refer to reality. But now we are living at the time of innovations! While understanding that this century-old dictum is negative for the chances of our work to be a success, we nevertheless decided not to 'loose' advanced physicists for the sake of their own interests. Thereto 10 practical examples are investigated and explained, the detailed 'hardware' of the work, other hundreds examples and routine comments are gathered at the References of the book and on site www.mathmodels.net. These new results are a humble initial contribution of author's to the 'coin-box' of the knowledge base for decision support to improve quality management, boost the efficiency and safety of different systems and/or reduction of nonproductive costs.

And now we can review briefly our own experience. There is an inconceivable ocean of practical problems which are subjects to the decision with use of system forecasting. Existing decisions from one area are far from being always appeared applicable in other areas. As a result in 9 cases from 10 it was necessary to create original mathematical models. Their quantity grew. Further, having analyzed problems and approaches to the decision of system engineering, there was clear uniformity of a train of thought of modern system analysts in various spheres. The logic scheme everywhere is identical: at first the set of destabilizing factors and threats against quality and-or safety is defined, then taking into account available resources the possible measures of neutralization should be chosen or developed. A vulnerability set of system comes to light. Technologies of system control and recovery of broken integrity should be implemented as counteraction against destabilizing factors and threats (there where expediently - continuous monitoring of conditions is used). Thus at every step of system life cycle the development of processes is supported by probabilistic forecasts, criteria of optimization are chosen in depending on the problem purposes. According to these rational decisions are made on the base of mathematical modelling and optimization.

Natural tests of road also are fraught with serious consequences. Therefore mathematical modelling becomes more and more popular. For this reason the described universal scheme of the system analysis has laid down in a basis of the presented below models methods and software tools. Modelling shouldn't be carried out only for modelling itself. If as a result of modelling we get only one value it is not quite clear how we should appreciate it (such a disadvantage is typical for simulation models, which require thousands executions of the same data, what is not made sometimes because of time-shortage; inverse problems solving is almost impossible for such kind of models). That is why the offered software tool suites, which uses only analytical models, is developed in such a way that it is possible in a split second to carry out computations, catch tendencies, reveal stability of processes in case of input data changes in the range –50% +200% and in a few minutes (!) to find admissible solutions of complex inverse synthesis problems. As a result of the offered models use a system analyst gets an ability to sense not a computed point but the whole quality field, which may be appropriate to system at different scenarios of system operation and environment behavior. Why you should trust to evaluating results by the offered models? In other words how models adequacy is substantiated? Though any answer to these questions won't be irrefragable for a certain system we shall try to formulate our arguments (experience readers understand that any model needs in similar arguments).

Argument 1. The fact is that while shaping models all mathematical results are initially drawn in the integral form. As input data are somehow connected with time after choosing distribution functions characterizing these data there were selected the gamma – distribution and the Erlang's distribution. Mathematicians know that these distributions approximate sums of positively distributed random variables well. Every temporary data are as a matter of fact such a sum of compound time expenses. Studies of regularities (Feller, 1971) have shown that extremes are achieved on bounds of these distributions, i.e. of exponential and deterministic (discrete) distributions. Thus, real values will be somewhere between lower and upper estimations of the software tools, if computation results are presented by one curved line they are lower estimations. The results reflect pessimistic value for following using.

Argument 2. As a basis of our models we used the probability theory and the theory of regenerative processes (i.e. recurring processes). Proofs of basic theoretical results are received, for example, by (Gnedenko, 1973; Klimov, 1983). If to return in the 70-s of the last century we may remember the boom of mathematical modelling, defining calls flow reliable and time-probabilistic characteristics. The boom passed and appeared the reliability theory, the queuing theory and a variety of models, which proved themselves to be effective. There are created standards and other normative documents regulating system methodical evaluations on the basis of these models. Nowadays these models are widely used and trusted because they produce reliable results confirmed in the course of time. It is worth to remind that these created theories and models are based on the probability theory and the theory of regenerative processes. The models of subsections 3.2 – 3.4 are the classical models of the 70-s improved and developed to meet the requirements of the present time. The other models are created on the basis of the limit theorem for regenerative processes developed in the 70-80-s in Moscow State University on the faculty of computing mathematics and cybernetics.

Argument 3. Skilled analysts know that if a probabilistic analytical model is incorrect then if input data are changed in the range from -∞ to +∞ there are always errors appearing either in infraction the probability theory laws or in illogic of dependencies behavior (most probably on the bounds of possible values) or in impossibility of obtained effects physical explanation. Bounds of input data in the offered software tools are assigned in the range from -∞ to +∞ (more precisely from 10-8 milliseconds to 108 years). Three-year testing of models including beta testing by fifty different independent companies raise confidence in software tools algorithmic correctness.

Argument 4. As far as possible any designer tends to use several models of different authors. If results of different models use are not divergent a designer begins to trust not only to results but also to the models. Comparison of results of the presented software tools with results of other models use proved their high adequacy (concerning computations of reliability and time-probabilistic characteristics, the other models don't have analogues).

The offered software tools are an original Russian creation patented. They have been presented at seminars, symposiums, conferences, ISO/IEC working groups and other forums since 2000 in Russia, Australia, Canada, China, Finland, France, Germany, Kuwait, Luxembourg, Poland, Serbia, Ukraine, the USA, etc. The software tools were awarded by the Golden Medal of the International Innovation and Investment Salon and the International Exhibition "Intellectual Robots", acknowledged on the World's fair of information technologies CeBIT in Germany, noted by diplomas of the Hanover Industrial Exhibition and the Russian exhibitions of software. The offered technology of modelling through the Internet has been acknowledged as the best project-2007 by the National Association of Innovations and Developments of Information Technologies of Russia.

Having analyzed results of long-term our practice, we, authors, have noticed the following. Many scientific researches, practical investigations, implementations and recommendations based on use of our models, methods, and software tools were bringing increasingly deep satisfaction not only to ourselves but, most important:

- to developers, i. e. all of our colleagues involved in the works (since the obtained results can be proved step-by-step and their usefulness checked; forecasts were confirmed in time; and, respectively, the number of profitable orders has been growing),
- to customers (since we managed to convince them that the residual system risks may and should be mitigated proactively; and now they have scientific justification of the amount of investments adequate to achieved quality and safety levels that may be guaranteed for the allocated money),
- to users (the forecast made in time has mobilized them on the basis of the 'forewarned is forearmed' concept; using our recommendations, in utilization stage they can extract from the system the best effects, that were assumed in concept, design&development and support stages).

This work is purposed for systems analysts from customers, developers, users, as well as investigators and staff of quality and security management, experts of testing laboratories and certification bodies. It can be used in system life cycle to form system requirements, compare different processes, substantiate technical decisions, carrying out tests, adjust technological parameters, estimate quality and risks. The decisions, scientifically proved by the offered models and software tools, can provide purposeful essential improvement of quality and mitigation of risks and decrease expenses for created and operating systems. The spectrum of the explored systems is indeed broad; it includes systems operated by government agencies, manufacturing structures (including enterprises, oil-and-gas and transport facilities, and hazardous production systems), food storage, power generation, financial and business, aviation and space industry, emergency services, municipal economy, military, etc. Moreover, our assessments and forecasts are generated much faster, feature innovations, have invariably high quality and, most important, the expected effects may be easily interpreted (what specifically is the result and how it can be reached) regardless of whether it pertains to increase in gains or reduction in losses. Eventually, having gained experience and being sure that those instruments are of use, we decided to share our knowledge and skills for analyzing and optimizing system processes in quality management. It should be stressed from the very beginning that no one forces you to use these proposed models, methods, and software tools. Any author trusts primarily his/her own models and is suspicious about someone else's if uses them at all. From this perspective we also understand our colleagues from the writers' community, share their doubts and nevertheless invite them... Join us, the esteemed reader. The knowledge that you will gain after even brief acquaintance with the work or just browsing the book and then comprehending its content will not allow you to continue unsystematic life without forecasting quality and risks! You can easily verify this author's forecast.

2. Review of system processes to reveal general engineering problems that are due to be solved by the mathematical modeling

System analysis is an important science intensive process, which is connected with system concept, development, production, utilization and support. As a result of adequate system analysis we extend our knowledge about systems, obtained quality dependency on different system characteristics and about a degree of system purposes achievement. This knowledge allows a customer to formulate substantiated requirements and specifications, a developer - to implement them rationally without wasted expenses, a user – to use system potential in the most effective way. Let's review some system standards - ISO 9001 "Quality Management Systems - Requirements", ISO/IEC 15288 "System Engineering – System Life Cycle Processes", ISO/IEC 12207 "Information Technology - Software Life Cycle Processes", ISO/IEC 15504 "Information Technology –Process Assessment", ISO/IEC 17799 "Code of Practice for Information Security Management", IEC 60300 «Dependability Management», IEC 61508 "Functional safety of electrical/ electronic/ programmable electronic safety-related systems", CMMI "Capability Maturity Model Integration", "GOST RV 51987 "Information technology. Set of standards for automated system. The typical requirements and metrics for information systems operation quality. General provisions", some standards for use in the oil&gas industry (ISO 10418 "Basic surface safety systems", ISO 13702 "Control & mitigation of fire & explosion", ISO 14224 "Reliability/maintenance data", ISO 15544 "Emergency response", ISO 15663 "Life cycle costing", ISO 17776 "Assessment of hazardous situations" etc. - from the role of system analysis point of view. These are the representative part of the modern system and software engineering standards.

In compliance with ISO 9001 to all processes there can be applied methodology known as "Plan-Do-Check-Act" (PDCA). Plan: from system analysis point of view it means that all parties should understand in equal measure the essence of customer requirements, metrics and admissible level of goals achievement. Do: it implies that implemented processes meet customer requirement on admissible level. Check: there should be used methods and tools for evaluations. Act: used methods should allow appearing dependencies and determining adequately an effective way for expected improvement. For any improvement a documented procedure shall be established to define requirements for determining potential nonconformities and their causes, evaluating the need for action to prevent occurrence of nonconformities, determining and implementing action needed.

In compliance with the standard ISO/IEC 15288 system analysis actions are the main actions for achievement system purposes in life cycle including required propositions in Agreement, Enterprise, Project and Technical Processes. In compliance with the standards ISO/IEC 12207 system analysis problems are to be solved to meet system requirements with resources optimization. The standard ISO/IEC 17799 is used for providing information security purposes. This and others like standards in security area (for example, ISO/IEC 15443, ISO/IEC 13335 etc.) imply that high effectiveness of system protection measures should be evaluated and confirmed quantitatively. It means that any system security evaluations need in an adequate mathematical methodology. The standard IEC 60300 describes the approaches to the risk analysis of technological systems from system analysis point of view. The standard IEC 61508 includes Parts "Examples of methods for the determination of safety integrity levels" and "Overview of techniques and measures" that recommend to evaluate system risks. An application of CMMI allows selecting the order of improvement that best meets the organization's business objectives and mitigates the organization's areas of risk. And these results are also based on system analysis.

To understand the situation with requirements and applicable methods to analize and optimize system processes an existing practices for providing system quality and safety were reviewed. The integral results of safety analysis are presented on Figure 2.

Some situations with modelling of processes for system quality are more wide viewed in this book. According to applicable mathematical models everyone (majority) solves the problems "how can", we can resume: all organizations need quantitative estimations, but only some part from them uses modelling complexes; used models are highly specialized, input and calculated metrics are adhered strongly to specificity of systems; existing modelling complexes have been created within the limits of concrete order for the systems and as a rule are very expensive.

Thus the summary of the analysis of existing approaches is the next.

1. Analysis of quality and risks is carried out mainly at qualitative level with assessments "better or worse". Independent quantitative estimations at probability level are carried out for specially created models.
2. Generally risk estimations from one sphere do not use in other spheres because of methodologies for risk analysis are different, interpretations are not identical. The methods for quantitatively risk analysis and quality analysis (on probability level) are in creating stage yet. The terms "Acceptable quality" and "Admissible risk" in use should be defined on probability scale level only in dependence on corresponding methods. As consequence probability estimations are not comparable for different areas, experience from other spheres is missing, comparisons for systems from different areas, as a rule, are not used, as universal objective scale of measurement is not established yet.
3. In all cases effective risk management for any system is based on: a) uses of materials, resources, protective technologies with more best characteristics from the point of view of safety, including integrity recovery; b) rational use of situation analysis, effective ways of the control and monitoring of conditions

and operative recovery of integrity; c) rational use of measures for risk counteraction.

4. It does not allow to solve the main problems of a substantiation of system requirements to parameters of information gathering and analysis, control, monitoring and counteraction measures at restrictions, and also to confirm about efficiency of the prevent measures for providing quality and safety!

Note. System integrity is defined as such system state when system purposes are achieved with the required quality.

In general case system methods for analyzing and optimizing are founded completely on the mathematical modelling of system processes. We understand that any process is a repeated sequence of consuming time and resources for outcome receiving. In general case the moments for any activity beginning and ending are, in mathematical words, random events on time line. Moreover, there exists the general property of all process architectures. It is a repeated performance for majority of timed activities (evaluations, comparisons, selections, controls, analysis etc.) during system life cycle - for example see on Figure 3 the problems that are due to be solved by the mathematical modelling of processes according to ISO/IEC 15288.

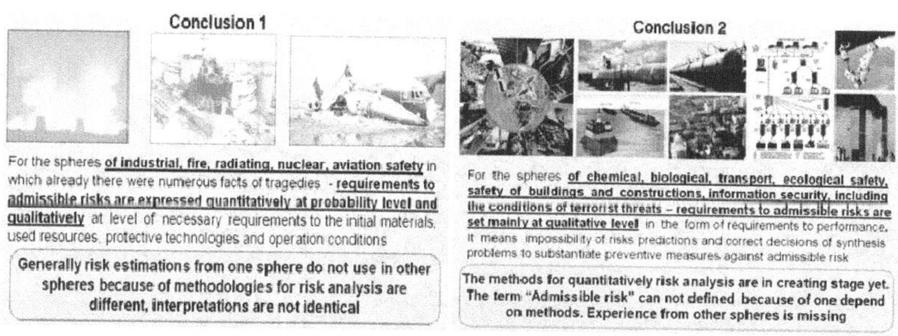

Figure 2. Conclusions of safety analysis

This work focuses on the way for extracting latent system effects from system processes by using universal metrics: probabilities of success or failure during a given period for an element, subsystem, system. Calculation of these metrics within the limits of the offered probability space built on the basis of the theory for random processes, will allow to predict outcomes on an uniform scale, quantitatively to prove levels of acceptable quality and admissible risks, to solve the problems of synthesis, answering preventive a question « What rational measures should lead to estimated effect without waste expenses, when, by which controllable and uncontrollable conditions and costs?».

Below we describe many-sided analysis of quality management. Thus we want to help the reader in solving problems of providing system effectiveness, which depends on both the reviewed system quality parameters and the parameters of pragmatic usefulness in a certain domain. They cover many important engineering problems in a systems life cycle - see Figure 4.

There exist different process-centered methods and integrated tool suites for systems analysis (see for instance Guide to the Software Engineering Body of Knowledge SWEBOK). In sections 3 and 4 we illustrate the original approaches based on mathematical modelling. Many analysis and synthesis problems and their solutions are demonstrated in section 5. However detailed mathematical definition for all problems is omitted not to overload a reader by complicated mathematical propositions, which require deep knowledge of the probability theory, theory of regenerative processes and mathematical analysis. You may find full mathematical models and their proofs (Martin (1972); Gnedenko (1973); Kleinrock (1976), Matweev & Ushakov (1984) etc.).

As a resume we can define the role of analysis and optimization system processes in compliance with modern engineering standards as decisive for rational reaching system operation quality. From analyst's point of view system analysis reduces system uncertainties and provides a quantitative basis to predict and choice in balancing business needs, quality, risks and specified requirements.

Figure 3. The problems that are due to be solved by mathematical modelling of processes

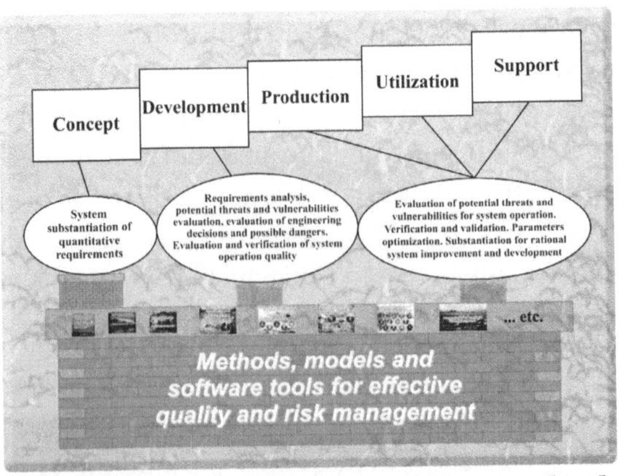

Figure 4. System engineering problems which are solved on the base of system analysis in quality management

As the first objects for demonstrating author's original approach to analyze and optimize system processes the information systems (IS) are selected. We will not attract readers' attention listing uncountable features of IS because the effect of their implementations is obvious in all spheres of human activities.

3. Models and software tools to analyze information system processes

3.1. General propositions

On information technology market there is offered a wide choice of engineering solutions that are able to satisfy functional requirements of a customer. You may choose an acceptable variant if you are guided by logical considerations. The thing is that you cannot be sure whether this variant is rational or not if to estimate it from the point of view of integral system operation goals achievement. The answer is likely to be positive if the technological solution is intended for an enterprise system, which goal is to get the highest benefit from goods manufacturing and sale. In this case the criterion for choosing a manufacturing computer system may be the one of upgrading goods quality and increasing the company profits under expenses limitations. And what will be the answer if to speak about IS, which production is output information? The criterion for providing IS high quality is the use of models and methods that allow to estimate, investigate and optimize processes of information gathering, processing, storage and producing. The basis for the functions performance is the integration of computers, software, communications and human capabilities. IS are the most important integral components of financial, transport, energy, customs, military and other SYSTEMS.

It is clear that requirements to IS operation depend on general SYSTEM purposes, use conditions (including potential threats), available resources, information sources facilities and communication requirements (see Figure 5). There is impossible to provide IS operation quality without the help of models and implementation tools. Its use allows to estimate the reliability and timeliness of information producing, the completeness, validity and confidentiality of the used information from users' point of view. This is the logical basis to create universal mathematical models and software tools which could estimate IS operation quality, compare various IS engineering projects, reveal "bottle-necks" and optimize the processes of information gathering, storage and processing. Such original mathematical models have been introduced in processes of IS development, use and maintenance.

The idea of estimating IS operation quality appeared as a result of studying potential threats to output information (see Figure 6). The results of their use to analyze technical solutions in processes of designing, developing, producing, using, supporting and certifying proved their effectiveness and multifunctional capabilities.

The main windows for choosing the mathematical models is illustrated on Figure 7. The modelling software tools complex CEISOQ+ is one of the few scientific and technical masterpieces, which satisfy most of the high requirements of the intellectual market. Moreover, this complex has appeared quite in time. The market requires the quantitative substantiation of engineering solutions and the IS quality validation. It is pleasant that CEISOQ+ developed by those who work in the field of defense reveals a new conceptual approach to quality.

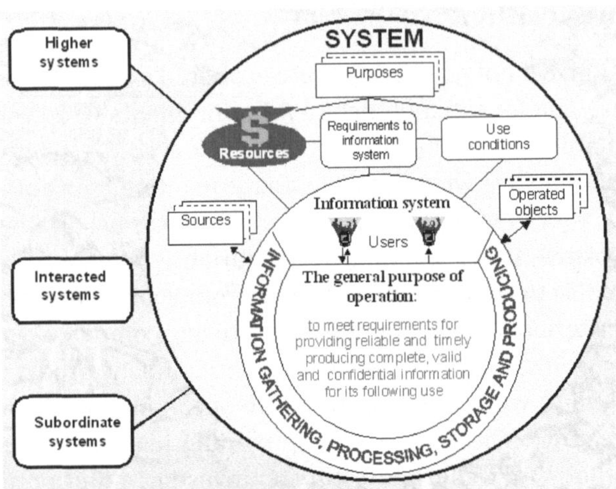

Figure 5. The place and the purpose of information system in a SYSTEM

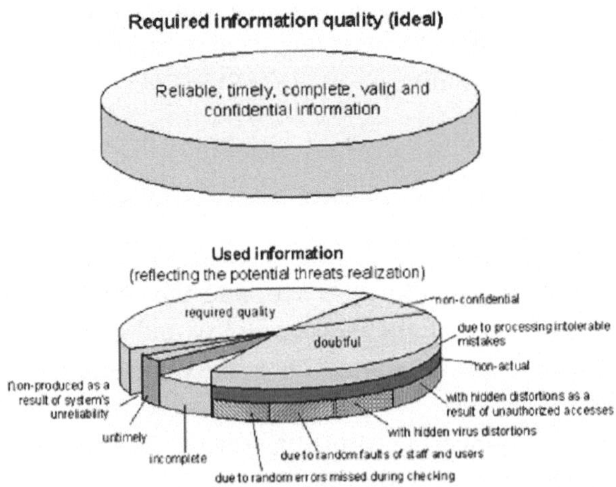

Figure 6. Potential threats to output information according to general purpose of IS operation

The created modelling Complex for Evaluation of Information Systems Operation Quality (CEISOQ+) allows to simplify and to spread the use of the next mathematical models: of functions performance by a system in conditions of unreliability of its components; complex of calls processing; of entering into IS current data concerning new objects of application domain; complex of information gathering from sources;

of information analysis; of dangerous influences on a protected system; of an unauthorized access to system resources.

The offered original mathematical models intended for estimating the level of the IS operation purpose achievement are supported by the created software tools CEISOQ+.

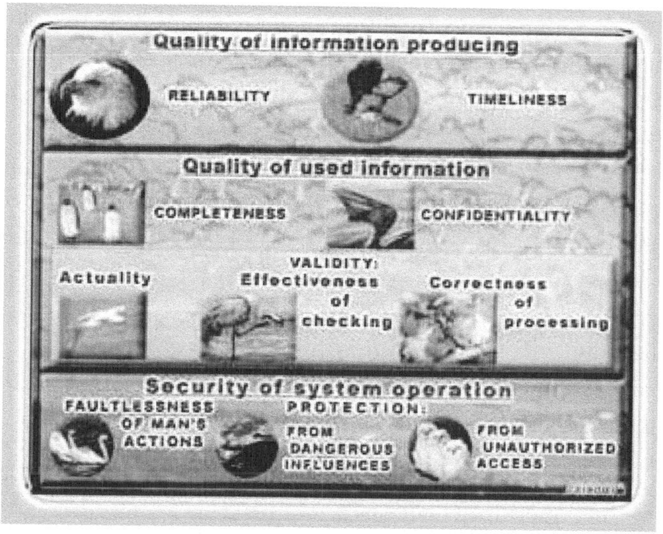

Figure 7. The main windows of software tools CEISOQ+

To make the understanding easier we didn't take into detail consideration that information quality depends on kind of input information, on functional tasks to be accomplished and on different users' requirements to conditions of IS operation. These dependencies were studied in a special complex IS operation quality investigation (Kostogryzov et al. (1994, 2000-2002)).

The software tools CEISOQ may be applied for solving such system problems appearing in an information systems life cycle as: substantiation of quantitative system requirements to hardware, software, users, staff, technologies; requirements analysis; estimation of project engineering decisions and possible danger; detection of bottle-necks; investigation of problems concerning potential threats to system operation and information security; testing, verification and validation of IS operation quality; rational optimization of IS technological parameters; substantiation of plans, projects and directions for effective system utilization, improvement and development.

Every system analyst (an IS customer, designer, developer, expert of testing laboratories and certification bodies etc.) may become a user of the software tools CEISOQ+. The CEISOQ+ may also be helpful in training programs for skilled specialists and educational programs of students studying information systems estimation.

The use of models and the software tools CEISOQ+ on different stages of an IS life cycle allows to answer the following questions: what quantitative requirements should be to hardware/software devices operation time between failures and to system repair time? which information operation processes should be duplicated and how? what processing devices and technologies should be chosen to achieve the necessary level of system throughput? what about the system tolerance to data flows changing? what data flows and functional tasks may be considered as the main causes of bottlenecks? what level of preparation, transfer and input productivity and what data gathering technologies can guarantee the completeness and actuality of information? which engineering solutions can provide the actuality of information? what about the quantitative level of information control quality? what qualification quantitative requirements should be for the staff and users? how dangerous are scenarios of environment influences and what protective technologies will provide the required security? how the use of protective technologies will influence on the IS operation quality characteristics? how the use of integrity diagnostics and security monitoring will worsen time-probabilistic characteristics of a system? what protection system effectiveness should be to prevent an unauthorized access? what about quantitative level of information security risks? etc.

This appendix is dedicated to building a probabilistic space (Ω, B, P) for the evaluation of system operation processes, where: Ω - is a limited space of elementary events; B – a class of all subspace of Ω-space, satisfied to the properties of σ-algebra; P – a probability measure on a space of elementary events Ω. Because, $\Omega=\{\omega_k\}$ is limited, there is enough to establish a reflection $w_k \rightarrow p_k = P(w_k)$ like that pk≥0 and $\sum_k p_k = 1$.

Such space (Ω, B, P) is proposed on the base of processes architectures formalization by the limited theorems for regenerative processes (Feller (1971), Gnedenko (1973), Klimov (1983), etc.) and also by using principal propositions of probability theory and theory for random processes. The proofs of the mathematical formulas used by the CEISOQ+, see (Kostogryzov et al. (1994, 2000-2002)).

3.2. Reliability of information producing

Problems of reliability have been solved already as related to technical means and systems but they are extremely urgent concerning. The reliability standards require acknowledgement of these values achievement. It is clear that without modelling acknowledgement of IS reliability, which consists of dozens territorially distributed software resources, may be obtained only as a result of its use. Such a use is a risk and every risk must be substantiated. Indeed there is no choice except modelling.

Modelling of functions performance reliability may be carried out with the help of the next model.

What about the logical idea for modelling processes from the point of view to provide reliability of information producing? From the point of formal reliability any system, subsystem or their components may be in "operable" or in "inoperable" condition during given period Θ. Let an operable condition is identified with the formulated condition "component provides reliable functions performance during period Θ ". Period, connected with system repairing after failure, is signed as "system does not provide reliable functions performance during period Θ". Then both mentioned conditions complete the set of elementary events for stochastic process $\xi_{rel.}(t)$ defining the system condition at the time t and functionality for period Θ after t, i.e.

$$\xi_{rel.}(t) = \begin{cases} \text{“system provides reliable functions performance during period}\,\theta\text{,”} \\ \text{if system is in operable condition before moment t and during period}\,\theta\,\text{begun at the moment}\,t; \\ \text{“system does not provide reliable functions performance during period}\,\theta\text{,”} \\ \text{if system is in inoperable condition at the moment}\,t\,\text{or a failure will be during period}\,\theta\,\text{begun at the moment}\,t. \end{cases}$$

The next variants are possible:

- a virtual moment t has overtaken the system in operable condition and there has not been a failure during period Θ (failure means change from operable in inoperable condition), in this case system operation is characterized by condition "system provides reliable functions performance during period Θ", i.e. the event of reliable functions performing is going on;
- a failure has happened during period Θ, in this case system operation is characterized by conditions "system does not provide reliable functions performance during period Θ";
- system is not capable for functions performing because one is in inoperable conditions at the moment t. Then it is going on the event "system does not provide reliable functions performance during period Θ".

The next statement 1 is proposed on the base of introduced formalization.

Statement 1. The limited probability of providing reliable function performance by system during the required time exists under the condition of existence for stationary probability distributions for considered characteristics and their independence. One is equal to

$$P_{rel} = \int_0^\infty \left\{ \int_t^\infty V(\tau - t)\,dN(\tau) \right\} dt \Big/ \int_0^\infty t\,d\left[N * W(t)\right], \tag{1}$$

where N(t) - is the probability distribution function (PDF) of time between neighboring failures (T_{MTBFnk} is the mean time); W(t) – is the PDF of repair time ($T_{rep.}$ is the mean time); V(t) – is the required period PDF of permanent system operation when reliability should be provided ($T_{req.}$ is the mean time);* - is the convolution sign.

The proof of this and others statements of the chapter 3 see (Kostogryzov et al. (1994, 2000-2002)) and site www.mathmodels.net.

Convolution of complex system framework into framework for one unit is implemented by usual methods (see, for example subchapter 4). The final clear analytical formulas for modelling are received by Lebesgue – integration (1) expression and convolution of complex system framework in to single-unit system.

The next variants are used in modelling of functions performance reliability: a) period Θ is strict deterministic and equals to $T_{req.}$ (discrete distribution V(t)); b) $V(t)=1-\exp(-t/T_{req.})$ when period Θ is exponential distributed (i.e. one is variable) and its mean is equal to T_{req} ; c) $W(t)=1-\exp(-t/T_{rep.})$. Input: n is the conditional number of a subsystem; k is the conditional number of a unit; TMTBF nk is the mean time between hardware/software failures for the k-th unit of the n-th subsystem;Trep. is the mean time of system repair after any unit failure. Customer requirements: Treq. is the mean required period of permanent IS operation when reliability should be provided; $P_{adm.}$ is the admissible probability of providing reliable functions performance by IS during the required time T_{req}..

With the subsystem "Reliability" of CEISOQ+ the next reliability metrics may be evaluated: $T_{MTBF\ n}$ – the mean time between failures of the n-th subsystem in an active redundancy mode; $T_{MTBF\ 1..n}$ - is the mean time between failures of a complex composed of 1, 2,..., n subsystems, each of which can perform its functions both in active and passive redundancy modes; $P_{rel\ n}$ – is the probability of reliable n-th subsystem functions performance during the period $T_{req.}$ both in active and passive redundancy modes; $P_{rel\ 1..n}$ - is the probability of reliable functions performance by a complex composed of 1, 2,..., n subsystems during the time $T_{req.}$ when redundant elements are used both in active and passive redundancy modes; P_{rel} – is the probability of reliable functions performance during the time $T_{req.}$ when redundant elements are used both in active and passive redundancy modes, $P_{rel} = P_{rel\ 1..N,}$ where N – is the number of subsystems in the modeled system; $K_{avail.}$ – is the system availability when redundant elements are used both in active and passive redundancy modes, $K_{avail.} = \lim_{T_{req.} \to 0} P_{rel.}(T_{req.})$, i.e. if to set very small $T_{req.}$ (for example 1 millisecond) the evaluated value P_{rel} approximates $K_{avail.}$.

3.3. Timeliness of information producing

Data circulated in a system, resources spent by process performer, queries for operator processing, output as a result of input flows for transforming into outputs may be formally calls for processing in a queue system. To estimate timeliness of required calls processing by process architectures let's examine existing approaches to their formalization. According to researches various methods of the queuing theory provide rather high degree of adequacy for calls processing modelling. There may be quite a few processing technologies: priority and unpriority processing by one or several servers, multiphase processing, time-sharing processing etc. Now there are some methodical approaches to estimation of some technologies under various conditions including the ones to analysis of computing systems and networks. As applied to queueing systems the term "processing technology" means the same with the term "processing mode, order or discipline", determining an order of call selection from a queue buffer for further processing. For example in accordance to information systems these calls are not only the ones on receiving of output documents but also on files transfer or information entering into a database, as well as technological calls on control of a computing process, access administration, information security providing.

It is proposed to formalize processes of users' information servicing as processing of Poisson flows by reliable singleserver or multiserver queuing system with a buffer of an infinite size. In practice process architectures for calls processing are formalized often as processing of Poisson calls flows by single-server or multi-server queuing system with a buffer of an infinite size. A supposition concerning Poisson calls flows may be substantiated by the fact that among Palma type flows a Poisson flow puts the queuing system in the most hard conditions and for queuing time metrics gives upper estimations. Moreover, calls flows of the same type as a rule constitute a compound flow from different sources. In practice, each flow intensity is very low in comparison with the compound flow. In such a situation theorem (Grigolionis (1963)) is applicable, according to which the compound flow is a Poisson flow. All the cited considerations as well as statistical researches results prove a possibility of assumption concerning Poisson calls flows. On the analogy with this a supposition concerning an exponential law of calls processing time distribution also allows to get pessimistic estimation of system response time.

There are several approaches to an analytical estimation of calls processing timeliness in queuing systems. The simplest is the one allowing a distribution function of system response on a call. It is necessary to note that an explicit distribution function may be got only for simplest systems without priorities, for example, for system $M/M/1/\infty$ (Gnedenko (1973)). There is another approach to estimation of systems for which distribution functions of system response time are expressed in terms of various Laplace-Stielies transformations (Gnedenko (1973)). For the wide range of priority systems $M/G/1/\infty$ with different processing technologies time-probabilistic characteristics are drawn in such a form. The expressions of joint distribution of a queue length and waiting queue time are drawn in the form of a functional dependency in terms of various Laplace-Stielties transformations and productive (generating) functions. They give an idea of mathematical complexity of models. In this case the desired probability may be computed on the basis of invert Laplace-Stielties transformations. Though there are some applied ways of such invert transformations practical computations require not only additional programming on a high level but also essential time expenses. Such conditions complicate a work of a system analyst. That's why in practice there often used approaches providing approximate estimations of the desired probability. The most popular way of approximate estimation consists in an approximation of a response distribution function with the help of the incomplete gamma-function. J.Martin's studies of some priority processing technologies proved rather high engineering accuracy of such an approximation (Martin (1972)). This approach is used by the CEISOQ+.

A supposition concerning infinite number of queuing buffer in practice means allotment for storage of calls, input and output data such system memory sizes that guarantee in case of right system use absence of information losses caused by possible buffer's overflow. Though last years we can trace a stable tendency of main storage and external storage memory size expansion together with its price reduction. Problems with lack of memory for information systems appear more seldom and it seems they won't cause any troubles in the nearest future. Taking into account all the abovementioned the supposition concerning infinite number of queuing buffer seems to be acceptable for many cases.

The core of formalization is: modelling by means of priority and unpriority queuing systems $M/G/1/\infty$ is possible (Gnedenko (1973); Kleinrock (1976), Matweev & Ushakov (1984) etc.). The offered models and software tools CEISOQ+ allow to estimate and to compare effectiveness of the next dispatcher technologies:

- technology 1 for apriority calls processing: in a consecutive order for single-tasking processing mode (regime "Singletasking"); in a time-sharing order for multitasking processing mode (regime "Multitasking");
- priority technologies of consecutive calls processing 2-5:
- technology 2 for calls processing with relative priorities in the order "first in - first out" FIFO;
- technology 3 for calls processing with absolute priorities in the order FIFO;
- technology 4 for batch calls processing (with relative priorities and in the order FIFO inside a batch) (Kostogryzov (1987));
- technology 5 is a combination of technologies 2, 3, 4 see (Kostogryzov (1987, 1992)).

In case of technology 1, single-tasking processing mode allows to process calls in the consecutive order FIFO. In case of multitasking processing mode if there are n calls they are all processed simultaneously but each call is processed n times as slower as it had been processed alone in the system. According to technology 2 calls with higher priority are processed earlier. If calls are of the same priority they are processed in the consecutive order. There is no interruption of begun call processing by another call of higher priority appeared. Unlike technology 2, technology 3 allows an interruption of processing if a priority of the coming call is higher than a priority of the processed call. Processing of interrupted calls continues from the interrupted place. In the case of technology 4, the first call, coming to an off-line system, forms the first batch. The next batch is formed by calls, which come during the previous batch processing time, and is processed immediately after all the calls of previous batch have been processed. In the processed batch all calls are processed according to technology 2 with the exception that the processing cannot be interrupted. Finally, for technology 5, all calls are divided into n groups. Calls of the g-th group have higher priority than calls of the e-th group if g<e (e, g = 1,..., n). Calls of one group have their own relative priorities that are actual only within this group.

Estimation of system operation time-probabilistic characteristics may be made with the help of the CEISOQ+ subsystem "TIMELINESS" (Kostogryzov et al. (2000-2002)). The models use allows to choose between the calls timeliness criterions: the mean processing time criterion 1; the probability criterion 2 of well-timed processing.

Criterion 1. An output information of the i–th type is considered to be well-timed according to the criterion of calls mean processing time if response time $T_{full\ i}$ is no less than required admissible time $T_{req.i}$: $T_{full\ i} \leq T_{req.i}$

Criterion 2. An output information of the i–th type is considered to be well-timed according to the probabilistic criterion if $P_{tim.i} = P(t_{full\ i} \leq T_{req.i.}) \geq P_{req.i}$, where $t_{full\ i}$ is the processing time, including queueing time and run time, $T_{full\ i}$ is the mean response time.

Note. The CEISOQ+ use proved the revealed analytical regularity for Technology 4: ratio of mean waiting time in a calls queue of low priority to mean waiting time of high priority calls doesn't exceed 3 units no matter what the system throughput is. At the same time for Technologies 2 and 3 this ratio may be measured in dozens or hundreds (other things being equal it is much greater for Technology 3 than for Technology 2). This very regularity is used for increasing the processing effectiveness owing to a combination of Technologies 2, 3 and 4 (see Technology 5).

The CEISOQ+ subsystem "TIMELINESS" use allows to estimate the next metrics: the mean wait time in a queue $T_{queue\ i}$; the mean processing time, including the wait time (it names also the mean response time) $T_{full\ i}$; the probability of well-timed processing during the required term $T_{req.r}$ ($P_{tim.i}$); the relative portion of all well-timed processed calls (S); the relative portion of well-timed processed calls of those types for which the customer requirements are met (C). For all technologies the probability function of well-timed calls processing is approximated by incomplete gamma-distribution:

$$P_{tim.i} = P\left(t_{full.i} \le T_{req.i}\right) = \frac{\int_0^{\gamma_i^2 T_{req.i}/T_{full.i}} t^{\gamma_i - 1} e^{-t} dt}{\int_0^{\infty} t^{\gamma_i - 1} e^{-t} dt}, \quad where \gamma_i = \frac{T_{full.i}}{\sqrt{T_{full.i2} - T_{full.i}^2}}.$$

$$S = \frac{\sum_{i=1}^{I} \lambda_i P_{tim.i}}{\sum_{i=1}^{I} \lambda_i}, C = \frac{\sum_{i=1}^{I} \lambda_i P_{tim.i}\left[Ind(\alpha_1) + Ind(\alpha_2)\right]}{\sum_{i=1}^{I} \lambda_i}, \quad Ind(\alpha) = \begin{cases} 0, \text{ if } \alpha = true \\ 1, \text{ if } \alpha = false \end{cases}'$$

a_1=(there is used criterion 1 and $T_{full.i}$£$T_{req.i}$); a_2=(there is used criterion 2 and $P_{tim.i} \ge P_{req.i}$.

3.4. Completeness of output information

A system will work in user's interest only after necessary initial input data concerning objects of its application domain have been entered. In the operational process there may be entered 2 types of current information sources:

- information concerning new objects which is firstly entered into a system;
- information concerning objects, which has already been stored in a system and is purposed for updating.

From the moment of information of the 1st appearance till the moment of its entering into a system is considered as incomplete as respects this information. In reality this information exists, characterizes states of new objects, which must be registered by system, but is not reflected in the system and therefore can't be taken into account by a system because he doesn't know about it. Concerning this information we may speak about its completeness only after its operational representation in system, after which a formal state of incompleteness disappears. Estimation of completeness level may be carried out with the help of the model of entering into system current data concerning new objects of application domain (below described), the CEISOQ+ subsystem "Completeness".

For both types of information the next question is reasonable: how actual is the information represented in the system at the moment of its use? The answer may be found in the next subchapter 3.5, where the models complex of information gathering from sources and the subsystem "Actuality" of the CEISOQ+ are used. In this chapter we pay attention to modelling process architectures for providing completeness of entering into system current data concerning new objects of application domain, i.e. of the first type of information. It is important to note that a theoretical solution of applied analysis and synthesis problems does not mean simplicity of its practical implementation. Though problems of data transfer and input into a system present no difficulties, problems of required information gathering and identification are still a stumbling block. As a result many systems are obliged to operate in conditions of information incompleteness because a scientific conception of required information gathering system characteristics does not provide practical possibilities of its implementation.

An analysis of system operation reveals that solving some problems it is often necessary to account a variety of objects and events, which occurrence is of a stochastic character. There may be considered some tasks of airport system, aircraft global positioning receiver system, tasks of reconnaissance, tracking of an area state in conditions of radioactive contamination, loads accounting by the customs etc. we shall consider output information complete if it represents all real objects and events necessary for system staff to perform their functions. In an automatic control system there is always represented only complete output information. All information circulating in an automatic control system is strictly determined and processed automatically, i.e. occurrence of new objects influencing on technological operations is eliminated. And vice versa any system is always operating in information incompleteness according to terrorist threats conditions. At the same time, we'll distinguish completeness and validity of represented information: the completeness concerns only that objects which appear, and validity concerns both new and stored information. In consequence, information may be complete, but not actual.

The essence of information incompleteness influence on decision-making consists in the impossibility of registration of all objects and events (OE), describing formal state of reality and influencing on decisions. As a logic result of decision-making may turn out to be inadequate to the situation, i.e. a decision turns out to be incorrect. A system contains information about states of all real objects and coincides if the number of occupied server for system $M/G/\infty$ is equal to zero.

Let's assume that appearance of new objects or events essential for a solution of a specific task occurs at random moments (we shall call them "causing"). Periods between these moments are also random, their duration is distributed by the exponential law with the parameter l. In a causing moment with the probability q_m there appear m new objects and events $\sum_m q_m = 1$. A generating function of appearing objects and events number at a causing moment we shall notate as $\Phi(z)$. In practice appearance of several objects and events is explained by their common origin: for example, as a result of a catastrophe on a chemical plant appears a set of zones of ecological contamination. After their appearance there is organized a message preparing during mean time w_I with a distribution function $B_1(t)$. Then the message is transferred for its loading into an IS during mean time d_i with a distribution function $B_2(t)$. There may be a delay in receiving of the message (for example, for a visual check of the data), then the message is loaded into an is within b_i with a distribution function $B_3(t)$. Thus, the loaded information allows to use classic results for queuing systems with an infinite number of servers $M/G/\infty$.

The probability that IS contains information about states of all real object and coincides with the probability that number of occupied server for system $M/G/\infty$ is equal to zero. It mails calculated by formula (Matweev & Ushakov (1984)). If with the probability q_m m new objects appear in random intervals exponentially distributed with parameter l, then the found probability:

$$P_{comp.} = exp\left\{-\lambda\int_0^\infty\left[1-\Phi\left(B(t)\right)\right]dt\right\}, \qquad (2)$$

where $\Phi(z) = \sum_{m>0} q_m z^m$ is productive (generating) function; B(t) – is the PDF of time for new information revealing and preparing $B_1(t)$, transfer $B_2(t)$ and entering into IS $B_3(t)$. $B(t)=B_1*B_2*B_3(t)$.

The next variants are used by the CEISOQ+: $\Phi(z)=z$; B(t) is exponentially distributed.

The final clear analytical formulas for modelling are received by integration (2) expression.

With the subsystem "Completeness" of the CEISOQ+ use the probability that IS contains information about states of all real object and coincides may be estimated.

3.5. Actuality of input information from using point of view

After information has been entered into a system it gains a property of actuality. It is clear that for real systems output information is received after it has been structured, formally processed and mixed with other information, which is gathered from different sources and is characterized by different significant state changes frequency of considered objects. Output information is a "mixture" of various input data elements with different actuality. As an analogue to actuality there may be product freshness by the moment of cooking. For example fresh fish has its useful life (a period between significant changes of consumer properties) equal to several hours, fruits life equals to several days, wine's – several years. Output information is also a product to be used by a man, that is why it should be "fresh" (in our terms – actual) for problems solving. As a stale ingredient spoils all food a non-actual part of output information may spoil information quality. The same information may be actual for solving one problem and non-actual for another one.

IS users usually use "anonymous" output documents received as a result of their calls processing by a computer. It would be ideal an output document were marked by the date by which the values included in it would have been actual. It would be completely similar with sold food on packages of which there are indicated its expiry date. However, in a reality it is wide from the truth. In practice the objects state changes are entered into an IS with some delay and for the different data delays are different. Moreover, the same information may be actual for a solution of one problem and turn out to be completely irrelevant for another one. In other words, in an output document under a date of its creation data of a different actuality degree may be represented. Thus, for problem solving there is used output information different from the real one because of its changes in the course of time. If this difference is essential, use of such information may cause errors. Therefore, substantiating engineering solutions of providing an actual condition of used information an engineer has to solve a problem of quantitative estimation of achieved actuality for different technologies of information gathering and bringing it to the notice of users.

Without any limitation we consider process architectures for items gathering on the examples of providing information actuality. According the offered below models complex of information gathering from sources the next statement for evaluation information actuality is proposed.

Statement 2. The limited probability of information actuality on the moment of its use exists under conditions of existence for stationary probability distribution for considered characteristics and their independence.

One is equal to:

a. for the mode D_1 when information is gathered in order "immediately after an essential object state change:

$$P_{act} = \frac{1}{\xi} \int_0^{\infty} B(t)[1-C(t)]dt, \tag{3}$$

b. for the mode D_2 when information is gathered without any dependencies on changes of objects current states (including regulated information gathering)

$$P_{act.} = \frac{1}{q} \int_0^{\infty} \left\{ [1-Q(t)][1-\int_0^{\infty} C(t+\tau)dB(\tau)] \right\} dt, \tag{4}$$

where C(t) is the PDF of time between essential changes of object states,FF– is the mean time; B(t) is the PDF of time for information gathering and preparing $B_1(t)$, transfer $B_2(t)$ and entering into IS $B_3(t)$; B(t)=B1*B2*B3(t); Q(t) is the PDF of time interval between information updating, q is the mean time (only for mode D_2).

The final clear analytical formulas for modelling are received by Lebesque-integration of (3) and (4) expressions.

Introduction of admissible limits of item suitable values changes is connected with a concept of an essential change of real characteristics of objects and events. We'll call a change of an object's characteristic essential for a solution of a certain problem, if well-timed representation of faultless information about this change to a user influences logic or result of the solution. Fully similar situation is peculiar not only to information process architectures but also to many other gathered items (inputs, system elements or components for a project in life cycle, required products or system operation outputs for an user, for instance, information, etc.).

With the subsystem "Actuality" of the CEISOQ+ use the probability of information actuality on the moment of its use may be estimated.

3.6. Information foultlessness after checking

Problems of item content analysis are everywhere in system life cycle. It may be nondestructive defects control for some objects safety checking, documentation or drawings checking, hardware or software testing against potential errors, information analysis for making decision etc. In any case there exist some objects, may be latent or suspicious for revealing and their following analysis. It is clear that more often item content analysis quality depends on system's application domain and used analysis methods. In a general case methods of analysis and decision-making may contain elements of both creative work and guessing. Nonetheless, any analysis is based on logical positions. Logic implies argumentation based on essential information use. The way of logical information use is an algorithm of given information analysis. In practice this algorithm is implemented by either a man or an applied software. Both of them we shall mean under the term "analyst". The cited sequence of positions concerning logicality of item analysis algorithm allowed us to formalize a process architectures for item content analysis according to the offered model. To apply this concept to information architectures, let's assume that in a system there are provided gathered information completeness and actuality and there is confidence in software/hardware tools correct operation. Is it enough for providing validity of output information? No, it is far from being enough. The person, from his/her date of birth, lives in conditions of information incompleteness, that is why modern IS are oriented on all possible ways of gathering complete information. One of the modern IS advantages is that they are developed to solve principal problems of information actuality owing to quick consumers informing. If it is not possible to use "the freshest" information then a man may use less fresh information to estimate the current and predict future states of considered objects. Incompleteness and non-actuality of information are the unavoidable properties of natural human environment. The main danger is in insufficient effectiveness of information gathering.

The problem concerning information faultlessness is completely different. In practice it is very difficult to provide information faultlessness. The fact lies not only in technological complexity but also in the term "error" and in man's physical inability of not making errors. So let's review the term "error" and the term "distortion" which is close to the first one. Despite seeming simplicity these terms concerning information circulating in an IS are not fully the same. A syntax error in spelling, which does not influence sense of input or output documents, does not have a deteriorative influence upon information validity. Moreover, if quantitative deviation of considered objects characteristics is considered significant if it is more than 10% (for instance) an appeared deviation in parts of percent cannot be characterized as an error. It is only degree of information correctness, which is taken into account by those who make decisions. From the other hand the term "to distort" is interpreted as "to show smth in the false color". The term "distortion" itself means inadmissibility of further use. Below we shall use the term "error" for accidental data deviations. In cases implying premeditation we shall use the term "distortion". To define these terms for IS let's assume that information gathering and processing is carried out instantly and during storage before its use accounted objects do not change their states (if they change them the information is instantly updated). Then under an information error or distortion in such idealized IS we mean such data changes which in case of correct information processing may cause paralysis and/or changes in results of system operation. Thus use of these terms we shall connect with information quality inadmissibility for its further efficient use. In fact data updating does not happen instantly but after a certain period of information transferring. In this case to errors and distortions are added non-actuality and other natural and artificial influences deteriorating information quality.

This subchapter is dedicated to studying faultlessness estimated with the help of the next model of items analysis. The core of modelling is illustrated by Figure 8 (for information checking application domain).

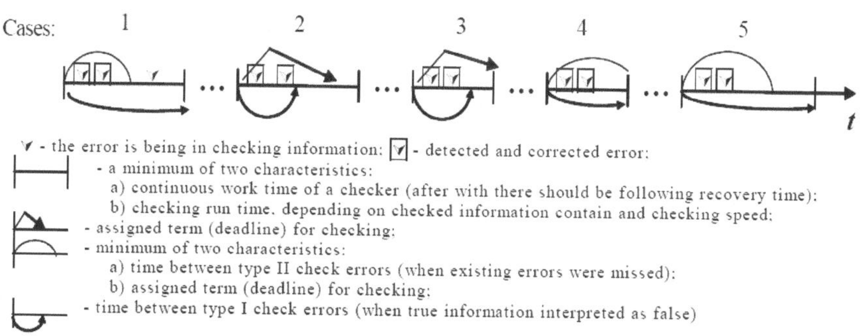

Figure 8. The illustration of processes for information analysis

On an example for visual checking, the cases 1, 2, 3 characterize presence if only of one errors, the cases 4, 5 and any other characterize that faultlessness after checking is provided.

It is not always possible in practice to draw a logical bound for admissible deviations. Often even an insignificant deviation may become important for system functions performance. At the same time it is undisputable that required data faultlessness should be set taking into consideration effects of system operation, damage caused by errors and also errors preventing and negative consequences elimination expenses. If to put an end to the above mentioned it turns out that for a user understanding of an "error" depends on information purpose and methods of its processing. Information processing implies: syntax and semantic information control to detect and eliminate errors; various processes (generalization, arithmetical and logical operations etc.) and as a final result – pragmatic information filtering and analysis and logical making decision with the purpose of its further use.

In real systems we always have to solve a problem of a balance between input information content and quality of its processing within the assigned period. If an input information content is big the number of errors increases what may cause a control action time waste. To be on schedule it is necessary to optimize the information content and to develop more rational information processing and representing technologies.

The model is used for evaluation of information foultlessness after checking and information processing correctness.

Definition 1. Information after checking is considered as faultless if all data errors were detected and corrected and no new errors were made. Definition 2. Information processing is considered as correct analysis if all essential information was faultless analyzed and no algorithmic errors was made.

There are possible the four variants of correlations between the characteristics.

Variant 1. An assigned term for analysis is no less than the real analysis time ($T_{real} \leq T_{req.}$) and the content of analyzed information is such small that it is required only one continuous analyst's work period ($T_{real} \leq T_{cont.}$).

There is proposed the next Statement 3.

Statement 3. Under the condition of independence for considered characteristics the probability of information faultlessness (for problems of checking) or processing correctness (for problems of analysis) during the required term is equal to:

$$P_{after(1)}(V,\mu,\nu,n,T_{MTBF},T_{cont.},T_{req.}) = \left[1-\hat{N}(V/\nu)\right]\left\{\int_0^{V/\nu} dA(\tau)[1-M(V/\nu-\tau)] + \int_{V/\nu}^{\infty} dA(t)\right\} \quad (5)$$

where N(t) is the PDF of time between type I analysis errors, η^{-1} is the mean time, for example, $N(t) = 1 - \exp(-t\times\eta)$; M(t) is the PDF of time between the neighboring errors in checked information, for example $M(t) = 1 - \exp(-t\times\mu\times\nu)$; A(t) is the PDF of analyzed type II errors, T_{MTBF} is the mean time; μ is the relative fraction of errors in information content (destined for problems of checking) or the relative fraction of information essential for analysis (destined for problems of analysis); $T_{real} = V/\nu$ - is the real time for complete information analysis; V – is a content of analyzed information; ν - is an analyzed speed; $T_{cont.}$ - is time of continuous analyst's work; $T_{req.}$ - is an assigned term (deadline) for analysis.

V, ν, $T_{cont.}$ and $T_{req.}$ are assigned as deterministic values. The probability that there are not errors without checking is $P_{no}(V) = e^{-\mu V}$.

The final clear analytical formulas for modelling are received by Lebesque-integration of (5) expression.

Variant 2. An assigned term for analysis is no less than the real analysis time (i.e. $T_{real} \leq \ T_{req.}$). But the content of analyzed information is comparatively large, i.e. $T_{real} > T_{cont.}$.

Statement 4. Under the condition of independence for considered characteristics the probability of information faultlessness (for problems of checking) or processing correctness (for problems of analysis) during the required term may be estimated by following:

$$P_{after(2)} = \left\{ P_{after(1)} \left(V_{part(2)}, , , n, T_{MTBF}, T_{cont.}, \tau_{part(2)} \right) \right\}^{N}, \qquad (6)$$

where $N = V/(\nu T_{cont.}), V_{part(2)} = V/N, \tau_{part(2)} \ = \ T_{req.}/N.$

Variant 3. An assigned term for analysis is less than the real analysis time ($T_{real} > T_{req.}$) and the content of analyzed information is such small that it is required only one continuous analyst's work period(Treal≡Tcont.).Treal≡Tcont...

Statement 5. Under the condition of independence of considered characteristics the probability of information faultlessness (for problems of checking) or processing correctness (for problems of analysis) during the required term may be estimated by following:

$$P_{after(3)} = \left(V_{part(3)} / V \right) \times P_{after(1)} \left(V_{part(3)}, , , n, T_{MTBF}, T_{cont.}, T_{req.} \right) + \left[\left(V - V_{part(3)} \right) / V \right] \times P_{without}, \qquad (7)$$

where $V_{part(3)} = T_{req.}, P_{without} = e^{-\left(V - V_{part(3)}\right)}$.

Variant 4. An assigned term for analysis is no less than the real analysis time (i.e. $T_{real} > T_{req.}$). But the content of analyzed information is comparatively large, i.e. $T_{real} > T_{cont.}$.

Statement 6. Under the condition of independence of considered characteristics the probability of information faultlessness (for problems of checking) or processing correctness (for problems of analysis) during the required term may be estimated by following:

$$P_{after(4)} = \begin{cases} \left[\frac{V_{part(4)}}{V}\right] P_{after(1)}\left(V_{part(4)},,,,T_{MTBF},T_{cont},T_{req}\right) \\ + [(V - V_{part(4)}) / V] e^{-\left(V - V_{part(4)}\right)}, if\ T_{real} \le T_{cont}; \\ \left[\frac{V_{part(4)}}{V}\right] \left\{P_{after(1)}\left(V_{part(4.2)},,,,T_{MTBF},T_{cont},_{part(4.2)}\right)\right\}^N + \\ + \left[V - \frac{V_{part(4)}}{V}\right] e^{-\left(V - V_{part(4)}\right)}, if\ T_{real} > T_{cont}. \end{cases} \quad (8)$$

where $V_{part(4)} = T_{req.}, V_{part(4.2)} = \frac{V_{part(4)}}{N}, t_{part(4.2)} = T_{req} / N, N = \frac{V_{part(4)}}{Tcont}$.

The fraction of errors in information after checking equals to $\mu_{after} = \mu \times \left(1 - P_{after}\right)$.

The final clear analytical formulas for modelling are received by integration (5) and using (6)-(8).

With the subsystem "Effectiveness of checking" of CEISOQ+ use the probability of errors absence after checking and the fraction of errors in information after checking may be estimated.

3.7. Correctness of information processing

The mathematical model of items analysis (subchapter 3.6) is recommended to be used also for estimating correctness of information processing. With the subsystem "Correctness of processing" of the CEISOQ+ the probability of correct analysis results obtaining may be estimated.

3.8. Faultless operation of staff and users

A man is an unavoidable element of an IS as its user and staff, providing system's functionality. How does this element influence achievement of system operation purposes? To answer this question the CEISOQ+ subsystem "Faultlessness man's actions" may be used. As the subsystem is completely analogous to the subsystem "RELIABILITY" (see subchapter 3.2) we'll use the minimum problem definitions and illustrating examples. The next metrics may be estimated: the mean time between errors for a complex composed of functional groups; the probability faultless operation of a complex composed of functional groups 1,...,n during the period $T_{req.}$

3.9. Protection against dangerous influences

Nowadays at system development and utilization an essential part of funds is spent on providing system protection from various dangerous influences able to violate system integrity including information integrity. Under information integrity we mean such information state which provides the required operation quality of a used IS.

Such dangerous influences on IS are program defects events, virus influences, influences of software bugs, violators' influences, terrorists attacks (in the information field), psychological influences on men by means of ordered radio and TV programs etc.

There are examined two technologies of providing protection from dangerous influences: proactive diagnostic of system integrity (technology 1) and security monitoring when system integrity is checked at every shift change of operators (technology 2).

Technology 1 is based on proactive diagnostics of system integrity. Diagnostics are carried out periodically. It is assumed that except diagnostics means there are also included means of necessary integrity recovery after revealing of danger sources penetration into a system or consequences of negative influences. Integrity violations detecting is possible only as a result of diagnostics, after which system recovery is started. Dangerous influences on system are acted step-by step: at first a danger source penetrates into a system and then after its activation begins to influence. System integrity is not considered to be violated before a penetrated danger source is activated. A danger is considered to be realized only after a danger source has influenced on a system. If to compare an IS with a man technology 1 reminds a periodical diagnostics of a man's health state. If diagnostics results have revealed symptoms of health worsening a man is cured (integrity is recovered). Between diagnostics an infection penetrated into a man's body brings a man into an unhealthy state (a dangerous influence is realized). The essence of protecting process architecture for the first technology is illustrated by Figure 9. The cases 1, 4 illustrate dangerous influences. The cases 2, 3, 5 illustrate secure system operation during a period T_{req}.

Note. It is supposed that used diagnostic tools allow to provide necessary system integrity recovery after revealing of danger sources penetration into a system or consequences of negative influences.

Technology 2, unlike the previous one, implies that operators alternating each other trace system integrity between diagnostics. In case of detecting a danger source an operator is supposed to remove it recovering system integrity (ways of danger sources removing are analogous to the ways of technology 1. A penetration of a danger source into a system and its activation is possible only if an operator makes an error. Faultless operator's actions provide a neutralization of a danger source trying to penetrate into a system. When operators alternate a complex diagnostics is held. A penetration of a danger source is possible only if an operator makes an error but a dangerous influence occurs if the danger is activated before the next diagnostic. Otherwise the source will be detected and neutralized. Thus in comparison with a man technology 2 reminds a continuous staying in a hospital when between rare diagnostics a patient is permanently under medical observation of operator. A dangerous infection penetrates into a man's body only because of a doctor's fault while it may be discovered later as a result of either an exacerbation of a latent illness or the next diagnostic.

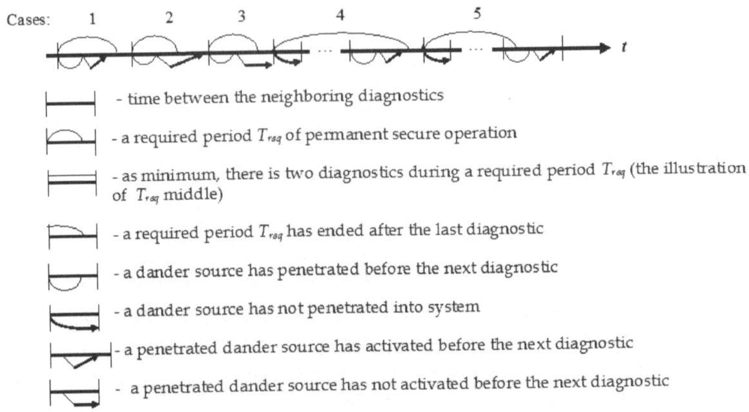

Figure 9. The illustration of processes for protecting system resources against dangerous influences by technology 1

For all technologies availability of means of danger sources total-lot detecting and existence of ways of violated system integrity total-lot recovery may seem to be a very high requirement. Nonetheless, a system which can't check and recover its integrity is a very vulnerable and knowingly doomed system.

With the subsystem "Protection from dangerous influences" of CEISOQ+ the probability of secure system operation within the assigned period may be estimated as a result of use the next mathematical models.

There are possible the next variants for technology 1: variant 1 – the assigned period T_{req} is less than established period between neighboring diagnostics ($T_{req} < T_{betw.}+T_{diag}$); variant 2 – the assigned period T_{req} is more than or equals to established period between neighboring diagnostics(Treq≥ Tbetw.+Tdiag). Here $T_{betw.}$ – is the time between the end of diagnostic and the beginning of the next diagnostic, T_{diag} – is the diagnostic time.

Statement 7. Under the condition of independence of considered characteristics the probability of dangerous influence absence for variant 1 is equal to

$$P_{infl.(1)}\left(T_{req}\right)=1-\Omega_{penetr} * \Omega_{activ}\left(T_{req}\right), \qquad (9)$$

where χ $\Omega_{penetr}(t)$ – is the PDF of time between neighboring influences for penetrating a danger source, for example$\Omega_{penetr}(t) = 1-e^{-st}$, s - is the frequency of influences for penetrating; $\Omega_{activ}(t)$ – is the PDF of activation time of a penetrated danger source, for exampleχfor example$\Omega_{activ}(t)=1-e^{-t/b}$, b – is the mean activation time; T_{req} – is the required period of permanent secure system operation.

Statement 8. Under the condition of independence for considered characteristics the probability of dangerous influence absence for variant 2 is equal to

$$P_{infl.(2)} = \frac{N(T_{betw.}+T_{diag.})}{T_{req.}} \cdot P_{infl.(1)}^{N}(T_{betw.}+T_{diag.}) + \frac{T_{req.}-N(T_{betw.}+T_{diag.})}{T_{req.}} P_{infl.(1)}(T_{rmn}), \quad (10)$$

where $N=[T_{req.}/(T_{betw}+T_{diag})]$ – is the integer part, the remainder time $T_{rmn} = T_{req}-N(T_{betw}+T_{diag})$.

Statement 9. Under the condition of independence for considered characteristics the probability of dangerous influence absence for variant 1 is equal to

$$P_{inf.(1)}(T_{req.}) = 1 - \int_0^{T_{req.}} dA(\tau) \int_0^{T_{req.}-\tau} d\Omega_{penetr.} * \Omega_{act.}(\theta). \quad (11)$$

Here $\Omega_{penetr}(t)$ – is the PDF of time between neighboring influences for penetrating a danger source, $\Omega_{penetr.}(t)=1-e^{-st}$, s - is the frequency of influences for penetrating; $\Omega_{activ}(t)$ – is the PDF of activation time of a penetrated danger source, $\Omega_{activ}(t)=1-e^{-t/b}$, b – is the mean activation time; $T_{betw.}$ – is the time between the end of diagnostic and the beginning of the next diagnostic ($T_{betw.}$=const); A(t) is the PDF of time between operator's error, T_{MTBF} is the mean time, $A(t)=1-\exp(-t/T_{MTBF})$. T_{diag} – is the diagnostic time ($T_{diag.}$=const); T_{req} – is the required period of permanent secure system operation.

Statement 10. Under the condition of independence of considered characteristics the probability of dangerous influence absence for variant 2 is equal to

$$P_{inf.(2)}(T_{req.}) = \frac{N(T_{betw.}+T_{diag.})}{T_{req.}} \cdot P_{wholly}^{N} + \frac{T_{rmn}}{T_{req.}} \cdot P_{infl.(1)}(T_{rmn}), \quad (12)$$

P_{wholly} – is the probability of dangerous influence absence within the assigned period $T_{req.}$:

$$P_{wholly} = 1 - \int_0^{T_{betw.}+T_{req.}} dA(\tau) \int_0^{T_{betw.}+T_{req.}-\tau} d\Omega_{penetr.} * \Omega_{activ.}(\theta), \quad (13)$$

and $P_{infl.(1)}(T)$ is defined above, but one is calculated not for all period T_{req}, only for the remainder time $T_{rmn} = T_{req}-N(T_{betw}+T_{diag})$.

The final clear analytical formulas for modelling by the CEISOQ+ are received after Lebesque-integration of (11), (13) expressions with due regard to Statements (7)-(10).

3.10. Protection from an unauthorized access

At all times a particular attention was paid to the problem of effective system resources (facilities, valuables, information, software etc.) protection from an unauthorized access (UAA). None clever solutions didn't guarantee complete protection from UAA to complex systems. As we have made sure there is also impossible to provide total-lot system reliability, information timeliness, actuality, faultlessness, correctness, and system protection from dangerous influences. Now we shall pay some attention to common regulations of providing protection from UAA in applications to IS.

Results of UAA may be the next: a dangerous influence on a secure system operation(on a subsystem of access control, a subsystem of account, a subsystem of integrity providing); a physical influence on system items (destroying, power supply failures, interceptions of electromagnetic radiation); an unauthorized withdrawing, acquaintance, use or dangerous influences on stored, processed, transferred and represented information (theft, fraud, insertion of spurious information, deleting, i.e. any violation of information integrity); an unauthorized use or change of system content, structure and functionality (including changes of configuration parameters, an introduction of bugs, viruses); hardware/software failures and malfunction; violating of network interconnection etc.

It should be formulated system security policy. The protected batch should create a virtual system operation environment, model potential threats, reveal vulnerabilities of a system, estimate potential risks and damage, compare expenses on the whole system operation and the subsystem of providing IS security. However we should remember that security provision mustn't hinder from real objectives achievement for the sake of which these systems are created.

To understand ways of overcoming protective barriers by a violator it is worth citing some examples of bottlenecks in existing information security systems:

- authenticating users a security service doesn't always have an ability to make sure of a user's authenticity. The particular ways of authentication (on the basis of a fingerprint or an eye retina analysis) have not been widely distributed yet;

- ❖ access delimitation to computer resources is not insuperable. The majority of systems does not support a mandatory access control, a cryptography information protection is not always introduced (sometimes such disregard of a cryptography information protection is justified if to take into account a required IS throughput but in other cases it is unjustified), a password access to the most relevant resources is not executed etc.;
- ❖ many protective systems do not prevent an unauthorized start of executable software files including a remote start of access procedures to resources of other computers;

- protection from viruses and bugs is still problematical (see the examples of this chapter);
- speed of crypto-transformations is not high enough, what often causes users' refusals of encoding;
- a control of a protection system operation correctness is quite often ignored, signaling functions are not performed;
- a required security of a network transfer is not provided (authenticity, capabilities of interception, insertion of spurious message are not checked and there is no keys distribution between network nodes);
- functions of substantiated information redundancy are often not realized. For example, a used background redundancy does not provide an information recovery owing to both failures of soft/hardware means and unauthorized actions;
- there are possible errors of network administrator (configuration, access to network resources and recovery control) and in control of a protection system operation correctness;
- weaknesses of used a cryptographic algorithms (absence of short and trivial passwords check, use of secret functions of overcoming cryptographic protection etc.);
- unfounded periodicity and order of tuned parameters changes (identifiers and authorities of the users, passwords and key information, frequency of resources integrity control etc);
- failures to meet requirements to the protocol of information intercommunication, correctness and completeness (there are possible interception of the transferred data, their thefts, changes and readdressing, unauthorized mailing on behalf of another user, a denial of data authenticity, etc). For example, an interception is possible if a violator synthetically connects to an unauthorized router network with readdressing of a messages flow;
- an IS elements malfunction (failures or an essential slowdown of executed operations), which can be as a result of a network overload, critical data deleting and performance of critical functions etc.

Thus the above-mentioned positions of information protection and overcoming of protective measures are to promote a better understanding of process architectures for protecting system resources from an UAA.

If to represent such a system from the point of view of its operation logics the process architecture for system resource protection is a complex of sequential obstacle barriers. A user is taught how to overcome these barriers to get an access to resources. A violator has to overcome these barriers trying to find system vulnerabilities (see Figure 10). A security service controls system operation thus reducing system vulnerability.

A probabilistic space for estimating of influences absence after UAA is created in the supposition that in a system there are realized protective measures from a potential violator. To get access to resources stored in a system there is a set of barriers known to an authorized user. A violator is a tool or person who does or doesn't know how protective barriers work. Somehow breaking an algorithm of barriers overcoming a violator may easily get access to system resources. We examine the most difficult mode of security system operation in conditions of the constant threat of its breaking. A violator is able to penetrate into a system only if: 1) he finds out how that part of the protection system works which is needed for his/her purposes achievement; 2) he gets access to information and/or software resources before this protective system will be changed (in this case the violator will have to overcome the new barriers). In modelling actions of "clever" violator equipped with high-tech means of system breaking are described by a greater speed of barriers overcoming.

The formalized process architectures for protecting system resources is fair for security estimating without and with considering objective period when resources value is high (see Figure 10b)). Often in practice this period is essentially limited. As an Air Transport System example there may be information resources used for performance of one passenger transportation or a certain task. After the task fulfillment the objective value of these resources comes to zero (is actual only archive value). The second example concerns a flight control system of an aircraft which operating lever are located in a cockpit. From the point of view of a flight security in conditions of terrorist threats a period of objective value of these resources coincides with a flight time or if an airplane is high-jacked it coincides with the period of its capture. As the third example there may be cash financial or gold-value resources in a bank after their receipt for storage. In this case from the point of view of the bank's security system the period of objective value of resources coincides with the period of their keeping in depositories of bank.

As a rule information has a certain period of its objective value (POV), which influences on the system protection from UAA. In the offered subchapter on the basis of use the model of unauthorized access to system resources taking into account POV of information resources, there is estimated confidentiality of used information. Estimations are based on a use of the CEISOQ+ subsystem "CONFIDENTIALITY". Thus the period of objective information confidentiality in a system is POV concerning information resources.

Thus, a period of objective value (POV) of a resource is time appropriate for the resource after expiring of which the resource loses its value and objectively does not need any protection from UAA. For the present model taking into account POV resources are considered to be protected from UAA if as a result of UAA after their POV has finished there is no penetration to them. The core of more general process architecture with due regard to resources value for one barrier is illustrated by Figure 10b)). Unauthorized access during objective period, when resources value is high, went through barrier in cases 1 and 4. Resources are protected in other cases 2,3,5.

With the subsystem "Protection from unauthorized access" of CEISOQ+ the probability of providing system protection from UAA by means of barriers 1st, 2nd..., mth and by means of all barriers may be estimated as a result of use the next mathematical model.

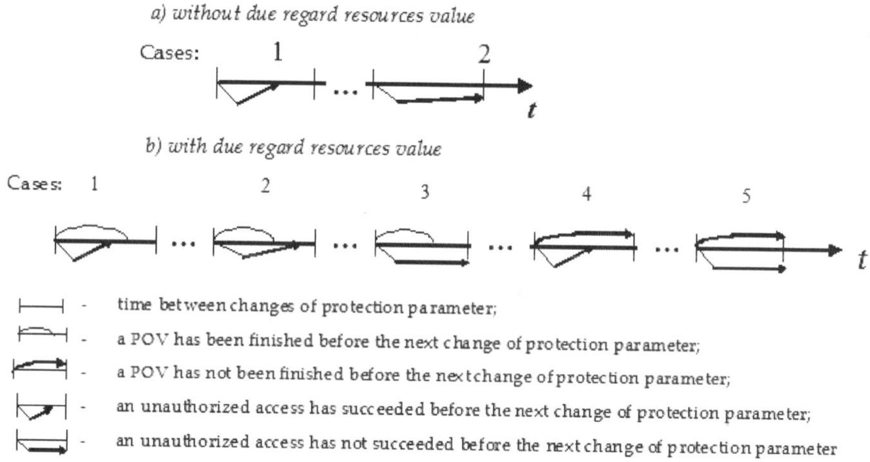

Figure 10. The illustration of processes for protecting system in application to a) UAA and b) information confidentially for an one barrier

As the model without considering objective period when resources value is high there is proposed the next Statement 11 for evaluation system resources protection against unauthorized access.

Statement 11. The limited probability of system protection against unauthorized access exists under the condition of existence for stationary probability distributions for considered characteristics and their independence.

One is equal to

$$P_{prot} = 1 - \prod_{m=1}^{M} P_{over\ m}, \tag{14}$$

where M is the conditional number of a barriers against an unauthorized access; $P_{over\ m}$ – is the probability of overcoming the m-th barrier by violator,

$$P_{over\,m} = \frac{1}{f_m}\int_0^{\infty}[1 - F_m(t)]dU_m(t)dt; \qquad (15)$$

where $F_m(t)$ is the PDF of time between changes of the m-th barrier with regulated parameter, f_m is the mean time;

$U_m(t)$ is the PDF of possible time of overcoming the m-th barrier, u_m – the mean time of a barrier overcoming.

The final clear analytical formulas for modelling by the subsystem "Protection from unauthorized access" of CEISOQ+ are received after Lebesque-integration of (15) expression with due regard to Statement (11).

3.11. Confidentiality of used information

The model taking into account a period of resources objective value is the next. With the subsystem "CONFIDENTIALITY" of the CEISOQ and CEISOQ+ the probabilities of providing information confidentiality by means of barriers 1st, 2nd..., mth from UAA and by means of all barriers may be estimated.

There is proposed the next Statement 12 for evaluation system protection against unauthorized access during objective period, when resources value is high.

Statement 12. The limited probability of system protection against unauthorized access during objective period, when resources value is high, exists under the condition of existence for stationary probability distributions for considered characteristics and their independence. One is equal to

$$P_{value} = 1 - \prod_{m=1}^{M} P_{over.m}, \qquad (16)$$

where M is the conditional number of a barriers against an unauthorized access; $P_{over\,m}$ – is the risk of overcoming the m-th barrier by violator during objective period when resources value is high,

$$P_{over} = \frac{1}{f_m}\int_0^{\infty}dt\int_t^{\infty}dF_m(\tau)\int_0^{t}dU_m(\theta)\left[1 - H(\theta)\right], \qquad (17)$$

where $F_m(t)$ is the PDF of time between changes of the m-th barrier parameters; $U_m(t)$ is the PDF of parameters decoding time of the m-th security system barrier, u_m – the mean time of a barrier overcoming; H(t) – is the PDF of objective period, when resources value is high.

The final clear analytical formulas for modelling by the subsystem "CONFIDENTIALITY"of CEISOQ+ are received after Lebesque-integration of (17) expression with due regard to Statement (12).

4. Models, software tools and methods to analyze and optimize system processes

4.1. General approach to mathematical modelling standard processes

The offered below mathematical models and supporting them dozens software tools complexes are focused on providing system standard requirements (ISO/IEC 15288 "System Engineering – System Life Cycle Processes", ISO 9001 "Quality Management Systems - Requirements" etc.) on the base of mathematical modelling random processes that exist for any complex system in its life cycle. The basic idea of the models develops the idea used for information systems - see subsection 3.1. At the beginning there were created the models of complex CEISOQ and CEISOQ+, after that - the other models.

The idea of mathematical modelling consists in the following. Any process represents set of the works which are carried out with any productivity at limitations for resources and conditions. This amount of works is characterized by expenses of resources (cost, material, human), accordingly works can be executed for different time with various quality. And conditions are characterized by set of the random factors influencing processes. It can be natural, technical, time, social factors, factors of the market and scientific and technical progress, say, all that is capable to affect processes. From the point of view of probability theory and the theory of regenerating processes it is possible to put formally, that all processes on macro- and micro-levels are cyclically repeated. If to assume, that number of recurrences of such processes is very large it is theoretically we can speak about probability of any events which can occur. The elementary example is a frequency loss of "heads" and «tails» at tossing up coins. If to enter conditions on a site (for example, on edge of gorge), on weather (a snow, a rain, a wind and so forth), on hardness of a ground it is possible to speak already not only about events of "heads" and «tails», and about other events falls, for example, that a wind will carry it away and coins will be lost. Actually course of complex system processes in life cycle from the mathematical point of view can be formalized absolutely similarly formalizations of tossing up coins process for the complicated conditions. In other words, the same as a matter of fact mathematical models can appear rather effective at their carry to other area of the practical applications. For example, the queueing theory which has arisen for calculations in a telephony, is used with success and for estimations at strikes on antiaircraft installations, and for estimations of time delays in networks of the computer systems, and for estimations of throughput of motorways etc.

In each of the offered models time characteristics of processes, frequency characteristics of any events and characteristics, connected in due course (for example, the set amount of works at known speed of their performance will give representation about mean time of performance of these works) are used as input. As final or intermediate result probabilities of "success" during a given time of forecasting or risks of failures as an addition to 1. They are used as evaluated output. So, at formalization of concept «customer satisfaction» estimations were under construction for the general case of following reasons. The customer expects performance of the certain amount of works with the acceptable quality and/or admissible risks for given time and money. In a reality the amount of works can appear other, the degree of quality essentially depends on applied technologies and management actions, time of performance and an expenses can undergo changes. As a result it is possible to speak about a degree of satisfaction in probability terms of performance of the set amount of works with the admissible quality for given time and money, and also about an expected and real part of the functional operations which are carried out with acceptable quality and/or admissible risks and expected and real expenses of the customer.

Thus the main proposition, implemented in the offered models, concludes the next: all amounts of works, characteristics of their performance, possible events and other inputs are interpreted as expense of time which can be reflected on a timeline. Probability metrics on the introduced limited space of elementary events are calculated by the rule of the probability theory (see section 3).

The basic ideas of correct integration of probability metrics are based on a combination and development of models, above presented, and consist in the following.

1st idea. As models are mathematical, the use of the same mathematical models is possible by a semantic redefinition of input and output of modelling. The idea is mentioned only for understanding the further logic in construction of modeled system, subsystems, elements and corresponding metrics on the basis of integrated modules.

2nd idea. For a complex estimation of the systems with parallel or consecutive structure existing models can be developed by usual methods of probability theory. For this purpose it is necessary to know a mean time between violations of integrity for each of element (similarly MTBF in reliability, but in application to violation of quality, safety etc.). Further taking into account idea 1 concept of a mean time between violations of an element integrity may be logically connected (for example, redefined) in concepts of a frequency of influences for penetrating into an element and a mean activation time of a penetrated danger source. The last concepts mean characteristics of threats.

Note. As logic element a subsystem, compound object or separate indicator from a complex of product indicators can be used.

Let's consider the elementary structure from two independent elements connected consecutively that means logic connection "AND" (Figure 11, left), or in parallel that means logic connection "OR" (Fig. 11, right).

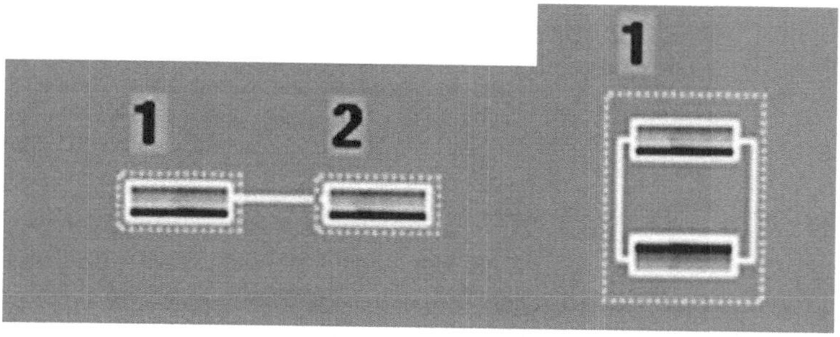

Figure 11. Illustration of system, combined from consecutively (left) or parallel (right) connected elements

Let's designate PDF of time between violations of i-th element integrity as $B_i(t) = P(\tau_i \leq t)$, then:

1. time between violations of integrity for system combined from consecutively connected independent elements is equal to a minimum from two times τ_i: failure of 1st or 2nd elements (i.e. the system goes into a state of violated integrity when either 1st, or 2nd element integrity will be violated). For this case the PDF of time between violations of system integrity is defined by expression

$B(t) = P(\min(\tau_1, \tau_2) \leq t) = 1 - P(\min(\tau_1, \tau_2) > t) = 1 - P(\tau_1 > t)P(\tau_2 > t) = 1 - [1 - B_1(t)][1 - B_2(t)]$. (18)

For exponential approximations: $B(t) = 1 - [1 - B_1(t)][1 - B_2(t)] = 1 - \exp(-t/T_{MTBF1})\exp(-t/T_{MTBF2})$.

Mean time between violations of integrity may be calculated by expression $T_{MTBF} = 1/(1/T_{MTBF1} + 1/T_{MTBF2})$;

1. time between violations of integrity for system combined from parallel connected independent elements (hot reservation) is equal to a maximum from two times τ_i: failure of 1st or 2nd elements (i.e. the system goes into a state of violated integrity when both 1st and 2nd element integrity will be violated). For this case the PDF of time between violations of system integrity is defined by expression

$B(t) = P(\max(\tau_1, \tau_2) \leq t) = P(\tau_1 \leq t)P(\tau_2 \leq t) = B_1(t)B_2(t)$. (19)

For exponential approximations: $B(t) = B_1(t)B_2(t) = [1 - \exp(-t/T_{MTBF1})][1 - \exp(-t/T_{MTBF2})]$. Mean time between violations of integrity may be calculated by expression $T_{MTBF} = T_{MTBF1} + T_{MTBF2} - 1/(1/T_{MTBF1} + 1/T_{MTBF2})$.

Applying recurrently expressions (18) – (19), it is possible to receive PDF of time between violations of integrity for any complex system with parallel and/or consecutive structure. The illustration of threats, periodic control, monitoring and recovery of integrity for combined subsystems of estimated system is reflected on Figure 12.

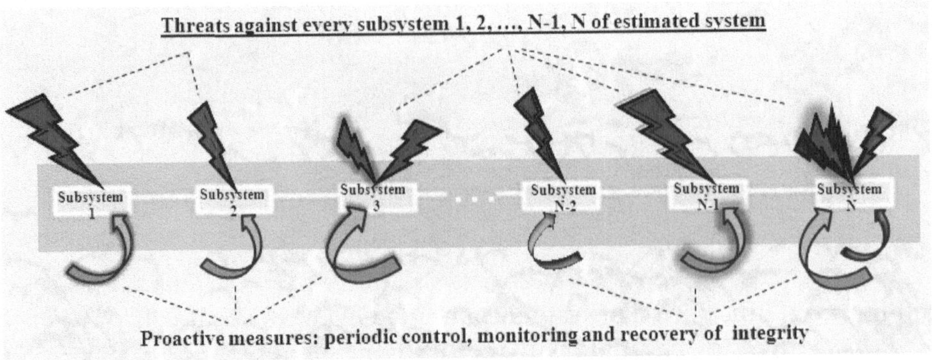

Figure 12. The illustration of threats, periodic control, monitoring and recovery of integrity for combined subsystems

3rd idea. Mean recovery time for system combined from consecutively connected independent elements may be calculated by expression $T_{rec.} = T_{rec.1}((1/T_{MTBF1})/(1/T_{MTBF1}+1/T_{MTBF2}))+T_{rec.2}((1/T_{MTBF2})/(1/T_{MTBF1}+1/T_{MTBF2}))$, for system combined from parallel connected independent elements $T_{rec.} = T_{rec.1}((1/T_{MTBF2})/(1/T_{MTBF1}+1/T_{MTBF2}))+T_{rec.2}((1/T_{MTBF1})/(1/T_{MTBF1}+1/T_{MTBF2}))$. Applying recurrently these expressions, it is possible to receive mean recovery time for any complex system with parallel and/or consecutive structure.

4th idea. If integrity violations are absent then diagnostic time for each element is equal on the average $T_{diag.}$. At the same time, if results of diagnostics require additional measures of integrity recovery this time increases. Thus mean time of diagnostics can be calculated iteratively with the given accuracy ε:

1-st iteration: Tdiag. (1) = $T_{diag.}$ that is given by input for modelling. I.e. for 1st iteration at detection of violation it is supposed instant recovery of integrity. Risk to lose required integrity $R^{(1)}$ is calculated (for example, by the models of subsection 3.9). Here recovery time is not considered; 2-nd iteration: Tdiag. (2) = Tdiag.(1) $(1 - R^{(1)}) + T_{rec.} R^{(1)}$, where $R^{(1)}$ is risk to lose required integrity for input Tdiag. (1). Optimistic risk to lose required integrity $R^{(2)}$ is calculated; …, n-th iteration is carried out after calculating risk $R^{(n-1)}$ for input $T_{diag.}^{(n-1)}$: $T_{diag.}^{(n)} = T_{diag.}^{(n-1)}(1 - R^{(n-1)}) + T_{rec.} R^{(n-1)}$, where $R^{(n-1)}$ is risk to lose required integrity for input $T_{diag.}^{(n-1)}$. Here recovery time is considering with the frequency aspiring to real, hence risk $R^{(n-1)})$ will aspire to the real.

The last iteration is when the given condition is satisfied: $R^{(n)} - R^{(n-1)} \le ε$.

5th idea. Existing models of section 3 are applicable to the system presented as one element. The main output of such system modelling is probability of providing system integrity or violation of system integrity during the given period of time. If a probability for all points $T_{given.}$ from 0 to ∞ will be calculated, a trajectory of the PDF for each combined element depending on threats, periodic control, monitoring and recovery of integrity is automatically synthesized. The known kind of this PDF allows to define mean time of providing integrity or between violations of system integrity for every system element by traditional methods of mathematical statistics. And taking into account ideas 2-4 it gives necessary initial input for integration.

Thus, applying ideas 1-5, there is possible an integration of metrics on the level of a PDF of time of providing system integrity or violation of system integrity. And it is the base for forecasting quality and risks.

Note. Ideas 2-5 are implemented in the supporting software tools (Kostogryzov, Nistratov at al. (2004-2011)) - see, for example, the "Complex for evaluating quality of production processes" (patented by Rospatent №2010614145) - Figure 13.

Thus models implement original author's mathematical methodology based on probability theory, theory for regenerating processes and methods for system analysis. An application of offered methodology uses to evaluate probabilities of "success", risks and related profitability and expenses. This helps to solve well-reasonly the next practical problems in system life cycle:

- ❖ analysis of system use expediency and profitability, selecting a suitable suppliers, substantiation of quality management systems for enterprises, substantiation of quantitative system requirements to hardware, software, users, staff, technologies;
- ❖ requirements analysis, evaluation of project engineering decisions, substantiation of plans, projects and directions for effective system utilization, improvement and development;
- ❖ evaluation of customer satisfaction in system design&development and possible dangers, detection of bottle-necks;
- ❖ investigation of problems concerning potential threats to system operation including protection against terrorists and information security;
- ❖ verification and validation system operation quality, investigation rational conditions for system use and ways for optimization etc.

The next complex for modelling of system life cycle processes "MODELLING OF PROCESSES" includes multi-functional software tools for evaluation of Agreement (models and software tools "Acquisition", "Supply"), Enterprise (models and software tools "Environment Management", "Investment Management", "Life Cycle Management", "Resource Management", "Quality Management"), Project (models and software tools "Project Planning", "Project Assessment", "Project Control", "decision-making", "risk management", "configuration management", "information management") and Technical Processes Modelling (models and software tools "Requirements Definition", "requirements analysis", "architectural design", "human factor ", "implementation", "integration", "verification", "transition", "validation", "operation", "maintenance", "disposal" tools) – see Figures 14-17 (one separate box is an implementation of one or more mathematical models) (Kostogryzov, Nistratov at al. (2004-2011)).

Figure 13 Subsystems of the "Complex for evaluating quality of production processes"

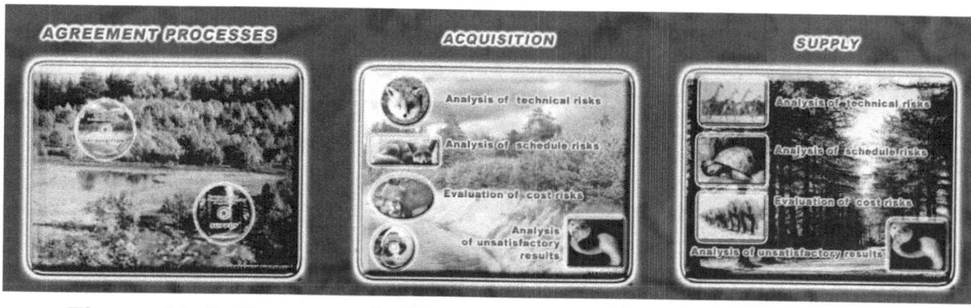

Figure 14. Software tools for evaluation of Agreement Processes

Figure 15. Software tools for evaluation of Enterprise Processes

Figure 16. Software tools for evaluation of Project Processes

Figure 17. Software tools for evaluation of Technical Processes

4.2. The formal statement of problems for system analysis and optimization

According to ISO 9000 management is defined as coordinated activities to direct and control an organization. In general case control is considered as the process of comparing actual performance with planned performance, analyzing variances, evaluating possible alternatives, and taking appropriate corrective action as needed (PMBOK). From system analysis point of view the main function of management is a purposeful change of a condition of process, object or system. Thus the process, object or system considered as managed if among all changes there is available one by means of which the purpose can be achieved. Management is based on a choice of one of set of any alternatives. Rational management is the management leading achievement of the purpose by criterion of an extremum (a minimum or a maximum) the chosen parameter at a set of limitations. Classical examples of rational management generally are maximization of a prize (profit, a degree of quality or safety, etc.) at limitations on expenses or minimization of expenses at limitations on a admissible level of quality and-or safety.

It is clear, that in life cycle of systems criteria and limitations vary. We shall consider briefly an essence popular today «process approach» for design, development and improvement of systems management quality according to ISO 9001. For successful operation the organization should define and carry out management of the interconnected kinds of activities. The activity using resources and performed with the purpose of transformation inputs in purposeful outputs, actually represents process. Thus, "process approach» means the application of system processes in view of their identification and interaction. The model of system management quality, based on «process approach», is directed finally to meet customer satisfaction. For rational management of processes it is necessary to know and predict their behaviour at various influences. For this purpose it is offered to use the mathematical models including models offered in this book. The metrics entered in these models, or their combination may be used as criteria metrics. Actually they are the quantitative measure (criterion function) describing degree of achievement of a purpose in view of management at various stages of system life cycle. For the enterprise there is important, for example, to optimize system management quality. A maximum of the probability of qualified work performance (i.e. in time and without any defects) or the probability of successful life-cycle processes running on condition that the competitiveness of each product type is retained can be used as criterion with corresponding limitations. For security services it is necessary to provide safety of object, process or system up to the mark. In this case the criterion of a minimum of expenses at limitations on an admissible risk level of dangerous influence on system contrary to counteraction measures or a minimum of risk of dangerous influence at limitations on expenses are possible. For the customer and the developer of the project the end result is important. In this case the criterion of a maximum of a relative quantity of system functions which are carried out with admissible quality or a relative level of customer satisfaction can be used as the integrated measure. The statement of problems for system analysis includes definition of conditions, threats and estimation a level of critical measures.

Thus the final choice of integrated measures is allocated on a payoff to the customer in view of specificity of created or maintained system. As probability parameters give higher guarantees in estimations of a degree of achieving purposes in comparison with average value at a choice it is recommended to use probability (i.e. on a degree of system quality operation - probability of providing admissible function performance quality during the given period of time) as the cores. And evaluated time characteristics (for example the mean time between violations of admissible system operation quality) are offered as auxiliary.

For example, there are applicable the next general formal statements of problems for system optimization:

1. on the stages of system concept, development, production and support:

system parameters, software, technical and management measures (Q) are the most rational for the given period if on them the minimum of expenses ($Z_{dev.}$) for creation of system is reached:

$$Z_{dev.}(Q_{rational}) = \min_{Q} Z_{dev.}(Q),$$

at limitations on probability of an admissible level of quality $P_{quality}(Q) \geq P_{adm.}$ and expenses for operation

$C_{oper.}(Q) \leq C_{adm.}$ and under other development, operation or maintenance conditions;

2. on operation stage :

system parameters, software, technical and management measures (Q) are the most rational for the given period of operation if on them the maximum of probability of providing admissible system operation quality is reached:

$$P_{quality}(Q_{rational}) = \max_{Q} P_{quality}(Q),$$

at limitations on probability of an admissible level of quality $P_{quality}(Q) \geq P_{adm.}$ and expenses for operation

$C_{oper.}(Q) \leq C_{adm.}$ and under other operation or maintenance conditions.

Of course these statements may be transformed into problems of expenses or risk minimization in different limitations. System parameters, software, technical and management measures (Q) is a rule a vector of input – see examples. There may be combination of these formal statements in system life cycle.

The order for use the developed classical formal approach to analyze and optimize system processes in quality management is illustrated by Figure 18. When analyst use this approach he'd like for several minutes to formalize a problem, perform mathematical modeling, analyze system processes in different conditions, choose the most rational variant and prepare analytical report. Such possibilities exist: an analyst should perform mathematical modelling by the Internet versions of the offered models – see Figure 19. He prepares input and receives analytical report in Word or pdf-file about 50-100 sheets as a result of interaction. This report will be formed automatically and include a formalization of analyst's problem, input, results of mathematical modeling in pictures (as demonstrated above in examples), analysis of system processes behaviour for different conditions, choice of the most rational variant and recommendations." It means that any analyst, understanding the used mathematical model, can receive during 1-3 minutes scientifically proved analytical report after interaction with an Internet version of model.

This report may be used for making decision and developing his independent report with additional materials. It is virtual outsourcing of high system analysis on the base of the offered mathematical models. The purpose is to give to analysts an opportunity of accessible and cheap high technology of studying standard processes in life cycle of estimated systems. This work has begun, the first models are accessible (see www.mathmodels.net).

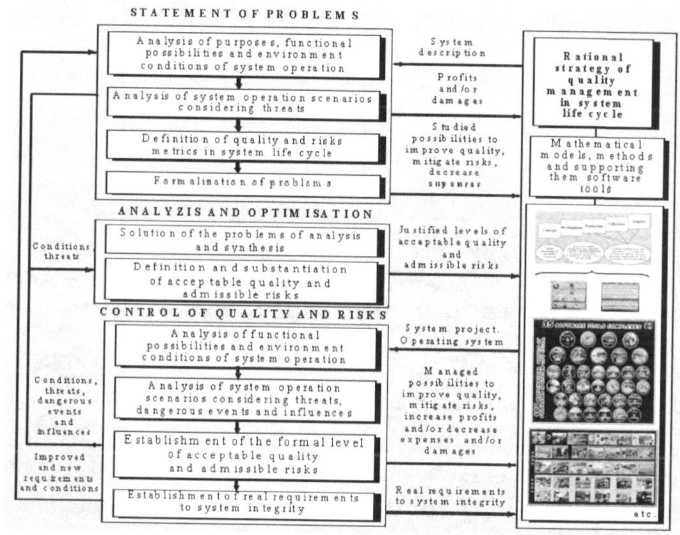

Figure 18.The approach to analyze and optimize system processes

The presented software tools complexes allow to solve problems for system analysis and optimization. Expected pragmatic effect from their application is the next: it is possible to provide essential system quality rise and/or avoid wasted expenses in system life cycle on the base of modelling system processes by the offered mathematical models.

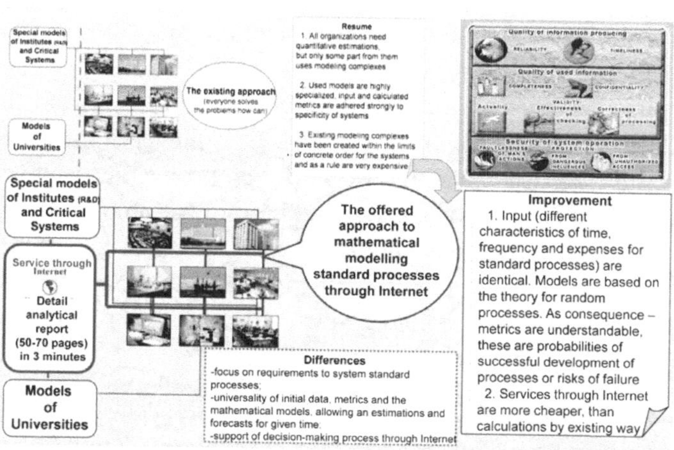

Figure 19. Mathematical modelling by the Internet versions of the offered models

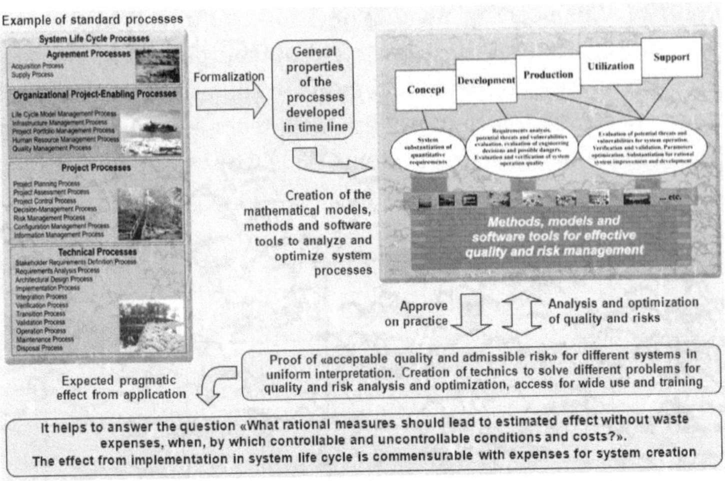

Figure 20. The offered way is the use of created methods to analyze and optimize system processes

Thereby necessary attributes of the offered innovative approach to control of system processes in quality management are above formed. Traditional approaches consist as a matter of fact in a pragmatical filtration of the information. In the decisions the responsible person, making decision, is guided firstly by the own experience and the knowledge and the advices of those persons of a command to whom trusts. Intuitively forming ideas which seem correct, this person chooses only that information which proves idea. The denying information is often ignored and more rare – leads to change of initial idea. This approach can be explained from the facts that at absence or limitation of used models it is difficult to investigate at once many ideas for given time. The presented models, methods and software tools, reducing long time of modelling (from several days, weeks and months to few seconds and minutes) change this situation cardinally.

The offered innovative approach is at the beginning substantiation of the system requirements, purposefully capable to lead to a success. Further, the responsible person, equipped by a set of necessary mathematical models and their software tools possibilities to forecasting quality and risks, is powered for generation of the proved ideas and effective decisions. These decisions are physically clear because of using accessible and operative analysis and optimization of processes in system life cycle. The offered approach allows to go «from a pragmatical filtration of information to generation of the proved ideas and effective decisions». The effect from implementation in system life cycle is commensurable with expenses for its creation (see Figure 20 and www.mathmodels.net).

We will demonstrate usability, universality and efficiency of the offered models, methods and software tools on the examples of their application for the analysis of "human factor», information actuality in commerce, errors during a use of SCADA-systems, efficiency of non-destroying control, preservation of foods quality, fire extinguishing, reliability of engineering equipment for enterprise object, flights safety in conditions of terrorist threats, information security, and also to the forecasts of risks for complex multipurpose systems.

5. Examples

Example 1 («Human factor»). Modern enterprises total tens – hundreds various workers. To solve a given functional enterprise problem there are required, as a rule, efforts of several specialists. For example, information gathering and control, its security providing, database and computing process administration, maintenance of computer equipment and information use are performed by different people. It is clear that their qualifications must be very high. Let's examine an example when it is not so. The reader may remember situations from his life.

Let the problem solution depends on joint but independent actions of 5 people. Let each of 4 specialists make 1 error a month and the 5th inexperienced person makes 1 error a day. System recovery time after an error equals to 30 minutes. It is required to evaluate faultlessness of such group's actions within a week.

The solution is based on the use of the CEISOQ+ subsystem «Faultlessness of man's actions» (see model in subsection 3.2). Integral computation results reveal that the probability of faultless joint actions of the first 4 skilled specialists within a 40-hours workweek equals to 0.80 but the low-quality work of the 5th unexperienced member mocks the whole group work. Indeed, the probability of faultless actions decreases to 0.15 (see Figure 21). Thus the computed results prove quantitatively the importance of thorough specialists training because a man is the main system bottleneck. It is impossible to detect all the system defects, but in some cases there is no full protection from "a fool". The quality management acts very wisely. As a rule an instructions with a training database and introduced assessment of users' readiness for a work with the real system is used. The proposed methods allow to estimate achieved levels of such system readiness.

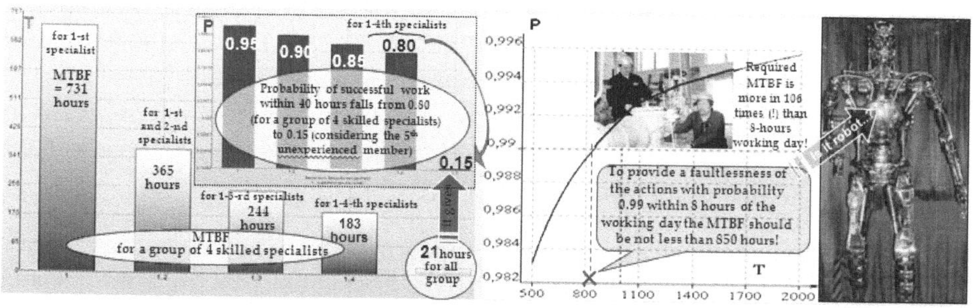

Figure 21. An estimation of human factor, examples 1-2

The question is lawful - what MTBF an worker should possess to provide a faultlessness of the actions with probability 0.99 within 8 hours of the working day? According to calculations the MTBF not less than 850 working hours is acceptable

– see Figure 21 right. It is more than 8-hours working day in 106 times (!), i.e. 4 months are necessary to work without errors, as the robot.

Example 2 (A role of information actuality in commerce). Nowadays the product market is being changed into an electronic one. What level of information actuality is peculiar to the successful companies within the possibilities of quality management and information technologies?

Solution. We'll try to answer this question by the subsystem "Actuality" of CEISOQ+ using as an example the worldwide known retail outlets network "Wall Mart", distributed in the USA, Canada, Mexico etc. (see model in subsection 3.5). Let's define an information aspect which makes for the company's success.

To increase productivity of each worker salesclerks were equipped with manual bar-code readers. Information contained in a bar-code is shown on a display. A worker can get a retrospective picture of products saling within a day, a week and several weeks. Moreover, on each article there are data about its quantity in the shop floor, in stock and how much has been ordered, i.e. everything what may be necessary for ordering. Let's evaluate actuality of information used by this system. It is logical to admit that significant changes happen not more frequent than 2 times per working day. There may be a leap in demand and supply, a change of goods quantity ordered by a customer, force majeur. Due to immediate information gathering by salesclerks (with the help of bar-code readers it takes 3 seconds), its transfer by satellite communications (10 seconds) and entering into a database (1 second) actuality of information in this network is not less than 0.992 (see variants i=1-4 on Figure22). It means that for successful company the probability to use actual information against non-actual one is more in 124 times (0.992/0.008=124)!

Correct use of this information turns out to be very effective. Using information read from a bar-code, which is transferred to the headquarters, they may order lots of goods, distribute them into their outlets and not worry about goods warehousing. For comparison, other shops, where usual bar-code readers are used and information is updated hourly, use information which actuality equals to 0.3-0.7 (see variants i=5-8 on Fig.23). Thus at the moment of use information can be as true as false. On the resulted figures it is possible to feel information roots of perfection quality management. According to "precedent" principle the achievable level 0.992 of information actuality can be defined as admissible.

Example 3 (Errors during a use of SCADA system). The control towers use SCADA system (Supervisory Control And Data Acquisition) for making decision. The data gathering and processing activities are modeled to evaluate the risk of misinterpreting of potentially dangerous events in control towers. Wrong interpretation may be caused by errors of dispatcher personnel, which can miss important information or turn harmless information into dangerous one, fails of

SCADA system. Let's consider a control station receiving information from the SCADA system for following processing. The information flow is measured in some conventional units and the information flow is of 100 units per hour. The total information contains not more than 1% of data related to potentially dangerous events. Taking into account automatic data analysis we suppose the speed of event interpretation to be near 30 sec per information unit. In this case 100 information units will be processed during 50 min. At that the frequency of errors for the whole dispatcher shift on duty, including fails of the SCADA system itself is about 1 error per year according to statistical data. The task is to estimate the risk of misinterpreting events on the control station for a time period of 1 hour, during one dispatcher shift turn of 8 hours, 1 month, 1 year, and 10 years.

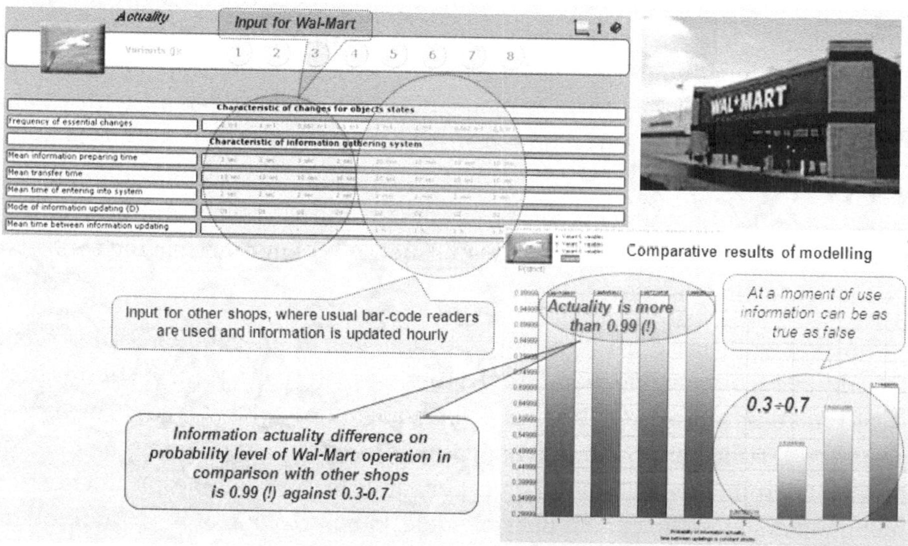

Figure 22. Input for CEISOQ+ and comparative results of modelling

The solution is based on the use of the subsystem «Risk evaluation. Risk of inadequate interpretation of events» of the software tools "Complex for evaluating quality of production processes" (see model in subsection 3.6). The analysis of modelling shows (see Figure 23) that for short time periods such as one shift turn or even for a month the risk of mistaken analytical conclusion is small enough (0.00076 and 0.07 accordingly). But when the time period grows the risk increases and becomes 0.565 for a year and almost unity (0.9998) during time period of 10 years. This means that during a month the probability for errors of dispatcher personal or SCADA system fails to occur is very small and their operation will be almost faultless. But for a more long time period such as a year is considered 1-2 errors of dispatcher personal or system SCADA fails will occur for certain. Considering high reliability of SCADA system and according to "precedent" principle the achievable level 0.07 for the risk of mistaken analytical conclusion during a month can be defined as acceptable.

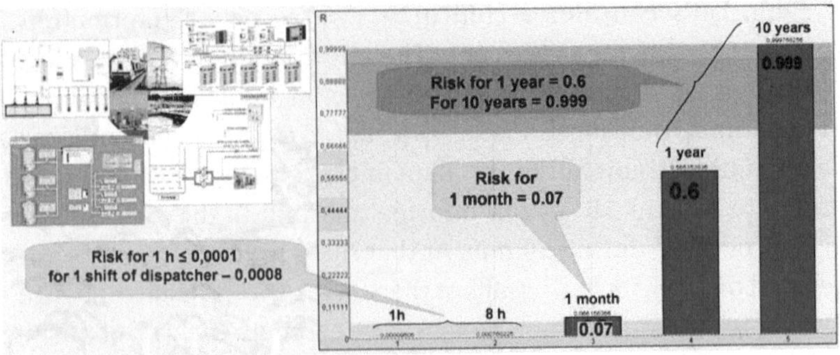

Figure 23. Some results of modelling a SCADA-system

Example 4 (Efficiency of non-destroying control). Let's consider two competing enterprises which are suppliers of pipes for transportation of production and guided in their quality management system by various technical politics. The first of these enterprises, guided by an innovative way of development with rational application of modern information technologies, effectively uses (as believed) existing innovations for quality and risks management. The second company uses cheaper and out-of-date technologies, keeping competitiveness on the market at the expense of it. At the enterprises various methods of non-destroying control are applied to revealing defects.

The first enterprise acquires input production from suppliers after quality control by all recommended methods of non-destroying control (acoustic, magnetic, optical, radiating, radio wave, thermal, electromagnetic etc.) that is confirmed by test reports and certificates on ISO 9001 and on output production. As a result for total controllable production in 100000 units per a month (for example, production tons, running meters etc.) the part of possible defects before control is 5%, a frequency of errors during the control is no more than 2 defects in a year (these are the latent defects not revealed by existing methods or passed at the control).

The second enterprise is satisfied by certificate on ISO 9001. And only radio wave method of non-destroying control is used by the suppliers. It allows to reveal such defects, as stratifications and deviations on a thickness in metal products (i.e. no more than 10 % of possible defects). At the expense of it the part of possible defects before the control is already 20 %, moreover, at the control defects of moulding (slag and flux inclusions, shrinkable bowls, gas bubbles, cracks, etc.), defects of processing by pressure (internal and superficial cracks, ruptures, tempers, dents, etc.), defects of heat treatment (overheats, hardening and hydrogen cracks, etc.) are missed. Totally about 30 defects per a year are possible.

Omitting questions of profits, we will compare technical politics of these enterprises by a risk of mistaken analytical conclusion within a month.

The solution is based on the same software tools as for example 3, but difference is the next: according to the 1-st idea of subsection 4.1 instead of metric "Risk of inadequate interpretation of events" we use metric "Risk of mistaken analytical conclusions". Input and results of modelling are on Figure 24.

The comparative analysis of the received dependences has shown:

- the risk of mistaken analytical conclusions for 1st first enterprise is 0.15, and for 2nd one – 0.92 (!);
- if the volume of controllable production is changed from 50000 to 200000 units per a month the risk increases for 1st enterprise from 0.08 to 0.58, and for 2nd one – from 0.71 to 0.96;
- the increase in a part of possible defects twice essentially does not influence value of risk, i.e. efficiency of applied technologies of the control depends essentially on other parameters, in particular from frequency of possible errors;
- if frequency of possible errors increases twice than the risk increases for 1st enterprise from 0.08 to 0.28, and for 2nd one – from 0.71 to 0.99.

Conclusion: For 1st enterprise the risk of mistaken analytical conclusions at level 0.15 after the control within a month can be recognized as acceptable. The 2nd enterprise supplies frankly defected production (probability nearby 0.9) that will negatively affect further at system operation.

Example 5 (Preservation of foods quality). We will demonstrate foods quality management on an example of probabilistic analysis of processes that are peculiar for grain storage. Quality of the grain supplied on longtime storage, decreases because of influences of dangerous biological, chemical and physical factors. Let's estimate the possible period before such moment of time when storing grain begin to loss required quality, and also expediency of introduction of continuous monitoring of grain quality.

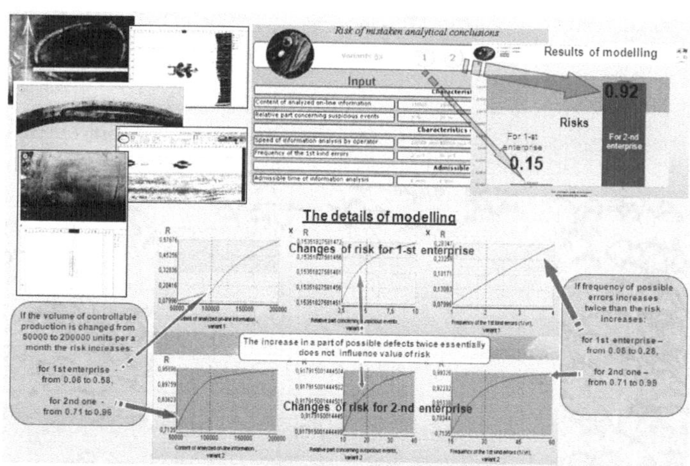

Figure 24. Comparative estimation of efficiency of quality management for enterprises which are suppliers of pipes

The solution is based on the use of the subsystem «Risk evaluation. Risk of uncontrollable development of situations» of the software tools "Complex for evaluating quality of production processes" (see model in subsection 3.9). The list of dangerous factors (threats), controllable parameters and proactive actions at grain storage in real conditions is resulted in table 1 (Machikhina et al. (2007)).

The cleared, dry and non-contaminated grain may be stored lost-free some years. However, the insects which are present in granaries and round them, occupy grain and breed. For example, every 2 months rice weevil increases in the number at 15-45 times at temperature from 20°C to 25°C. If in batch of wheat in weight 1000 tons contamination reaches 16 bugs on 1 kg of grain, losses are expected more than 5 %. The grain polluted by wreckers and products of their vital functions (excrements, dead bodies, uric acid, etc.), becomes toxic. It cannot be used for the food purposes. Therefore we will consider security of grain from insects, believing within the example, that exactly the main dangers are from them.

Table 1. The list of dangerous factors, controllable parameters and proactive measures at grain storage

Dangerous factors (threats)	Controllable parameters	Proactive measures
Biological: –microorganisms; –contamination of grain stocks by insects	Grain, spoilt as a result of self-warming and growing mouldy. Insects and pincers, a dung of rodents.	Observance of requirements of the standard documentation on grain storage. Complex of practical and exterminating measures against insects.
Chemical: - mycotoxins; - products of fats oxidation in grain (free fat acids, aldehydes, ketones, peroxides); - harmful products of vital functions of grain wreckers; - pesticides	The content of the spoilt and damaged grains as a result of microbiological spoiling. Organoleptic indicators (colour, a smell), and also the content of the beaten and brought down grains. Total density of pollution by live and dead wreckers, no more than 15 copies /kg. Residual quantities.	Observance of the general sanitary norms. Observance of regulations for pesticides use and terms of grain endurance after processing. Decrease of storage temperature to low positive temperatures of air. Observance of the instruction for pest control. Observance of requirements to grain after desinsection.
Physical: - extraneous subjects, casual and weed impurity; - grain temperature and humidity	Rough, large and casual impurity. Stable temperatureand humidity	Grain clearing on separators. Regular cooling of grain to low positive temperature (no more 10°C). Observance of the requirements of the general technological regulations

Let's a frequency of latent occurrence of critical situations during hot months is often not less than 1 time a day (i.e. every day at air temperature above 12°C infection or the further damage of grain is possible). Our consideration: at 12-15°C a duration of insects development (for example, weevil) is 141-376 days, and in a laying from 300 to 600 eggs a cycle of development is 1.5-2 months. In the conditions of cooling of grain below a temperature threshold of insects development (more low than 10.2°C) their pairing, eggs putting off and development of all stages stop. Insects become inactive and almost do not eat. Long stay of insects at such temperature leads to their slow extinction. Besides, humidity maintenance at a level of 13%-15% also promotes extinction of insects.

Thus, input for modelling is defined: frequency of latent occurrence of critical situations – from 1 time a day to 1 time a week; mean time of danger source activation – 1.5 months; time between diagnostics of system integrity (analysis of temperature and humidity) – 1 hour; duration of diagnostic, including recovery time – 1 hour.

It is enough to predict a risk of uncontrollable development of situations with grain storage. The results of modelling for the period from 1 year to 6 years have shown the following.

If a frequency of latent occurrence of critical situations is 1-2 times a day, risk of uncontrollable development of situations within a year will grow from 0.28 to 0.47, and during 2-years period it can exceed 0.5 – see Figure 25 left.

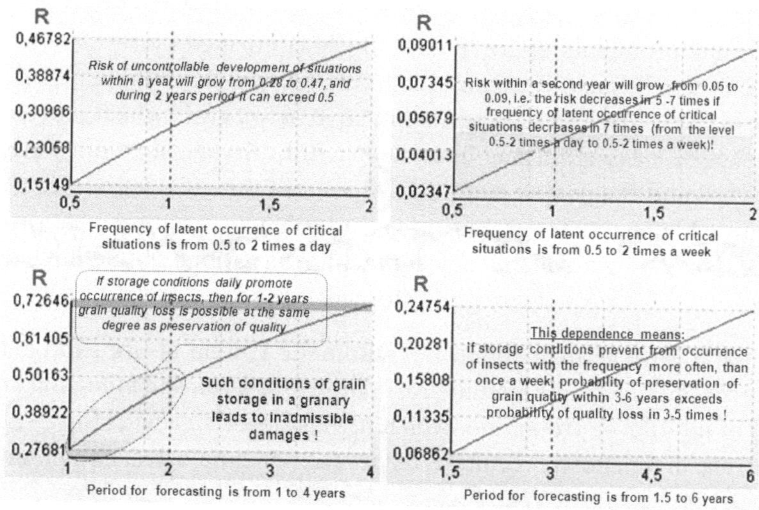

Figure 25. Some detail results of modelling and analysing

These results can be interpreted so: if storage conditions daily promote occurrence of insects, then for a 1-2 years grain quality loss is possible at the same degree as preservation of quality. Thus the next conclusion is right: the accepted conditions of grain storage in a granary leads to inadmissible damages. For prevention such danger scenario the following basic requirements (Machikhina et al. (2007)) should be performed: a smell unusual for grain should not be felt; isolation from dampness and from penetration of subsoil waters should be provided; grain-elevator should not have unfixed vertical and horizontal joints; doors should be densely closed, floors and walls should be smooth, without cracks, roofs – in a serviceable condition; fixtures should be protected by protective caps with grids; inlet of active ventilation should be densely closed preventing a penetration of an atmospheric precipitation, etc.

Performance of these requirements conducts to decrease a frequency of latent occurrence of critical situations in granaries. Further we will answer the question – what about a risk in conditions of more rare occurrences of critical situations? And, on the contrary, what the level of a frequency of latent occurrence of critical situations can be considered as admissible for granaries?

Results of modelling show: if frequency of latent occurrence of critical situations will be 1-2 times a week, risk of uncontrollable development of situations within a year will grow from 0.05 to 0.09, i.e. the risk decreases in 5-7 times! (against the level from 0.28 to 0.47), and within 6 years risk will make 0.25-0.43 (it is better, than risk within a year when frequency of latent occurrence of critical situations is 1-2 times a day!) – see Fig. 24 right. These results can be interpreted so: if storage conditions prevent from occurrence of insects with the frequency more often, than once a week, probability of preservation of grain quality within 3-6 years exceeds probability of quality loss in 3-5 times!

The results of modelling are quantitatively confirmed by results of long-term researches of the Russian Research Institute of Grain (Machikhina et al. (2007)). According to these researches experimental batches of grain wheat met to standard requirements of class grain has been kept within 6 years without deterioration in dry, cleared and the cooled condition. Moreover, the received values of risk can define admissible quality for grain storage. Indeed, new recommended result is: the acceptable risk of uncontrollable development of situations should not exceed 0.10 for 1 year and 0.25 for 6 years of grain storage.

Example 6 (Fire extinguishing). An automatic system of fire extinguishing for an enterprise of dangerous manufacture operates, as a rule, on following principles: provision of multilevel protection, which highest level means a stop of all servers operation; use of diagnostic results of devices and technological equipment.

The next measures are carried out for system availability to provide operation and fault tolerance: reservation of input for signals to acting; duplication of data transfer for switching-off equipment; consideration of switching-off only at the command of the safety officer (from the button); the voltage control in chains for executive mechanisms; implementation of intellectual devices with self-diagnostics; reservation of power supplies; reservation of safety control and emergency stop in conditions of failure of the basic system means.

To avoid false operation after detecting a fire-dangerous situation, the automatic system of fire extinguishing starts with delay 0,5 seconds. Control from the panel of the safety officer is blocked for the period of operating the automatic system of fire extinguishing. Duration of diagnostics with possible actions of fire-prevention protection is about 8.5 seconds. Control comes back to safety officer after end of automatic system act.

The solution is based on the use of the subsystem «Risk evaluation. Risk of uncontrollable development of situations» of the software tools "Complex for evaluating quality of production processes" (see model in subsection 3.9). But according to the 1-st idea of subsection 4.1 instead of metric "Risk of uncontrollable development of situations" we use metric "Risk of occurrence an emergency". Analysis of real situations allowed to form approximately the next input for modelling: frequency of occurrence of a danger source = 1 time a day, activation time of a danger source = 1 minute, the period between integrity diagnostics = 0.5c, duration of diagnostics with performance of actions of fire-prevention protection = 8.5c, MTBF for system = 2000 hours (it is commensurable with MTBF for complex technical systems and also with the period between maintenance service). Mean time to system recovery is about 1 hour.

Results of modelling show the next (see Figure 26). At the expense of automatic monitoring and fire-prevention protection the risk of occurrence an emergency within a year equals to 0. 065, and within 2th years is nearby 0.125. The mean time between possible emergencies will be about 131590 hours (this does not mean, that such successful system operation time is peculiar to the equipment. This figure characterizes effectiveness of the whole technology of the control, monitoring and integrity recovery in the given conditions of threats).

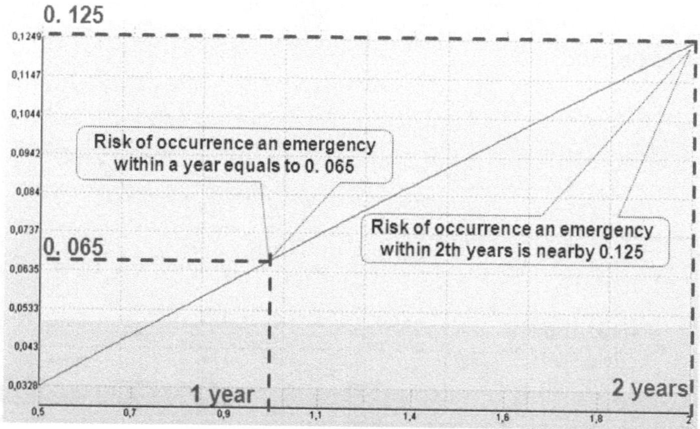

Figure 26. Dependence of risk from the forecasting period

As modern automatic systems of fire extinguishing are an example an effective utilization of information technologies implemented into various industrial systems, the reached level of risk (not above 0.065 within a year) can be de facto recognized as admissible according to "precedent" principle. At the same time, the risk of occurrence an emergency within 3th years will already exceed 0.6. This means, that at daily threats of a fire within the next 3-5 years at least one potentially emergency will be real. And moreover it can't be prevented by the operating automatic system. Here the additional measures of fire-prevention protection (including forces from the state fire service) should be provided.

Example 7 (Reliability of engineering equipment for enterprise objects). Prediction of operation reliability of computer-aided engineering equipment against usual non-automated engineering equipment is needed for the stages "Concept" and "Development" within quality management. Let the estimated object (for instance, the center of information processing and storage) includes power supply subsystem, an air conditioning subsystem, supported by 2 sources of an uninterrupted supply and a server, supported by 1 source of an uninterrupted supply and disks for information storage, supported also by 2 sources of an uninterrupted supply. In turn, the power supply subsystem includes the switchboards, supporting by 2 sources of an uninterrupted supply. All listed above engineering equipment is supported by 2 engine-generating installations.

The solution is based on the use of the subsystem «Prediction of integral quality» of the software tools "Complex for evaluating quality of production processes" (see combination of models from subsections 3.2 and 3.9 according to proposition of subsection 4.1). Within the example two subsystems are allocated (see Figure 27): a subsystem 1 – the city power supply formalized as basic and reserve subsystems; a subsystem 2 – an object fragment.

It is supposed, that operation reliability of the object is provided, if "AND" in 1st subsystem "AND" in 2nd subsystem there will be no power supply infringements during predicted term.

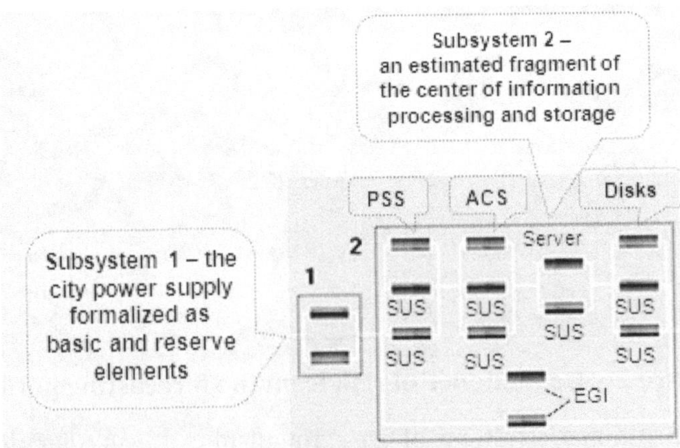

Figure 27. Logic model of the object for modelling (PSS - power supply subsystem, ACS - air conditioning subsystem, SUS - source of an uninterrupted supply, EGI - engine-generating installation)

Results of modelling are reflected by Figure 28. The analysis shows, that, at estimated technology of the control, monitoring and integrity recovery the MTBF for computer-aided engineering equipment will equal to 42219 hours. The probability of reliable object operation within a year equals to 0.828. In turn, for usual non-automated engineering equipment (there is no the monitoring implemented for computer-aided engineering equipment) efficiency characterized by estimations on Figure 28 below.

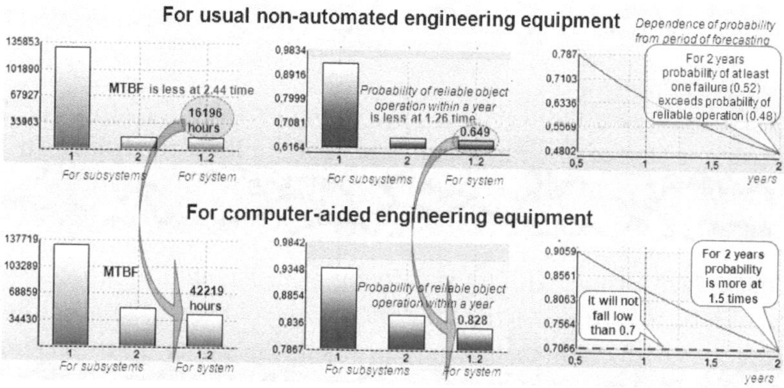

Figure 28. Results of modelling for example 7

For usual non-automated engineering equipment the MTBF will make 16196 hours (it is at 2.44 time less, than for computer-aided engineering equipment that uses monitoring), and the probability of reliable object operation within a year equals to 0.649 (at 1.26 time less, than for computer-aided engineering equipment). Moreover, without automation for 2 years the probability of at least one failure (0.52) exceeds probability of reliable operation (0.48). Against this the probability of reliable object operation within 2 years for computer-aided engineering equipment is more at 1.5 times and will not fall low than 0.7.

Example 8 (Flights safety in conditions of terrorist threats). We understand that a system component of the global terrorism problems can't be fully studied within any monograph. Nevertheless, we'll offer an approach, which allows to estimate quantitatively and compare some organizational and technical ways of its solution within quality management (safety aspect). From the modelling point of view a flying airplane is a protected system operating in conditions of threats to its integrity during the flight. We'll try to answer the next questions: "How effective was the existing before 09/11 system of flights safety provision in Russia and the USA from the point of view of opposing to terrorists?" and "How this level of the safety may be increased and by what measures?"

The answers are based on the use of subsystems «Risk evaluation. Risk of uncontrollable development of situations» and «Risk evaluation. Efficiency of protection barriers» of the software tools "Complex for evaluating quality of production processes" (see models from subsections 3.9 and 3.10).

Note. The basic results of an example 8 have been received in a week after events of 09/11, and presented for working groups on system and software engineering WG7 and WG10 SC7 JTC1 ISO/IEC in Moscow in October 2001г.

For answering the first question "How effective was the existing before 09/11 system of flights safety provision?" comparative analysis is based on the use of subsystems «Risk evaluation. Risk of uncontrollable development of situations» and «Risk evaluation. Efficiency of protection barriers» of the software tools "Complex for evaluating quality of production processes" (see models from subsections 3.9 and 3.10).

To gather necessary input data for modelling let's recall pictures of the some acts of terrorism. One of these is a highjacking of the Russian airliner Tu-154 (the company "Vnukovo" airlines) on March 15, 2001. And it is a terrorist attack on the USA committed on September 11 with the help of several passenger airliners.

The passenger airliner Tu-154 was flying from Istanbul to Moscow with 162 passengers on board. Three terrorists armed with cold steel captured the airliner and threatening with a bomb blowing-up made the pilots fly to Medina (Saudi Arabia). All terrorist attempts to break open the door to the cockpit failed. The pilots not controlled by the terrorists explained the situation on board, terrorists' maneuvers, necessary details concerning the airliner arrangement before the start of a rescue operation. Moreover, they secretly communicated with stewardesses situated in the plane cabin. On March 16 Arabian troops of special purposes made an attempt to capture the airliner. A Russian stewardess, Julia Fomina, who was fatally wounded during that storm, opened a ramp. At the cost of her life she rescued lives of the passengers. From the moment of highjacking till the moment of capturing there passed about 24 hours.

In September an unprecedented attack was committed on the USA. That attack killed thousands of people. Two skyscrapers of the World Trade Center were rammed by two passenger airliners "Boeing-767" and "Boeing-757" (the company American Airlines), captured by terrorists on their flights from Boston to Los Angeles (92 people on board) and from Washington to Los Angeles (64 people on board). The Pentagon was attacked by " Boeing-76" (the company "United Airlines") flying from Newark (New Jersey) to San Francisco (45 people on board).

Now we go to modelling of unauthorized access to airliner resources. From the point of view of terrorists opposing formalization the existing system of security provision represents a sequence of technological barriers, which should be overcome. What are the barriers?

For the existing before 09/11 safety system it is: the 1st barrier is pass and inter-object modes in aerodromes and centers of air traffic control; the 2nd barrier is a preflight examination and control of passengers and their luggage during the registration; the 3rd barrier is a preflight examination before boarding; the 4th barrier is a lock-up door to the cockpit; the 5th barrier is an on-line warning about a highjacking (this barrier is critical if terrorists try to hide the fact of highjacking). It is clear that the first three barriers if a passenger behaves well are conditional because terrorists reveal their criminal nature only on board an aircraft. Moreover, the character of the last terrorism acts proves that among terrorists there are trained executors. The terrorist actions are worked out in details.

Taking the above considerations into account we'll form input data for modelling. At first we'll discuss time of barriers overcoming. For a trained terrorist (not "wanted", having valid documents and luggage) both in Russia and in the USA mean time of the 1st barrier overcoming equals to 10 minutes necessary for identification (m=1). For an untrained terrorist the main task is not to be taken into those who are checked by security service of the aerodrome. Let only 0.5% of passengers be checked. This check may result in imprisonment during 10 days. This means that mean time of a barrier overcoming equals to ≈ 1.36 hours.

To evaluate input characteristics of the 2nd and the 3rd barriers we'll analyze the existing facts and specialists' reports. On one hand prevention of guns and explosives carrying through customs in the USA seems to be rather reliable. From the other hand carrying of penknives with blade length up to 8 centimeters had been officially allowed before September 11. On September 11 the terrorists were armed with knives for cutting of thick carton ("cutters"). Moreover, American specialists in terrorism-fighting cite facts when in 2000 employees of the USA Department of Transport decided to check 8 American airports for their vigilance. They could carry bags with guns in 68 cases of 100 ones. Finally in several shops of airports there were sold knives-souvenirs, which are brought right to the airline ladder, i.e. without any control. In Russia the situation was not better. It was worsened by the fact that in some airports modern systems of electronic examination are not used. Let's assume that a fraction of such airports mounts to 30%. The above-mentioned allows to state that for a trained terrorist overcoming of the 2nd and 3rd barriers in the USA takes about 2 hours (for each barrier) and in Russia – 1 hour. The same actions will take an untrained terrorist 10 days appeared as a result of his/her imprisonment. Then in the USA mean time of a barrier overcoming equals to ≈ 3.3 days and in Russia it equals to ≈ 2.6 days. Mean time of pass and examination in the airport is not less than a year before any essential change happens (usually before a next serious incident and start of an appropriate fight for providing airports security). The authors of the monograph know about real control service on local airlines of the USA and Russia not through hearsay. Thus the input data necessary for computations concerning the first three barriers may be considered to be formed.

The 4th and 5th barriers are the only barriers on board an airliner. A cockpit door in American Airlines "Boeing" is usual. It can be broken within a few minutes. This was done to rescue pilots in case of a catastrophe. For the same purpose some airliners take off and land with open doors. To make it clear let's set mean time of the "Boeing" 4th barrier overcoming equal to 15 minutes. A door of a Russian airliner is armored. Impossibility of such a door breaking within a few hours allowed avoiding more grave consequences on March 15. Nonetheless, according to the specialists' opinion it is not a great difficulty to blow it up or open it with the help of a fire extinguishing ax or a forcer. Let's assume that using additional improvised means it takes not more than 2 hours to overcome this barrier.

Russian aircraft are furnished with a special button of reporting about a highjacking. Not all foreign airliners are furnished with such a button and terrorists may cut off the communication with the Earth. According to specialists it is possible to escape radars by reducing height to its critical point and sharp changing of an airliner's course. On Earth it is possible to guess that an airliner is high-jacked only on the basis of indirect signs: a disappeared communication, a change of course, strange maneuvers. Sometimes passengers may use mobile phones what happened on September 11 in the USA. So, let's set time of preventing a warning about the highjacking equal to flight time.

Results of modelling are on Figure 29. An analysis of computation results reveals the following:

- both in Russia and the USA the existing systems of flights safety provision are very effective against inexperienced or untrained terrorists (the probability of security provision is not less than 0.99). It is achieved owing to preflight electronic examination and control of passengers and their luggage;
- the probability of flights safety provision in Russia and in the USA consisting in preventing of trained terrorists' penetration into a cockpit is practically the same: it equals to 0.52-0.53. In case of on-line warning about a highjacking and owing to this warning a possibility of essential opposing to terrorists this probability increases to 0.76. In Russia an armored door is the essential obstacle and in the USA it is a modern electronic examination system. According to the computations both in Russia and the USA the probability of terrorist's goals achievement in case of a thorough preliminary training is unacceptably high.

The drawn frightening figures (0.52-0.53) mean that the time of "single terrorists" has passed. They may act only on local airlines of developing countries where are no means of electronic examination and control of passengers. The computations allow with a high degree of confidence to come to the conclusion that all the taken place terrorist acts were committed after their thorough preliminary planning and preparing.

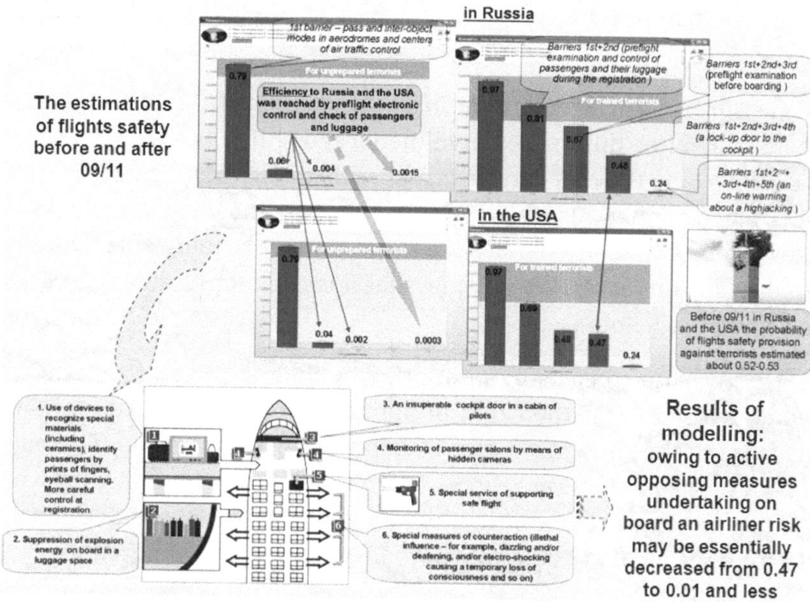

Figure 29. Results of modeling for example 8

Conclusion: in Russia and the USA the existing before September 11 systems of flights safety were ineffective against planned actions of trained terrorists; the bottleneck of flights security provision system were a weak protection of a cockpit and absence of active opposing measures on board an airplane.

Let's answer the second question "How the level of the safety may be increased and by what measures?"

It is possible to think up a set of such measures, however ways of their application should be carefully proved depending on the scenario of terrorists actions. We will stop only on some of the measures that already are implemented at the various airports.

The 1st measure consists in using devices for recognition the special materials (including ceramics), in an identification of passengers by prints of fingers, in eyeball scanning, in using the general databases of prospective criminals, in restrictions on hand luggage. We will designate a measure 1 as 6th barrier in addition to considered above. We will put the mean time of keeping effectiveness of 1st measure equals to 1 year (i.e. for overcoming this barrier it should be spent about 1 year). As effective devices appear annually, we will estimate time before the next adequate strengthening of a measure in 1 year.

The measure 2 allows to suppress an energy of explosion on board in a luggage space. It is 7th barrier. We will put the mean time of keeping effectiveness of 2nd measure equals to 5 hours (i.e. some effect can be achieved on the average within 5 hours, commensurable in due course flight). As annually there are more effective remedies. Time before the next adequate strengthening of a measure also will be estimated in 1 year.

The measure 3 is an armour door in a cabin of pilots (or two doors, the second door opens only after the first one will be locked) - it is 8th barrier. The armour door should become the real barrier insuperable to terrorists during all flight. It is necessary to notice, that this measure will not secure the members of crew serving passengers. Thus unlike 4th barrier the mean time of keeping effectiveness of 3rd measure logically increases. We will put, that it is commensurable with duration of flight and equals to 5 hours.

The 4th measure consists in monitoring behind passenger salon by means of videocameras. As soon as the cabin of pilots becomes unapproachable it can be transformed into the situation center of safety of passenger salon. Thereby before pilots, and also the land officers the real picture of an events opens. They will have access to complete and valid information on board. We will consider a monitoring on board from the auxiliary point of view for other additional measures.

The measure 5 is a special service of flight. At the same time the boomerang effect is possible - terrorists can detonate an explosive after having encountered resistance from special service. And if terrorists can disarm a specialist of special service, they will have an additional weapon. It is 9th barrier. We will put the mean time of keeping effectiveness of 5th measure equals to 1 hour (i.e. a specialist of special service can be detected in average for 1 hour and an effect of it is not clear). The period between strengthenings of special services we will estimate in 5 years.

The measure 6 is formed from special measures of counteraction (temporary depressurization of salon, not lethal influence). Really it is 10th barrier. For the explanatory of this measure we will consider some scenario reasons:

- as counter-attracting maneuver at average altitude the salon can be temporarily depressurized for a disorientation of terrorists and granting of the initiative to crew and special service of flight (that at a low altitude this measure may be inefficient, and at a high altitude it will quickly lead to irreparable consequences);
- terrorists are obliged to be active, for this purpose those from them which have found out itself obviously, are in standing position, passengers – in sedentary. The first problem of protection is to destroy these subjects of threats at least for some minutes. And means of not lethal influence should be used because of passengers can also be influenced simultaneously. Then 6th measure is capability of using means of dot not lethal influence on the revealed terrorists. It may be influences by lulling gas and-or short-term shocking influences (for instance, blinding and-or deafening and-or the influences of electroshock type leading to a temporary loss of consciousness). The ways of influence should be a little, because against one way a simple counteraction can be found (against

gas – a gas mask, against blinding – goggles and so on). Thereby some revealed terrorists can be practically neutralized.

As at salon there can be the accomplices capable to recapture after additional preparation, methods of compulsory keeping of suspicious passengers on the places before emergency landing should be made. It is one of versions within the limits of 6th measure (which can be used by the individual lulling influence and-or jammed fastening, etc. Considering possible variants, we will put, that the mean time of keeping effectiveness of 6th measure equals to 5 hours (commensurable in due course flight). The period between strengthenings of 6th measure we will estimate at 2 hours taking into account various possible variants.

All listed measures seem at first sight rather impressive, but how much they are effective? Really, their effectiveness should be proved quantitatively! This is a very complicated task. It is impossible to make natural experiments. We may only use mathematical models.

Analysis of results has shown, that after implementation of the described measures the integrated risk to lose complex safety of flight during 5 hours of flight against terrorist threats is equal to 0.000004. And if duration of threats will be increased to 5 days the risk raises from 0.000004 to 0.002. The last can be commented by the next interpretation: safety will be achieved in 998 cases from thousand hypothetical terrorist attacks. Even taking into account an essential error of initial scenarios and preconditions it is an obvious indicator of high efficiency of additional safety measures according to "precedent" principle! Still it is not a victory. It is clear that the first failures will make terrorists to analyze their causes and find new bottlenecks of the safety system thus continuing the counteraction. This counteraction will be ended when there are taken proactive measures which effectiveness is based on modelling.

Example 9 (Information security). In quality management measures of protection of valuable resources from an unauthorized access (UAA) should be provided. The most important for any enterprise are information and software resources of an IS. We will consider the approach to an estimation of IS security against UAA and information confidentiality. A resources protection from UAA is a sequence of barriers. If a violator overcomes these barriers he gets access to IS information and/or software resources. In the Table 2 there are shown supposed characteristics of barriers and mean time of their overcoming by a specially trained violator (real values of such characteristics may be drawn as a result of actual tests or use of models not included in the monograph). It is required to estimate IS protection against UAA.

Solution. We'll try to answer this question by the subsystems "Protection from unauthorized access" and "Confidentiality" of CEISOQ+. The analysis of computed dependencies (see Figure 30 left) shows the next. The barriers 1,2,3 will be overcome with the probability equal to 0.63. However, monthly password changing for barriers 4, 5, 6 allows to increase the protection probability from 0.37 to 0.94 but the level of IS protection (the first six barriers) is still low. The introducing of 7,8,9 barriers is useless because it does not practically increase the level of IS protection. The use of cryptography allows to increase the level of IS protection to 0.999. This is probability for all time of IS operation (i.e. about 20-30 years). It is possible to establish a conclusion, that with the use of cryptographic devices the achieved protection level exceeds similar level of quality and safety for processes from examples above. But according to "precedent" principle this level of protection can't be recommended as high customer requirements for every cases.

Table 2. Input for modelling

Barrier	The frequency of barrier parameter value changes	The mean time of the barrier overcoming	Possible way of the barrier overcoming
1. Guarded territory	Every 2 hours	30 min.	Unespied penetration on the territory
2. Admission system for coming into office	Once a day	10 min.	Documents forgery, fraud
3. Electronic key for powering the computer	Every 5 years (MTBF = 5 years)	1 week	Theft, collusion, forced confiscating
4. Password to login	Once a month	1 month	Collusion, forced extortion, spying, password decoding
5. Password for access to devices	Once a month	10 days	Collusion, forced extortion, spying, password decoding
6. Password for requesting information resources	Once a month	10 days	Collusion, forced extortion, spying, password decoding
7. Registered device for information recording	Once a year	1 day	Theft, collusion, forced confiscating
8. Confirmation of user authenticity during a computer session	Once a month	1 day	Collusion, forced extortion, spying
9. Television monitoring	Once a 5 years (MTBF = 5 years)	2 days	Collusion, disrepair imitation, force roller
10. Cryptosystem	1 key a month	2 years	Collusion, deciphering

Let's look on example condition more widely. The violator is interested in a certain IS resources during a certain period of time. This period is called the period of objective confidentiality. Unlike UAA information confidentiality should be provided within these lasting 7 days. Fig. 30 (right) shows how this period influences on protection:

- in comparison with the results above the use of the first 5 barriers provides confidentiality during 7 days on the level 0.98 which is more higher than protection from UAA by the 9 barriers (0.946 – see Fig. 30 left);
- the use of all the 10 barriers provides the required confidentiality on the level 0.99997. It eliminates the customer's risk in providing system protection. It explains the role of a considered period of objective confidentiality – its consideration allows to understand, that real protection of resources during 7 days is essentially higher - 0.99997 against 0.999!

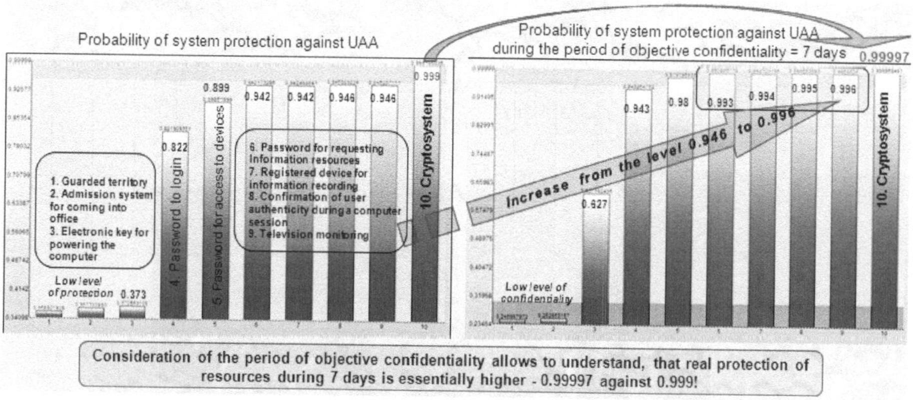

Figure 30. Comparison of protection levels

Example 10 (Forecasts of risks for complex multipurpose system). Let's consider a hypothetic multipurpose system which formally composed from functional system (similar, for instance, to commerce system, enterprise non-destroying control system or system of foods preservation from examples 2, 4, 5), gathering and data processing systems (similar to SCADA system from example 3), system of fire extinguishing (from example 6), system of engineering equipment for enterprise object (from example 7), information security system (from example 9). «The human factor» is considered in the parameters of control, monitoring and integrity recovery measures for corresponding elements. It is supposed, that a required integrity of system is not lost, if during given time a required integrity is not lost by all subsystems: "And" by 1st subsystem, "And" by 2nd subsystem, ... "And" by the last 6th subsystem (the logic illustrated by Fig. 12). It is required to estimate the measures of risk management, including the periodic control and, where it is possible, continuous monitoring of integrity of each components – see Figure 31.

The input for subsystem 1-6 is described in examples 2-7, 9. The general results of complex forecasting of risk are reflected by Figure 32. Analysis of results shows, that with using of measures of the periodic control and where it is possible, monitoring of elements operation, the integrated risk to lose integrity of system during operational 1 – 4 years is changing from 0.11 to 0.67.

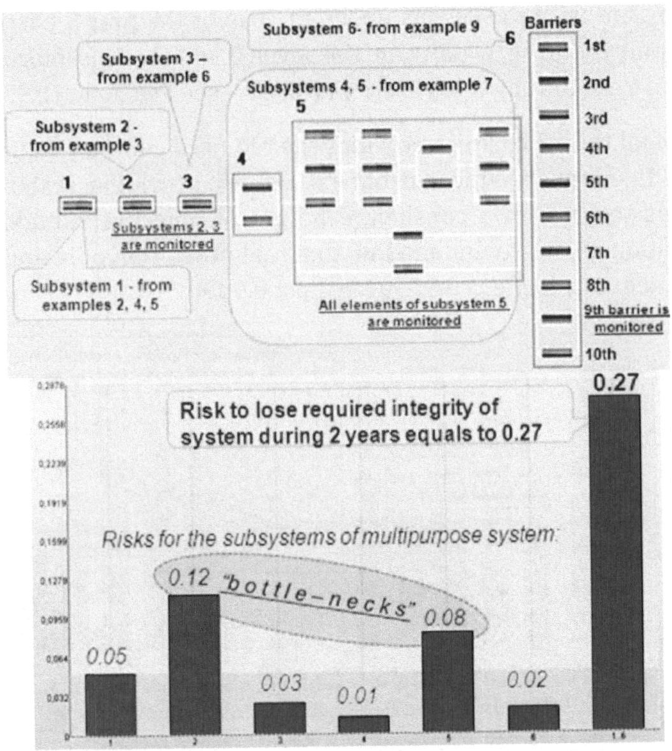

Figure 31. The formal scheme of multipurpose system for a complex risks evaluation

The general logic proposition is right for a given period of forecasting: as a rule, the risk to lose system integrity increases in depending on increasing time period. But there are the features demanding a logic explanation. Serrated and nonmonotonic character of dependence on Figure 32 is explained by the periodic diagnostics of elements, monitoring presence or absence and their quantitative values. Let›s remind: for every monitored element a penetration of a danger source and its activation is possible only if an operator-monitor makes an error but a dangerous influence occurs if the danger is activated before the next diagnostic. Otherwise the source will be detected and neutralized. Immediately after element diagnostic the risk decreases because during diagnostic all dangers are detected and neutralized and at the beginning of a period after diagnostic dangerous influences don't have enough time to accumulate and be activated. Nonetheless, there is a lack of protection accumulated for the previous full periods that's why the risk doesn't decrease to 0 for every element. By the middle of a period between neighboring diagnostics there is an increase of the calculated risk because new danger sources can begin to influence. Moreover, for the longer period of forecasting monitoring possibilities are weaken, thereby the moment of operator error comes nearer. And, if on timeline the following diagnostic does not come yet, risk increases. Similar effects paradoxes are explained – for example, that risk to lose integrity during

2.96 years (0.58) is more, than risk during more long time - 3.12 years, 58 days longer (0.57). One more effect of modelling: if to do forecasting not for 2.04 years, and for 2 weeks longer (2.08 years, i.e. 2% longer period) the expected risk to lose system integrity increases from 0.28 to 0.36. This is higher on 28 %! These results of modelling should serve as a substantiation for development of predicting counter-measures.

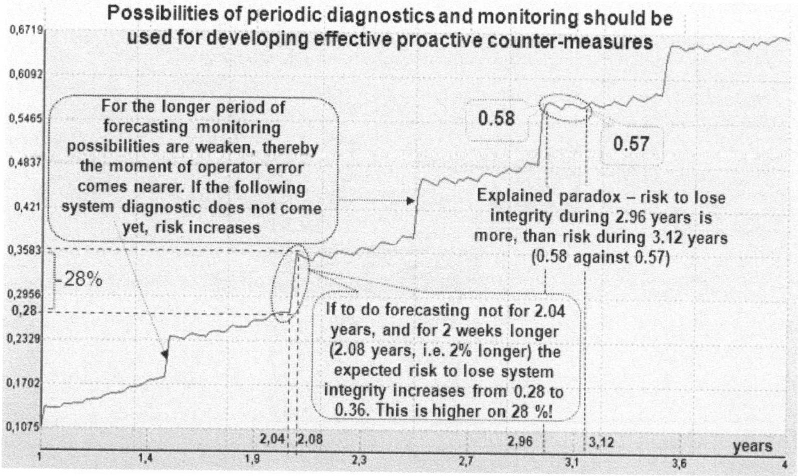

Figure 32. Integrated risk to lose integrity of system during operational 1 - 4 years

Indeed, on the basis of a rational choice of parametres for technologies of the control, monitoring and integrity recovery an optimization of processes offered in work is possible.

6. Conclusion

Rational management means wide use of existing models and software tools for decision-making in life cycle of systems. The criteria used for rational management are maximization of a prize (profit, a degree of quality or safety, etc.) at limits on expenses or minimization of expenses at limits on a comprehensible degree of quality and-or safety or their combination.

As a result of adequate modelling more deep and extend knowledge of system allows the customer to formulate well-reasoned system requirements. And it is rational to developer to execute them without excessive expenses of resources, and to the user – as much as possible effectively to implement in practice the incorporated power of system.The presented models, methods and software tools, allowing to forecast quality and risks according to system requirements of standards, are real levers to analyze and optimize system processes and improve quality management. The investigated practical examples demonstrated their functionality and possibilities to use "precedent principle» for definition the justified levels of acceptable quality and admissible risks. For complex systems the proposed results helps to answer the questions «What rational measures should lead to estimated effect without waste expenses, when, by which controllable and uncontrollable conditions and costs?» and allows to go «from a pragmatical filtration of information to generation of the proved ideas and effective decisions» (see Figure 33). The effect from implementation in system life cycle is commensurable with expenses for system creation.

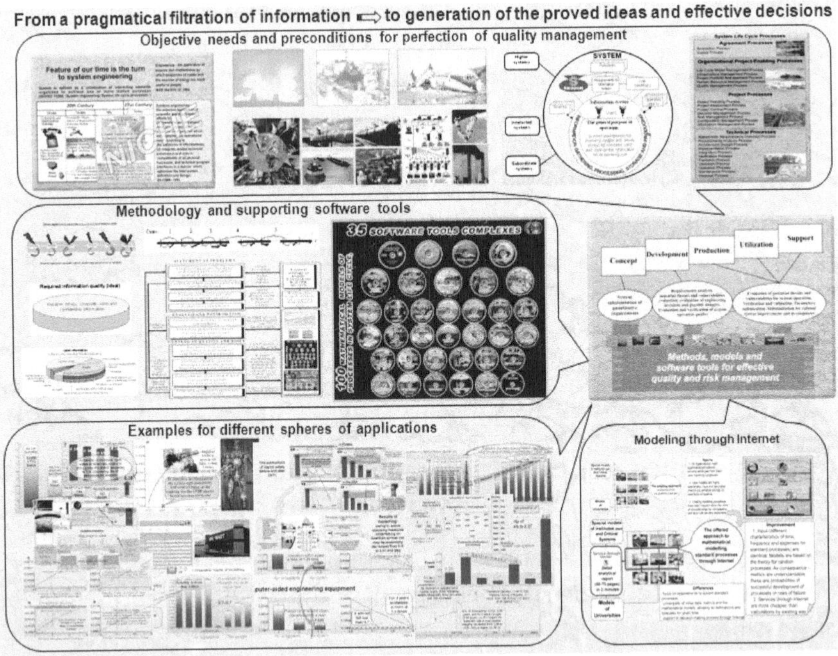

Figure 33. The proposed results helps to answer the questions «What rational measures should lead to estimated effect without waste expenses, when, by which controllable and uncontrollable conditions and costs?»

References

1. FellerW.1971An Introduction to Probability Theory and Its Applications. Vol. II, Willy, 1971.
2. Gnedenko B.V.1973et al. Priority queueing systems, MSU, Moscow, 4481973
3. Klimov G.P.1983Probability theory and mathematical statistics. MSU, Moscow, 3281983
4. MartinJ.1972System Analysis for Data Transmission. V. II, IBM System Research Institute. Prentice Hall, Inc., Englewood Cliffs, New Jersey, 1972.
5. KleinrockL.1976Queueing systems, 2Computer applications, John Wiley & Sons, New York, 1976.
6. Matweev V.F. & Ushakov V.G.1984Queuing systems. MSU, Moscow, 2421984
7. Kostogryzov A.I., Petuhov A.V. and Scherbina A.M.1994Foundations of evaluation,

providing and increasing output information quality for automatized system. Moscow: "Armament. Policy. Conversion", Moscow, 2781994

8. Kostogryzov A.I.2000Software Tools Complex for Evaluation of Information Systems Operation Quality (CEISOQ). Proceedings of the 34-th Annual Event (25-29 September 2000) of the Government Electronics and Information Association (GEIA), 2000 Engineering and Technical Management Symposium, USA, Dallas, 2000, 63
9. Bezkorovainy M.M., Kostogryzov A.I. and Lvov V.M.Modelling Software Complex for Evaluation of Information Systems Operation Quality CEISOQ. 150 problems of analysis and synthesis and examples for their solutions. Moscow: Armament. Policy.Conversion, 20012nd edition-2002), 303p., 5-89370-015-5
10. Kostogryzov A.I.Modelling Software Tools Complex for Evaluation of Information Systems Operation Quality (CEISOQ). Proceedings/International Workshop-Information assurance in computer networks: methods, models and arhitectures for Network Security. MMM ACNS 2001, St.Peterburg, Russia, May 21-23 2001, LNCS (200190101
11. GrigolionisV.About approximating stepped processes sum to Poisson processes. Probability theory and its applications, 81963
12. Kostogryzov A.I.1987Conditions for Efficient Batch Job Processing of Customers in Priority-Driven Computing Systems Where the Queueing Time Is Constrauned, «Avtomatika i telemehanika». 1987. №12. 158164
13. Kostogryzov A.I.1992Study of the Efficiency of Combinations of Different Disciplines of the Priority Service of Calls in the Computer Systems, «Kibernetika i sistemny analiz». 1992. №1. 128137
14. KostogryzovA.NistratovG.Standardizationmathematical.modellingrational. managementcertificationin.thefield.ofsystem.softwareengineering. (.standards1. mathematicalmodels. .softwaretools.morethan. .practicalexamples.Moscow: Armament.Policy.Conversion, 2004 (2-nd edition- 2005in Russian), 393p., 5-90231-305-8
15. Kostogryzov A.I., Nistratov G.A. 100 Mathematical Models of System Processes According International Standards Requirements.Transaction of the XXV International Seminar on Stability Problems for the Stochastic Models. Maiority, Italy, September 20-24,2005University of Solerno, Italy 196201
16. KostogryzovA.NistratovG.KleshchevN.Mathematical Models and Software Tools to Support an Assessment of Standard System Processes. Proceedings of the 6th International SPICE Conference on Process Assessment and Improvement (SPICE-2006Luxembourg, 2006, 6368
17. KostogryzovA.NistratovG.Mathematical Models and Software Tools for Analyzing System Quality and Risks according to standard requirements. Proceedings of the 6th International scientific school "Modelling and Analysis of safety and risk in complex systems" (MASR- 2006), SAINT-PETERSBURG, RUSSIA, July 4- 8, 2006

18. KostogryzovA. I.StepanovP. V.Innovative management of quality and risks in systems life cycle. Moscow: "Armament. Policy. Conversion", 2008p., 5-89370-012-0

19. Grigoriev L.I., Kershenbaum V.Ya. and Kostogryzov A.I.System foundations of the management of competitiveness in oil and gas complex. Moscow: National Institute of oil and gas, 2010p., 5-93157-086-1

20. KostogryzovA.et al.Mathematical models and applicable technologies to forecast, analyze and optimize quality and risks for complex systems, Proceedings of the 1st International Conference on Transportation Information and (SafetyICTIS 2011) June 30th ~July 2nd 2011Wuhan, China, 845854

21. MachikhinaL. I.AlexeevaL. V.L'vovaL. S.Scientific foundations of food grain safety during storage and processing Moscow, DeLi print, 2007p., 978-5-94343-140-1

8

Competence Education and Training for Quality

1. Introduction

The world global changes that are in their character historic and far-reaching are experienced at the end of the twentieth century. The phenomenon of globalization of world economy and world market has done to radically alter notions of time and distance. The world is becoming global and transnational systems of production / services in conditions of high mobility of people and capital.

Each country tends to its products and services reach the market freely in other countries, so that the world and national markets are increasingly overwhelmed with a large and diverse number of goods and services [1]. Removing technical and other barriers to trade created more opportunities for consumers and clients over the world. However, the increased flow of goods and services, is now more than ever, a need for increased consumer protection and consumer services. This is achieved by quality, to protect and connect people around the world. So to confirm and strongly affirm the assertion that no ideology has failed or will fail to unite the world, is only to have quality. The concept of quality includes, at the same time, all areas of human activity: quality products, quality management, quality management (government) and quality of the life of the people.

In the modern business world, globalization of markets, the quality is extremely important instrument for achieving competitive advantage of organizations. Success of the business is now determined by the ability of organizations to respond to the demands set by the market, and improving the quality of the business becomes an imperative of contemporary market trends. Improving the quality of business is long-term goal of all organizations that seek business excellence and achieving world-class products and services. The process of continuous quality improvement is based on the improvement of knowledge and productivity of all employees in the organization of individuals, especially those who are responsible for the growth of productivity.

2. Term of quality

The basis of the modern terms of quality "quale" is a Latin word that means what, how is something or someone. Of course, in some earlier languages of Latin, there are terms that describe exactly what it was something. The definition of the universal cosmic phenomenon and, despite the differences, they all end in terms of talking about the same. Long ago the brilliant Aristotle (384-322. Years BC) wrote about the quality "as the difference between individuals" in "Metaphysics".

The scientific approach to management occurs at the beginning of last century, in parallel with the work of Frederick Winslow Taylor from 1856 to 1915 i Henri Fayol (1841-1925). However, management appears much earlier and may be said to be as old as civilization and as the oldest types of organizations. The idea of management by Claude George, author of one of the most popular books on the history of management thought (Claude S. George Jr. The History of Management Thought, Prentice Hall, Englewood Cliffs, New Jersey, 1968.), dates back to the Sumerian civilization (5000 BC). The Sumerians lived in Mesopotamia, the area of present-day Iraq and Iran. They were in the beginning in clans, and then slavery society.

They have developed a high culture that was based on developed agriculture. It is believed that they achieved it by the construction of irrigation systems. Among the first they developed form of written records on cuneiform clay tablets. Then, the construction of the pyramids in Egypt during the Old Empire (2700 - 2200 BC). It is considered that the construction of the Cheops pyramid was more than 20 years. All this had a plan to organize, coordinate, and so on. Significant written records date from the old Babylonian state (2200 - 1700 BC), and for the first time in its written law regulating moral issues and the quality of work. As the law 229: "If mason builds a house for a man, and does not make it hard enough, and if it fell and caused the death of the owner - builder will be executed." There are still some of the current principles of war of the Chinese military leaders from 600 BC When it comes to ancient China we must not bypass the philosopher Confucius (ca. 552-479 BC).

Also, the Bible can be found elements of the vision and leadership. In the works of thinkers from ancient Greece can be found many wise thoughts on management, primarily state and army. There are Ksenofant (480-355 BC), his best pupil Socrates (469-399 BC). Socrates thought that someone who is unable to lead his own business, will not be able to keep any state. We mentioned the genius of Aristotle, who first mentioned the concept of quality - quality, and some of his ideas, which he gave in his major work "Politics" are current today, such as the:

- Specialization of labor,
- Grouping of jobs in the work unit,

- Leadership ("Who never learned to listen - can not be a good commander"),
- Centralization / decentralization,
- Synergy ("The whole is naturally superior to parts")

Aristotle's genius is not only in terms of quality but also to its versatility and legacy, which on very conservative estimates exceed 20,000 pages. It's not just the work but the whole library. For this we highlight the wealth "Metaphysics," which, according to many authors, is considered a major offense. The origin of the term metaphysics, the author dares to formal explanation. The phrase following the merger of the metaphysics of words μετα τα φιςνκα and haplologijom - omitting τα syllable that is repeated. There are similar examples in our language, such as - water carrier. In the section "Metaphysics" as item number 14 the heading TERM OF QUALITY. Before we move on to consideration of this concept, one more note, which refers to the fact that you should know that all of Aristotle's writings are divided into published and unpublished. The latter group includes Metaphysics. This means, in its reading and study we often have to ask whether we are dealing with authentic texts of Aristotle, as they are entered in the unpublished writings of his lecture for a narrower circle of older students who at least add some words of the teacher.

"The notion of quality. Quality is called, in one sense, the difference substances: for example, a man is a living being of certain quality because it is a biped, the quality of horses that quadruped; circle is a figure whose quality is no angles, all these things show that difference to the quality of substance really. In this first sense of quality is said to be a difference of substance. In another sense the quality mathematical meaning of real things, in this sense, the numbers have a certain quality: they are, for example, made up numbers, not numbers that denote not one size, but those whose size and weight are their image (those are numbers that are the product of two factors and the numbers that are the product of three factors), quality is what lies in the importance, in addition to its quantity: in fact, the substance of each issue is what exists once, so, for example, two is not six or three times one number, but one time since six times one is six Quality is also called properties of substances that move, as heat and cold, and whole hiatus, weight and lightness, and the other to determine the species, according to which, when changed, and the body said to suffer a change. On the other hand, in the field of quality are virtue and vice, and in general good and evil.

The meanings of quality could be largely reduced to two, of which is one the main. The first quality is the difference in the substance of things, and the quality of the numbers is a part of it, because it is the difference between substances, but substances that either are not movable. or are taken as moving. The second meaning involves determining the motion being taken as such, the difference of motion. Virtue and vice are to some extent among these forms: namely, they show the difference of movement and acting according to which the creatures that move or suffer the act good or evil, in fact, what you can not move or act in this way is right and what you can move or work in one, the opposite way is bad. Good and evil are expressed in particularly the quality of living beings, and of those most in those who are endowed with free choice "[2].

Concerning this passage from Aristotle's "Metaphysics" of quality (14-term of quality), it can be concluded that the quality, in fact, the quality. I refer primarily on product quality - quality ingredients. In addition to product quality, it could be an analogy with what we "can not move or act in this way is right and what you can not move or work in one, the opposite way is bad" and get to something that would be might recognize as the product of agreement and disagreement. Also, one could see traces of the procedures and codes of behavior through, for example, "Virtue and vice are to some extent among these forms (quality): namely, they show the difference of movement and acting according to which the creatures that move and act suffer good and evil." The presence of motion to describe the quality can be considered as a process.

In addition to the concept of quality (quality), Aristotle mentions the most important resource, which is knowledge and that in the first chapter of the same book (Metaphysics) on page 3: "All men are by nature tending to get to know, the proof is the joy caused by the experienced knowledge; namely, in spite of their benefits we like then the visual information more than any other. Because we appreciate the sight of all, so to speak, not just to be able to do, but even assuming that we do not want anything to do. The cause of this is that, of all our senses, sight is a sense by which we gain the most knowledge and discover a multitude of differences. " This is Aristotle's conception of knowledge when compared with today's is understanding of knowledge and with knowledge that the most acquired knowledge is visually, about 70%, it is not difficult to see his greatness. Here is one of today's understanding of knowledge:

"Knowledge that is formed as a world view or self-consciousness is expressed in the form of religion, philosophy, art, science, folk wisdom, that is in the form of conceptual and artistic consciousness. It occurs as an expression of individual and group consciousness and experience and has deep roots in the past. Knowledge has always been part of organizing society and its development. "[3]

Aristotle's genius in universal is reflected in the fact, which in a large scale could also serve as Catholic theologians type Aquinas and Islamic mystics to the Indian Ocean. There is still no universal definition of quality. To help understand the concept of quality, quality

experts answer what is the quality:

- Quality is not what most people think it is.
- Quality is noting new, what the majority of employees in jobs of the quality have already known and has attempted (and failed) to improve it.

- Quality is not just quality of products and services.
- Quality is not a "commodity" that can be purchased at the market. "[4]

Even the world of quality gurus have different definitions of quality. Guru means a respected teacher, spiritual leader, who in his field has not only made a great contribution and innovation, but also a large-scale revolution. People who have established themselves and profiled philosophical trends in quality, are the gurus of quality. Guru of quality, in addition to its basic meaning, means a person who, with their concept and approach to quality, significantly contributed characterizing a period of time.

Although all the quality gurus contributed significantly to the development and improvement of the quality we will name six of them: Edwards Deming, Joseph M. Juran, Ishikawa Kaoru, Genichi Taguchi, Armand Feigenbaum and Philip Crosby.

In addition to these quality gurus at this time without fail the name of Philip Kotler, who is the "Financial Times" in 2001. was ranked among the greatest management guru Peter Drucker by side. On several occasions he traveled throughout Europe, Asia and South America, holding consultations and lectures to many companies. Among his clients are some of the leading, global companies like IBM, Michelin, Bank of America, General Elrctrica and Motorola. He is the author of many papers and books, the most famous is the "Marketing Management", and now the current of his new book "Chaos Magick, management and marketing in turbulent times." According to Kotler, turbulence has two main effects. One vulnerability, for which the company must develop a defense shield. The other is an opportunity which should be used with a new model called the Control System of Chaos. This is an innovative model that minimizes the vulnerability and exploit opportunities, and thus creates a competitive advantage.

However, definitions of quality are given in international standards. The audit standards and improve the quality of definitions, ranging from ISO 8402:1986 (1994), Vocabulary ("set of characteristics that an entity has to satisfy all requests, and even anticipated the wishes and preferences of the customer") [5], then, ISO 9001:2000, Fundamentals and vocabulary (3.1.1Quality degree to which a set of inherent characteristics (3.5.1) fulfills the requirements (3.1.2)). and ISO 9001:9005, Fundamentals and vocabulary[6]

2.1. Basic quality elements

The basic elements of quality are: metrology, standardization, accreditation, certification and market control. Historically quality developed in accordance with the development of society. Analogously, evolved, also the quality elements.

2.1.1. Metrology

Measure some size means to compare it with a known size - standard. The science of measurement is metrology. Metrology is not reserved only for science, that is, scholars, it is vital for all people. Primarily for their health, consumer protection, the trust of customers / users, and to complete development and economic success of nations. The first recorded mention of metrology in ancient Egypt, and to measure the length. In addition, measures of length in the Bible - Old Testament reference to the weights - the weight measures and volume measures. Measures of length are calculated according to the human body: the finger width (1.875 cm), width of palm (7.5 cm), inch (22.5 cm) - width of the hand when the fingers are spread, from tip of the thumb to the tip of the little finger and elbow (45 cm) - from elbow to tip of middle finger. Most of measure of length was the Egyptian reed thar had six cubits, or twelve inches, 36 palms or 144 fingers. Bushel were given the names of the vessels that they received a certain amount agreed, noting that measures the volume of liquids and solids were different. For the liquid is 1 baht (22.9 liters) and ten times a homer, and the dry matter of a chief, five times a litek and 10 head of a homer ("ass load"). Measures for the mass of 1 tera (0.5 g), 10 times larger is 1 guard, 2 guard is a shekel, 50 shekels was a mine, mine 60 1 talant (30 kg).

Besides the scientific metrology, which deals with the organization, development and maintenance of standards, there are legal and industrial metrology categories. Legal Metrology deals with the accuracy of measuring instruments, and industrial metrology has to ensure proper application of them. For these reasons there must be a national organization for metrology, which is a regional member, for example, EUROMET - European Collaboration in Measurement Standards, and / or the World Organization for Metrology OIML.

The importance of metrology is described in the old motto of the old traders: The number and dimensions - my faith

2.1.2. Standardization

Activities aimed at determining the provisions for common and repeated use, in relation to actual or potential problems, are undertaken in order to achieve an optimal level of standardization of neatness. These activities are specifically related to the processes of formulating, issuing and implementation of standards. Standard is a document established by consensus and approved by a recognized body. Standards should be based on the consolidated results of science, technology and experience. Under by consensus in reaching the standards does not mean unanimity, but a general agreement characterized by the absence of categorical opposition to any substantial question of interested parties, made in the process which seeks to take into account the views of all stakeholders to harmonize all conflicting arguments. Depending on the degree of involvement, level of standardization and standards may be international, regional, national and local. However, standards can be made on other grounds, such as branch standards or standards of the company (internal), which can be applied in several countries. The importance of standardization, ie, the standard is to improve the benefits of processes, products and services for their purposes, prevention of barriers to trade and facilitating technological cooperation.

A prerequisite for the smooth trade and free flow of goods and services is the application of international standards. As defined by the WTO - World Trade Organization under the term "international standards" is ment the only standards that are developed on the basis of the international system of standards and conformity assessment, and that was obtained by consensus, voluntary and by the influence of the market. International Organization for Standardization's ISO-International Organization for Standardization, which in the foreword of its standards gives the way to prepare and adopt standard of: ISO (International Organization for Standardization) is a worldwide federation of national standards institutes (ISO member). The work of preparing International Standards is carried out in ISO technical committees. Each member of ISO, when the interest in the subject of some of the technical standards committee has the right to delegate representatives to the Committee. International organizations, governmental and non-governmental, in relation with ISO, also take part in the work. ISO collaborates closely with the International Electrotechnical Commission (IEC - International Electrotechnical Commission) with respect to all matters of electrical standardization.

2.1.3. Accreditation

The term accreditation is relatively common in everyday speech, and depending on the area in question is given a different meanings. For example, in journalism, getting right to the reporting of an event or place. Accreditation occurs in diplomacy where the competent state authority entitles a person to represent his country in another country. Often occurs in foreign trade, banking, and in correspondence. As can be seen, these are different areas of human activity on different instruments, but basically they all have a common factor in establishing trust.

At this time of globalization when the world trade in a rising market and a growing number of goods and services, there is a question of establishing a mechanism of trust between the supplier or service provider and customer or service user. Each customer or user wants the goods or services you purchase or use to be safe. that is not to be harmful to his health and the environment, and to meet their needs and expectations. However, every individual is unable to satisfy itself of the quality of goods or services, a time to gain confidence. To achieve this, he used different documents. These documents may be certificates which provide various certification authorities, certificates of inspection by the quality control organizations or test reports issued by testing laboratories. Also, here raises the question of confidence in these documents. In most cases the customer-user is unable to directly verify the validity of the certification bodies, inspection bodies, testing and metrology laboratories, but also in his behalf, the work national accreditation bodies, which determine the competence of these organizations. Accreditation body of the organization, which defined competency issues the decision on accreditation or accreditation certificate.

Thus, accreditation is formal recognition that a national body for accreditation after the procedure, which was by all compliant to international standards, confirming that the organization is competent to perform the defined scope accreditation. All these organizations are engaged in conformity assessment activities. Compliance is fulfilling the acquirements.

Depending on which jobs and organizations are involved is given the accreditation for the appropriate type of accreditation, and the types of accreditation and the scope of accreditation. Depending on the type of accreditation, these organizations can begin the process of accreditation:

- Laboratory testing
- Laboratory calibration (metrology lab),
- Control of the organization,
- Certification body of process for products and services,
- Certification body for quality management system,
- Certification body for management systems and environmental
- The organization for the certification of persons performing conformity assessment activities.

The introduction of standards and technical regulations in almost all commercial and industrial areas, significantly increases the number of products / services, processes, personnel and systems that may be subject to conformity assessment. For these reasons, the range of accreditation by an accreditation body performs is very extensive and covers mainly the following areas:

- Acoustics
- Agriculture
- Fire fighting and anti-explosive devices
- Chemistry and chemical products
- Electronics, IT, radio and telecommunications
- Construction
- Elect
- Mechanical Engineering and Materials
- Protecting the environment
- Food and Food Security

- Energy
- Medicine and drugs
- Textiles, rubber, plastics and packaging
- Business and Environment
- Protective devices and equipment
- Tobacco and tobacco products
- The objects of general us
- Non-destructive Testing
- Waste

Accredited organizations with the national accreditation body, which manages the accreditation system, make the accreditation system of the country. National accreditation systems attempt to integrate with regional accreditation system, as well as international accreditation system. To accomplish this goal, the national accreditation body for the activities under its jurisdiction shall adopt and implement a document governing the criteria, rules and procedures in accordance with the general requirements of a series of harmonized standards EN 45000, adopted by the European Organization for Standardization (CEN) and the series of international standard series 17000, adopted by the International Organization for Standardization (ISO) in collaboration of the International Electro-technical Commission (IEC).

European Cooperation for Accreditation (EA), at its General Assembly in June, 2002. in Bucharest established the general principles concerning the status of politics and national accreditation body. The adopted principles are binding for both existing members and new. Existing members of EA must agree with them, and new ones have to meet before accession. Principles of EA:

- The principle of national recognition. Each accreditation body can become a member of the EA, only if the country is recognized as a national body..This means that from one country may be delegated only of single accreditation body.
- The principle of profitability. Given that in one state can be only one accreditation body and thus is in a monopolistic position in order to preserve the independence and impartiality, the accreditation body must remain non-profit organization.
- The principle of financial independence. Accreditation Body must be able to assume their responsibilities for the service it provides in its entirety without limitation.

- ❖ The principle of orientation towards the user. Accreditation body, regardless of ownership, should conduct their activities in such a manner as to ensure quality services that meet market needs conformity assessment. It must also be provided with the influence of all stakeholders at all stages of accreditation.
- ❖ Principle is not concerned assessment of compliance. Accreditation body can not deal with the assessment of compliance in the area in which accredits other.
- ❖ The principle of non-competitiveness. Should avoid applying any other national accreditation bodies, as they infringed upon the independence and credibility of the member countries. But do not exclude the possibility that a foreign accreditation body provides its services in the areas for which national accreditation body has not yet qualified, but in agreement with the local accreditation body.
- ❖ The principle of supremacy. The national accreditation body shall conduct its activities in such a way that accreditation remains the highest level of control of conformity assessment activities. Means that there should be no possibility that someone at the administrative or other proceedings reversing a decision it has reached the final as the national accreditation body.

The national accreditation body is competent to determine the competence of organizations that carry out conformity assessment and accreditation competent body established by the Commission on the mechanism of EA peer evaluation (peer assessment / evaluation). Accreditation bodies which have successfully passed the assessment become members of the EA and are eligible to sign the Multilateral Agreement on Mutual Recognition. In this way, certificates, certificates of inspection and test reports are recognized in the EA member states. This eliminates additional conformity assessment and obtain conditions for the unimpeded flow of products / services. In addition to multilateral agreements within the EA members are bilateral agreements with accreditation bodies, which are not members of the EA.

2.1.4. Certification

Certification is the process by which accredited organizations issuing documents (reports, certificates) of Conformity confirms that a particular process, product / service, quality system and environmental protection system are in compliance with the requirements of relevant standards, technical and other regulations. In this way it allows the consumer, the customer / user reach compliance, safe, high quality and reliable products, processes and services. The link between accreditation and certification is shown in Scheme-1.

Accreditation laboratories perform testing of food, environmental parameters, electrical devices of pressure equipment, chemical and textile products, human clinical samples, metals and metal products, animal feed, pesticides, seeds, construction materials and others.

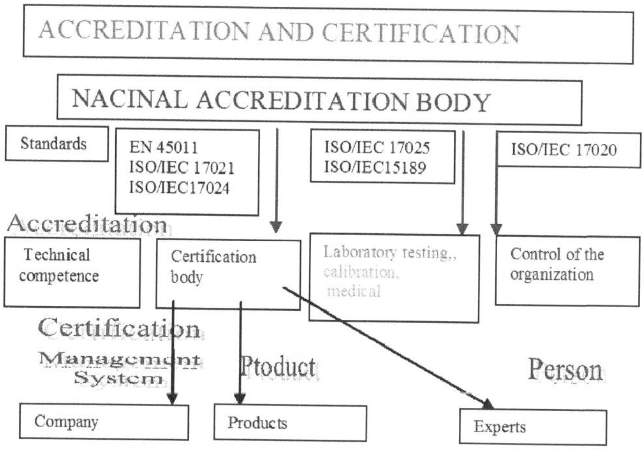

Figure 1 Schematic of accreditation and certification

Laboratories calibration performing calibration of measuring and test equipment.

Control bodies performing inspection of elevators and cranes, pressure equipment, electrical equipment and installation, quality furniturc, quality of food and water, textile products, controlling the field of motor vehicles and others.

The organizations accredited by accreditation bodies get a sign-a symbol of accreditation that can be used in their reports, certificates and other documents in accordance with the Regulations on the use of accreditation symbol.

2.1.5. Control of market

Placing on the market is the initial activity that makes the product available for use or distribution. Products sold to the European Union should be harmonized with the relevant Directive, ie Directive. Existing EU rules guarantee free movement of products and high level of protection for consumers and users with the basic principle that only safe products are placed on the market. Articles 28 and 29 of the EU Treaty on free movement of goods prohibits quantitative restrictions on imports, exports or goods in transit and all measures having effect equivalent to that of the Member States. These measures are not listed in the legislation, but are defined through the PRX is an important principle of mutual recognition:

"A product that is lawfully marketed in one Member State should be allowed the placing on the market of other Member States. Member States - the destination may be refused marketing is in its given form, unless it can show that it is completely necessary for protection, for example, public safety, health or the environment. In this case, the Member States - the destination must also show that its measure is such that at least restrain trade. " National legislation must be in accordance with this principle

2.2. Quality control world. A system of accreditation

Ensure quality control system in the world is shown in Figure-2. At the top is the WTO World Trade Organization (WTO), the International Accreditation Forum - IAF (International Accreditation Forum), and then follow the organization of regional and national level. There are four regional organizations for accreditation to the:

- EA-the European Cooperation for Accreditation
- RAS, Accreditation Asia, Australia and Canada
- IAAC, American accreditation
- SDCA, South African accreditation

At the national level, the national accreditation bodies that meet the general requirements of International Standard ISO / IEC Guide 61 earlier, i.e., ISO 17011, now and instructions EN, EA and IAF.

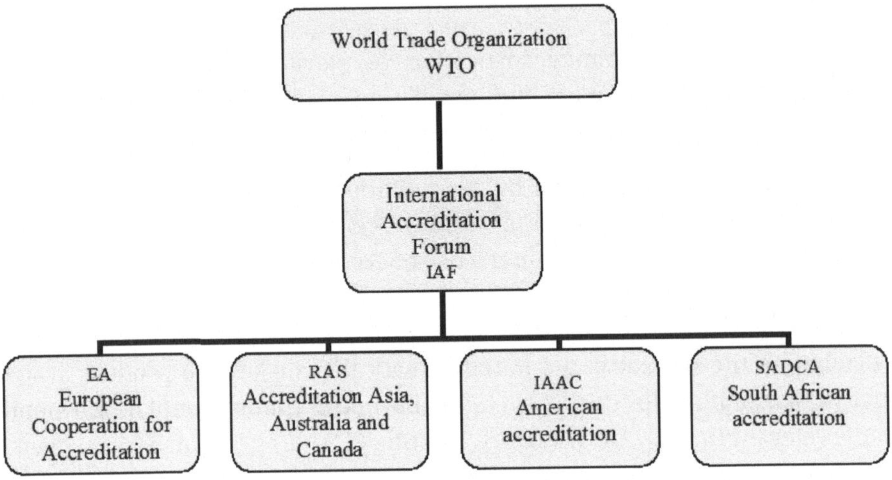

Figure 2. World Quality Control

Further quality control at the national level is done by certified bodies, inspection bodies, testing and metrology laboratories, the Institute for Standardization and Metrology, inspection systems and intellectual property protection.

3. Education and training for quality

A key resource is knowledge of modern business, and thus the improvement the quality of operations is based on the effective application of knowledge. Rapid technological changes are demanding higher and higher level of general knowledge, so that the level of development of a national economy is increasingly measured and brought into causal connection with the capability of creating and using knowledge. New type of professional skills necessary for the broader understanding of complex tasks, or successful completion of certain complex projects. For these reasons, the developed countries, to increase the participation of highly educated people in the employment structure. Human knowledge is treated as development potential, and is one of the basic elements of competitive advantage and national industry organizations, including the quality of products and services, as well as the quality of education and training for quality. In this sense, education and training for quality is a basic prerequisite for a successful build, implementation and promotion of the concept of quality management in every organization. Educated and professional people are a guarantee that the product or service will be good. Therefore, employees should not be viewed as a cost but as a value in which to invest.

One of the major problems facing the majority of local business organizations is the lack of application of the concept of quality management. Complete absence of change in the concept of quality management and its inadequate application resulted in the fact that there is a lack of competitive domestic organizations. At the conclusion of operations, especially in the case of businessmen from abroad, one of the first questions is all the more reports, "Are you accredited or certified by national accreditation bodies and accredited certification body." In other words, if you have established the concept of quality management.

Training and education are essential for the quality of any organization. All this begins from these. Training is needed for all employees, especially management-managers. There must be a permanent improvement of the knowledge life-long learing. For these reasons, every organization should adopt a program of education and training of management and employees (long and short). They are very distinctive and important role of training. The importance of education and training for quality is essential for the QMS, a primary and decisive role in improving the knowledge and skills of employees and management. The role of training is to enable employees to understand better the demands of their job.

4. Competence of education and training for quality

The need for education and training for quality is also expressed in the many standard requirements. Competence is the most important factor in the performance of any business, especially when it comes to education and training for quality. The basis for competence is knowledge. Knowledge is acquired through formal and informal education.

In a series standards ISO 9001:2000, ie, ISO 9001:2008, when the most significant innovation in education and training is their competence.

Competence can be mathematically presented through meetings or through cross-set of the three most important element in the education and training. These are: knowledge, the desire to work and work-place work. Graphic competence (K) is shown in Diagram-1.

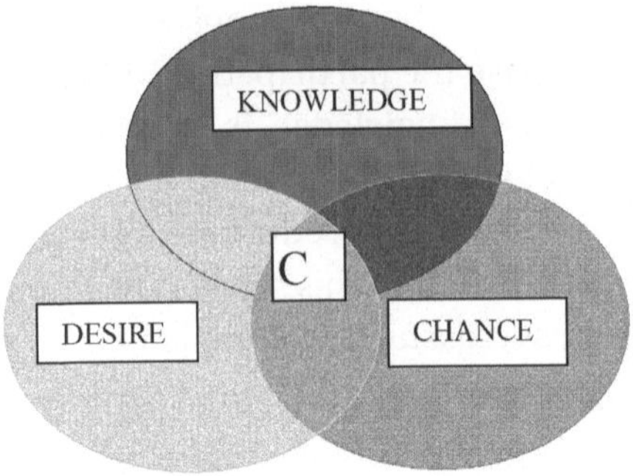

Diagram 1. Competence

DESIRE ∩ KNOWLEDGE ∩ CHANCE = COMPETENCE

There is International standard ISO 10015:1999 Quality Management – Guidelines for training. In this standard are given instructions and training cycle, as shown in Figure-3 However, a comprehensive evaluation of training can not be implemented until trained are not seen and tested it the workplace. Therefore, evaluation of training is done in two parts:

- Immediately after training,
- After observation and testing in the workplace.

Identification of training needs can be presented as overcoming what is missing, or the gap between what is happening and what we want to happen in the organization to quality products and meet customer requirements.

Knowledge (K) is one of the basic elements of social development and the institutions (I), technologies (T) and value system (VS) are the driving force-the power of society (PS). It can be displayed in the Diagram-2, by the sets.

$$I \subset K \wedge VS \subset K \wedge T \subset K \Rightarrow PS = K \cap I \cap VS \cap T$$

When it comes to education and training for the role of quality consultants and consulting firms are extremely important. The world is growing consulting industry, so that the sales potential of knowledge is increasing. The world arranged this issue in a way that they formed associations of consultancies. In Europe there is an association of national consulting house, according to which the members must meet, at a minimum, the common criteria

However, here arises the most important issue when it comes to training and education quality, and that is competence. Competence relates both to the trainer (trainer) for quality, and on all types of training. Obviously, it is necessary the need for management education and training for quality. For teachers is crucial to have a combination of theory and practice. For example, the assessor of the quality system that should have experience in introducing quality management system and he has to be theoretically prepared. In addition he has to possess personal qualities, in an order to understand and interesting way to transmit knowledge. Also it is very important, who determines the competence.

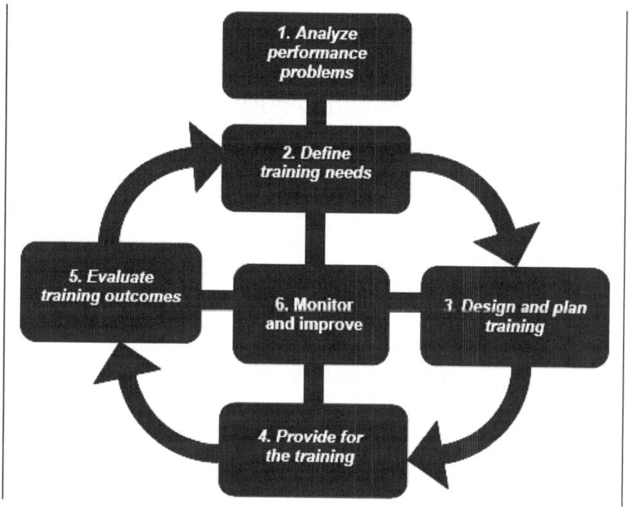

Figure 4. Training cycle

Competence is synonymous with modern business [7].

This issue is not regulated in many countries it is similar in our country, so it is possible that the quality of training providers not competent, and there is no possibility to verify it. To solve this problem there is thought for doing so as stipulated in the documents IRCA-International Register of Certificated Auditors. First the organizations that conduct training are accredted. Then, the contractor must have training for accreditation: the selected program, selected and expertly documented, the documentation and accredited personnel. Therefore you should create a national registration system of independent monitoring and training, or a license from the relevant organizations in implementing and monitoring the quality of training. What organizations could apply to international standards and to maintain a register of accredited training and education for quality?

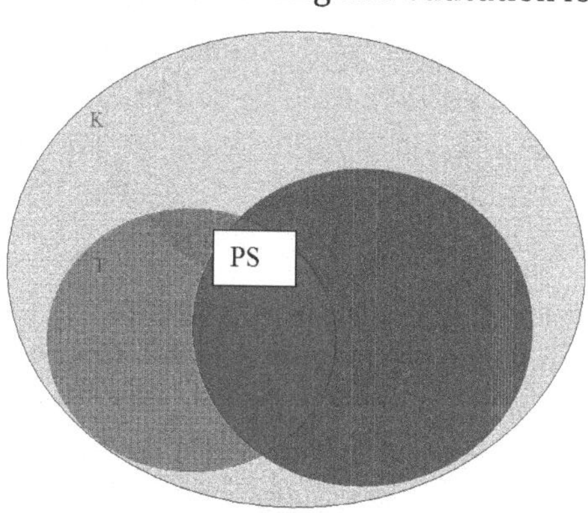

Diagram 2. Power of society

4.1. Research

Having in mind the problems exposed in this paper about the competence of education and training for quality in order to find possible solutions, the results of research. The research was carried out with the competent organizations and people, or accredited organizations and people who deal with quality issues. The study was conducted on two occasions with an interval of eight years.

When contacting the respondents it is noted that they all gladly accepted and wanted to explore that contribute to the improvement of quality and improving the accreditation system. However, it must be noted that there are answers to all questions in the survey form, except several of them on one issue, which is related to the cost of education and training for quality accreditation. The reason not answering this question, according to their explanation is because in that period there was a change of ownership, and in some leadership. Sample of the research was picked, because it included almost all the accredited organizations which employed people with rich experience in the areas of quality and accreditation, which they have gained in our country and / or abroad. Their composition is impressive, not only for its expertise, but also because the organizations where they work. For example, a dean or vice dean for science, factory manager, and the highest percentage was of accredited organizations 47.82% then 30.43% of quality managers and 13.04% of respondents.

The analysis of survey data using the scientific method came to very interesting results on training people for jobs in the field of quality.. They can be grouped into three parts. The largest portion, 70% are those whom formal education is not enabled for activities related to quality (NO). The reason they stated as that programs did not search this issue. The second group (DO) of 20%, are those with formal education partially equiped, and only in the part of the quality control of products within individual subjects. The third group (OK) of 10% are those who are trained and have studied this issue only at postgraduate studies. Which means, according to this research, formal education did not educate or it did not quality enough the students for work in the field of quality in addition to postgraduate studies. Basically they were trained and they acquired skills in non-formal education. Beginning self-training over, attending and participating in conferences and seminars, scientific meetings, then, attendance and completion of local and foreign courses and schools in this area they acquired skills. It is clear that in all that invested much effort, time and money and it is a high capital value.

A repeated survey after eight years shows that the situation is more favorable in terms of formal education. This applies primarily to higher education. Training in the field of quality can be both at the postgraduate studies and at the undergraduate level, where there are scientific fields and disciplines such as management development, quality management and the similar. A large contribution to this improvement has contributed to the Bologna Declaration. One of three basic elements of the Bologna

Declaration is just quality. Thus in many countries on almost all the colleges there is the subject of quality management. In secondary education was also made progress. In some educational profiles of there is the system of quality as optional or this field is studied in the frame of some other subject.

The results surveyed indicate that there have been improvements in training activities related to quality through formal education, but that does not mean that there is still much to improve. First of all, we need to improve the quality of research in the field through all forms of formal education, including preschool education. As for the competence of education and training for quality that is taught in formal education I was formally established as there are national Commission for the Accreditation Program, and the staff is not defined. Given that more and more intertwined formal and informal education and that the limit is decreasing, the common problems are such as competence education and training for quality. This is particularly true for countries in transition that are in the form of donations flooded with various trainings and with education and training for quality of questionable competence. To solve this problem, it is necessary to have management training and education for quality

4.2. Education management and training for quality

Education management and training is required and there must be an integral part of quality management systems namely the management.

To address the aforementioned problems, you should first improve the quality of management at the state level. It is essential that the state establishes a quality of infrastructure in accordance with international, regional and other standards, regulations and directives. It is essential for the existence of institutions: metrology, standardization, accreditation, conformity assessment and market surveillance. It is common to all these institutions of quality the they are nationally recognized organizations such as that in strictly defined rules and standards become members of regional and international organizations.

Continuous improvement of the accreditation system is particularly important. The results of other studies and the results of this research have helped to develop a model of continuous improvement of the national accreditation system. Both studies were done in the form of expert interviews with the application of scientific methods, techniques and procedures.

Education management and training for quality should be in accordance with the new philosophy of quality management, which is based on principles. The principle of orientation towards the customer-user may be considered or treated as a business philosophy, but philosophy which is transformed into action.

In order to improve education management and training for quality, competence and thus to improve the education and training for quality it is possible to define the model. The model should be in accordance with the new philosophy of quality management, based on its fundamental principles, particularly on the process and the principle of the system with constant improvement. Principle directions to customers / users can be treated as a business philosophy, but philosophy is transformed into action. The model consists of two groups of factors that need to be optimized to achieve the developmental effects. The first three factors are factors that are the result of policy at the state level. By their nature they are opting for the fact that they have a significant impact on the quality of education. The second group includes factors that directly influence the direction of development of quality education and training for quality at the level of the national economy. They come from all interested parties and by their character are focussed

Decisive factors:

- ❖ State policy in the field of formal education.
- ❖ The role of state institutions in the field of stimulating education for quality.
- ❖ Development and organization of institutions of informal education.
- ❖ Transfer of knowledge from science to economy.
- ❖ The level of economic development.

The addressing factors:

- ❖ The degree of development of quality awareness to the community level.
- ❖ The degree of development of awareness of quality at the level of business organizations.
- ❖ The necessity of internationalization of business.
- ❖ The role of consumers in developing a climate for quality improvement.
- ❖ The role of media in the popularization and promotion of quality.

Quality education should be considered in all stages of formal education, from primary, secondary and higher schools. In elementary school, it could be done within a subject, for example, good manners, civic education, or similar subjects. In high school should be introduced a new subject-management of quality in all educational backgrounds, and in some would be the easiest way to introduce it instead the work organization. In the process of faculty quality is significantly affected and it is well on the way that all the faculties have the subjects of quality, and there are also studies in the field of quality.

Competence is the most important question in formal as well as in informal education. Competence relates both to the trainer, and all kinds of programs. For teachers is crucial to have a combination of theory and practice, for example that the teacher is the assessor of the quality system that the teacher has experience in quality system and to have theory. In addition the teacher should have to personal qualities, in order to understand and in an interesting way to transmit knowledge. Besides this there should be a national organization to establish competence - accreditation of education and training in quality as the register of accredited training and education, as well as its contractors. As possible organizations could be the excisting national accreditation body or the National Association for quality or National Association of consulting companies.

The world is growing consulting industry. Potential sales skills are growing. They arranged the matter so that they formed a national association of consulting companies. In Europe there is a Federation of European associations of consulting organizations - FEACO, which consists of national associations of consulting firms. According to this association the national associations may require different conditions for membership, but must keep the following common criteria as a minimum:

- Independence.
- Expertise.
- Ethics.
- Experience.
- Qualifications.
- Verification.

5. Conclusion

- Quality management is necessary for continuous operation
- Management of continuous education and training for quality is needed, as a process that is integrated into quality management, i.e., management operations, in order to constantly improve the desires of the customer-user
- Education and training for quality should be competent and that is acquired during formal and informal education,
- The state should create conditions for normal operation of the economy, primarily by improving the quality of their management level.
- Investment in education and training investment in capital goods.
- Better cooperation with business organizations, scientific institutions,
- Effective application of knowledge,

This is a never-ending process that requires constant review and continuous improvement.

References

1. D.Djordjevic, Quality as a factor in the competitiveness of domestic firms, Belgrade, 2003
2. AristotleMetaphizikaCulture.Belgrade, 1960
3. Z. Avramovic, Is a knowledge society a new type of society, the IER, 2007
4. D. Stanivukovic, Quality and standards- The development and trends, Research Technology Center, Novi Sad, 1996.
5. ISO 8402:19861994Vocabulary
6. ISO2000Fundamentals and vocabulary
7. MaistorovicV.ModelT. Q. M.BusinessPolicy.Belgrade. 2000

9

The Integration of TQM and Six-Sigma

1. Introduction

Since the 1980s, several important quality management systems, or programs, such as ISO 9000, TQM, Six-Sigma program, Reengineering, and Toyota production system (or lean production), have been launched. Most of these quality imperatives have been widely adopted by industries around the world. All the firms expect good results from the implementation of these quality programs. But the prerequisite is that the employees are familiarized with the quality systems and know how to implement the related practices as a firm plan to adopt these quality systems. In order to help the industries, we will describe the meanings of 'quality,' the evolution of quality management, and the content and practices of some important quality imperatives.

Usually, some firms will adopt several quality programs simultaneously. If a firm implements several quality programs separately, the employees, especially the managers and staff, will encounter some trouble. Among the quality management imperatives, the TQM and Six-Sigma program are widely adopted by the industries around the world; many organizations even implement both of these quality management systems. In order to implement these two quality management programs effectively, it is necessary to integrate TQM with the Six-Sigma program, or even with other quality practices. After the Malcolm Baldrige National Quality Award (MBNQA) and European Quality Award (EQA) were launched, many organizations consider MBNQA and EQA as the 'business excellence model' and use these systems and the related evaluation items to perform self-assessment. Based on the integrated model of TQM and Six-Sigma, and referring to the constructs of MBNQA and EQA, a holistic business excellence model can be developed.

2. What is 'quality'

There are many scholars and practitioners who have given definitions of 'quality.' In this section, we will mention several representative examples. Edward defined 'quality' as the capacity of a product or service to satisfy the consumer requirements in [1]. Usually the consumer's wants are complex and multi-faceted, thus it may not always be satisfied in a particular way. Juran defined quality as being 'fitness for purpose of use,..., it is judged by the users, not the manufacturers, or the merchants' in [2]. Juran also asserted that each product/service has multiple quality characteristics, which can be divided into two kinds: the features desired by customers, and the freedom from deficiencies. Thus Crosby defined quality as 'conformance to customers' requirements' from the viewpoint of the customers, he also emphasized the ideal of 'zero defects' or 'meeting all the specifications of product/service all the time' in [3].

The definitions mentioned above are not mutually exclusive, they are almost the same. There are several researchers who have given similar definitions, for examples, see [4, 5, 6]. Japanese Industrial Standard (JIS Z8101) and International Standard Organization (ISO 8402-1986) give the same definition of 'quality' as the totality of features and characteristics of a product/service which determines the ability to satisfy the customers' needs and expectations in [7]. Thus, the providers of products/services need to determine the specifications upon these features and characteristics which can meet the customers' requirements and expectations.

There are some critical concepts of quality to be emphasized. Japanese quality philosophy is 'zero defects - doing it right the first time.' It means that quality is the result of doing the right thing and doing the thing right the first time, 'doing the right thing' is to meet the customers' needs and expectations, and doing the thing right' is to follow the standards of the totality of quality. The definition of quality by Crosby has the same concept. Deming's quality concept is customer-focused; he emphasized that quality is only assessed by customers; the quality is surpassing customers' needs and expectations throughout the lifetime of product/service in [5, 8].

We can summary the meanings of quality as follows.

1. Quality is conforming to the standards and specifications of a product/service.
2. Quality is zero defects or meeting the specifications 100%.
3. Quality means that product/service possesses the fitness for purpose of use based on its functions.
4. Quality is the ability of a product/service to meet the customer's needs and expectations.
5. Quality is assessed by customer only borne upon the critical features and characteristics of a product/service considered by customer.

6. Quality is determined by the deviation of the measures of quality characteristics of a product.
7. Quality is customer satisfaction.

3. The evolution of quality management

Quality, price, product function, delivery, and reliability are the competitive aspects for any industries, of which quality has become the most important one in [9] since customers only buy the goods with accepted quality. In order to assure the delivery of good quality products to customers, industries have adopted many actions to control the quality of the products during the manufacturing process. These actions are somewhat different due to the change of the quality concept. In the beginning, the major quality concepts were product-focused and manufacturing-focused and then changed to user-focused, customer-focused, and value-focused. The evolution of quality management is coincidental with the change of quality-focused, which consists of several stages.

3.1. Inspection quality control (IQC), since 1910~

Ford Company created the assembly line in 1913 due to the influence of the scientific management of Frederick W. Taylor. The implementation of the assembly line led Ford to reduce manufacturing costs significantly. Therefore the assembly line and the resulting volume production became very popular among the manufacturing industries. But it caused the issue of quality control. In this period, inspection activities were formally recognized as the popular control of product quality in [10]. In most manufactures, the foremen are responsible for the inspection works. Thus, it is also called foreman quality control.

Engineers and management level design the standards of the quality upon the critical attributes of the product, and set up the process standards and the related task specifications. Workers are requested to perform the tasks according to the standards and specifications. The inspectors will check the dimensions and characteristics of products, detect the errors and failures, and take the necessary steps to improve the quality.

3.2. Statistical process control (SPC), since 1930~

Inspection quality control is costly since it fails to effectively control the process quality. Walter Shewhart thus created the quality control tool 'control chart' as he had worked in Bell Labs as a quality control inspector in [11]. He suggested using a sampling inspection method instead of 100 percent inspection to reduce the amount of inspection, due to his study of chronic variation of production. The control chart is used to monitor the quality performance of the process by using the sampling methods upon the critical aspects of the process and the attributes of the product in [10].

Since many statistics tools are used in the statistical process control, we also call the quality control method 'statistical quality control (SQC)' Using sampling inspection will cause fewer defective products to be shipped and result in some extra costs, but Shewhart argued that if the missed number of defects is small, then the savings in inspection costs make it worthwhile in [11].

3.3. Total quality control (TQC), since 1950~

Starting in the early 1950s, J. M. Juran propounded the concept of quality costs. He addresses the economics of quality in the book 'Quality Control Handbook' in 1951 in [9]. It is often that the losses due to defects were more than the costs of quality control. Thus the model of 'costs of quality,' which is subdivided into prevention, appraisal, internal failure and external failure costs, is proposed. The way of SPC can't effectively reduce the quality costs, especially the costs caused by internal failures and external failures.

Armand Feigenbaum joined General Electric since 1944 in [9]. He used the statistical techniques to improve the product quality while he was working in the jet engine factory. But Feigenbaum also used the concept of cost-of-quality and adopted a user-based approach to quality. He thought that this approach requires the management and employees to have an understanding of what quality means and its relation to the company's benefits. He emphasized that quality assurance cannot be achieved by the control just on production process. Thus he propounded the concept of Total Quality Control in 1956 in [12]. This means that the quality is determined at all stages of the whole product lifetime, and all the functions are included in the quality control. The quality activities start with the product design, incoming quality approval, and continue through production control, product reliability, inventory, delivery, and customer service. Actually, Feigenbaum's quality concept and ideas are similar to those described by Deming, Juran, and Crosby in [12].

3.4. Company-wide quality control (CWQC), since 1970~

After World War II, the Union of Japanese Scientists and Engineers (JUSE) was formed in 1946. Its members were constituted of scholars, engineers, and government officials in [13]. They devoted themselves to improving Japanese productivity and product quality in order to enter the foreign markets, especially the American market. In 1950, JUSE invited Deming to Japan to introduce the quality concepts and statistical quality methods to the top managers of Japanese industries in [11]. Juran also visited Japan in 1954 and instilled the concepts of quality control, costs of quality, and the strategic role of management in the quality activities for the Japanese industries in [11]. The concept and approach of TQC were introduced to Japan during 1960. JUSE synthesized the concepts, principles, and approaches of statistical process control and total quality control.

During this period, Japanese industries realized the concepts of TQC. All the departments and employees, from the operators, first-line supervisors, engineers, managers, and top managements, participated in the quality programs and activities. Thus, we called this Japanese TQC company-wide quality control (CWQC). Japanese industries emphasized the education and training of quality for all employees and the cultivation of quality culture intensively. Kaoru Ishikawa, a pioneer in quality control in Japan, advocated the use of statistical methods. But his largest contribution was to promote the realization of total quality and continuous improvement. He contrived the Quality Control Circle (QCC) activity, and used the seven QC tools and improvement tools to apply the QCC improvement activities in [9].

3.5. Total Quality Management (TQM), since 1985~

The realization of CWQC led Japanese industries to possess core competitiveness and quickly move into western markets that were once dominated by western companies by providing the customers with high quality products at lower prices in [14]. The western firms, especially the American companies, encountered serious global competition from Japanese and Asiatic competitors. The western companies saw their shares eroded by foreign competitors. This situation caused American and western industries to benchmark Japanese CWQC performance and learned the management of quality control from Japan. As a result, total quality management (TQM) was developed and widely adopted by the industries around the world. The industries considered TQM as a powerful tool that can be used to regain the competitive advantage.

The development of TQM was also influenced by the western quality gurus: Deming, Juran, Feigenbaum, and Crosby in [15]. TQM is thus an integrated model of management philosophy, quality concept, and set of practices. However, to implement the TQM successfully it is necessary to integrate the so-called 'hard side' of the system (that is, the technical aspects of quality control) with the 'soft side' of the program (that is, the aspects associated with 'quality concept, culture, and people factors') in [16]. Statistical methods, quality control tools, process standardization, and improvement are the elements of 'hard side,' and quality concept, employees' participation, education and training, and quality culture are included in the 'soft side.'

From the mid-1990s onward, several important quality programs were being launched. Besides the development of TQM, the ISO system and Six-Sigma program, which was initiated by Motorola, were started in 1987. Until now, ISO system has had three revisions in 1994, 2000, and 2008 respectively. The Six-Sigma program was being widely imitated by GE in 1995 in [17], while most were copying from Motorola. The successful implementation of Six-Sigma by Motorola and GE caused this improvement methodology to become popularly adopted by the industries around the world.

3.6. Business Excellence Model, since 2000~

The rapid development and application of technology and internet have caused significant changes in market environments in [18, 19] and, consequently, in business management in [15]. In particular, the effects of the borderless global economy are clearly evident in virtually every aspect of business activity in [20]. The increased competitive pressure from both domestic and forei gn competitors has forced businesses to pursue speed, innovation, quality, and value in [21, 15]. In the past two decades, the industries adopted several strategic actions: Total Quality Management (TQM), ISO system, Reengineering, Six-Sigma program, Toyota production system (TPS), etc. in [22, 15]. But in today's world of serious competition, implementing these actions may not be enough to possess the competitiveness.

The enterprises need to develop their core competencies and core capabilities in order to excel at the contrivance of core competitiveness and then develop the innovative business model in [23-27]. The integrated system of these critical ingredients, in order to pursue the long-term high profits and development, can be called a business excellence model. But there is no coincidence of the formal 'business excellence model.' Several scholars and practitioners consider the model of Malcolm Baldrige National Quality Award (MBNQA) or the model of European Quality Award (EQA) as the 'business excellence model' in [9]. Kanji developed a business excellence model that was suitable for organizations that incorporate the critical success factors of TQM in [28]. Based on this business-excellence model, Kanji then developed a 'business scorecard' in [28]. Kanji & $S_{\hat{A}}$ later developed a 'Kanji business excellence measurement system' by integrating the business excellence model and Kanji's business scorecard in [29]. Yang also developed an integrated model of a business excellence system in [30].

4. The development and implementation of TQM

TQM began in the mid-1980s and was based on benchmarking and learning from Japanese CWQC. In the beginning, there was a lack of consensus on the content and practices of TQM. But several gurus, like Deming, Juran and Ishikawa have contributed much to the development of TQM, especially the Deming 14 points and Juran quality trilogy in [31, 9, 32]. Additionally, the characteristics of CWQC also affected the content of TQM.

4.1. The fundamental concepts of TQM

Now we state the concepts, practices, and characteristics as follows.

4.1.1. Deming 14 points:

1. Create constant purpose toward quality improvement of products and service.
2. Adopt the new concept of 'zero defect' that we no longer accept the commonly accepted levels of delays, mistakes, and defective products.

3. Stop the dependence on mass inspection of quality control to achieve the quality assurance; instead, set up the built-in quality system in the production processes.
4. Cease the practice of material purchases based on the decision of the price alone.
5. Use statistical methods to find the root causes of the problems and ultimately eliminate these problems.
6. Institute modern methods and systems of employees' on-job training.
7. Execute new methods of leadership for the supervision of workers.
8. Drive out fear, so that every employee can work effectively.
9. Break down barriers between departments; instead, team-work can be realized.
10. Eliminate slogans and the exhortations by numerical goals for the workforce; instead, encourage employees to challenge high levels of quality and productivity.
11. Eliminate only work quotas without accounting quality and remove the obstacles that prevent employees from achieving their challenge.
12. Remove barriers that rob people of their pride of workmanship.
13. Develop and execute a complete program of education and training for all employees.
14. Perform all above actions and push for continuous improvement.

4.1.2. Juran quality trilogy

Juran divided quality management system into three stages, which are

1. Quality planning:

The firms first identify the focused customers and their needs and set up the goals to satisfy customers and achieve excellent business results based on the development of new products and strategic processes. This planning stage also attempts to eliminate problems which may become chronic as the process was designed that way.

2. Quality control:

The firms need to establish a control system to monitor the quality, evaluate the process performance, and compare the operating results with the goals. It is also critical to discover the problems, especially the chronic problems.

3. Quality improvement:

In this stage the firms will identify the improvement projects and teams and analyze the root causes and eliminate them. After the problems are solved, the firms will standardize the new process and establish the mechanisms to control the new process in order to assure the quality.

4.1.3. Characteristics of CWQC:

1. Customer-focused and quality-first.
2. Full participation and teamwork.
3. Education and training of quality for all employees.
4. Cultivation of quality culture.
5. 'Continuous improvement' is the key quality activity.
6. Concept and realization of 'zero defect.'
7. Realization of 'do the right thing first time.'
8. Everyone is responsible for the quality.
9. Emphasizing on the prevention activities and quality assurance.

4.2. The content and framework of TQM

During this period, the ISO 9000 quality system was launched and Motorola implemented Six-Sigma improvement projects in 1987. The USA also started the Malcolm Baldrige National Quality Award (MBNQA) in 1987, which was based on the referring to the Japanese Deming Award. After MBNQA launched, many countries also developed their national quality awards based on the MBNQA system. The development of TQM is displayed in Figure 1.

Additionally, many researchers and experts on quality management have been eager to study the essentials of TQM. The development and implementation of TQM today has become a very consistent consensus on the content in [33, 34, 15] as follows:

1. Customer focus – To understand the requirement of customers proactively, take proper actions to fulfill the customers' needs, and the aim to satisfy customers.
2. Continuous improvement – To discover problems, analyze the critical root causes, and eliminate those barriers completely.
3. Employees' participation – Every employee is accountable with one's responsibility for quality, and also everyone needs to be involved and commit oneself to every quality activity.
4. Teamwork–It is necessary to overcome sectionalism and to realize the teamwork and cooperation for improving quality and embark on quality activities.

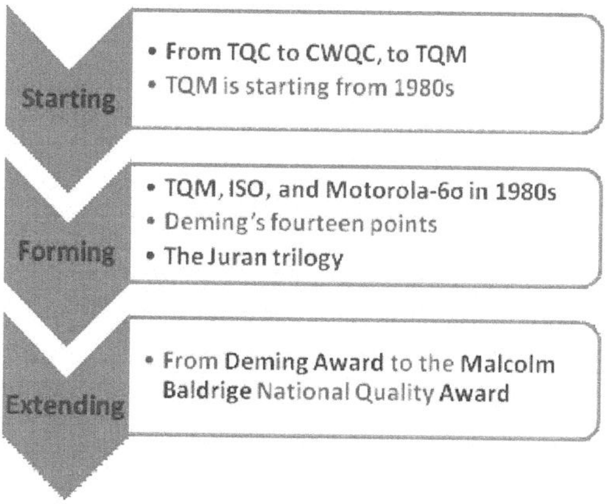

Figure 1. The development of TQM

1. Process focus – Standardizing the processes and taking proper quality control in the key steps of the operation procedures to prevent any defects occurring in processes.
2. Systemization – For bettering the prevention and control of quality, all the quality activities should be conducted and implemented systematically.
3. Empowerment – It is critical that every employee can be autonomous to do the right thing the first time in order to get good quality performance. Therefore, it is necessary to empower the employees.
4. Leadership – During the implementation process of TQM, the top management should play a key role. The top management should be a coach, to teach and influence the subordinates.
5. Management by facts – For the sake of quick decision and solving problems, it is necessary to use numerical methods and statistical tools effectively. It is also essential to develop the quality information system and powerfully apply this system.
6. Training and education – Japanese industries emphasize the training and education for the employees, which is focused on the quality concepts and the improvement tool, and the implementation of quality practices. Thus, employee training and education is the fundamental activity for the adoption of TQM.
7. Corporate quality culture – In order to successfully perform the above imperatives, the top management needs to cultivate the organization quality culture, and all the employees can maintain it forever.

Based on these imperatives, we can develop the framework of the implementation of TQM, see Figure 2.

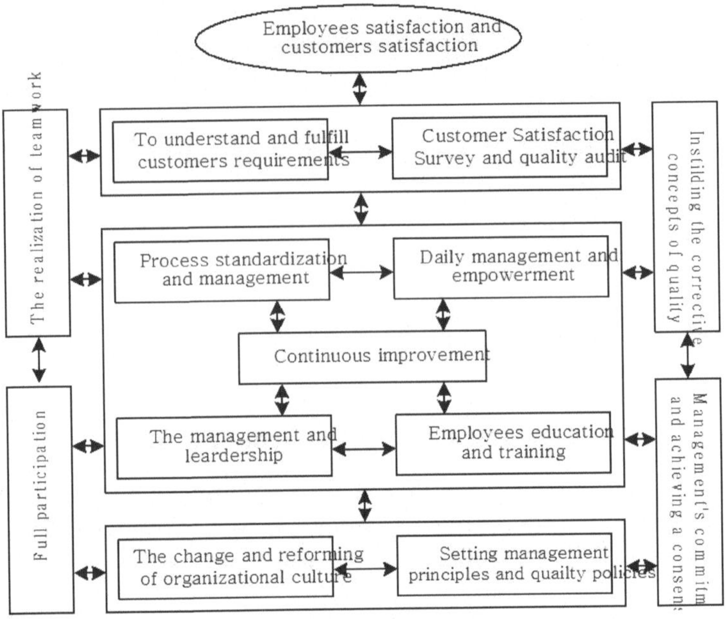

Figure 2. The framework of the implementation of TQM

5. The development and implementation system of Six-Sigma program

The Six Sigma program was first espoused by Motorola in the mid-1980s. The Six Sigma architects at Motorola produced results far more rapidly and effectively. The successful implementation of the Six Sigma program in Motorola led to several famous companies following Motorola in successfully implementing the Six Sigma program in [17]. In this section, we first introduce the development of Six-Sigma program.

5.1. The development of Six-Sigma program

By the end of the 1970s, Japanese industries possessed strong competitiveness; their competitiveness was based on the ability to develop the core competencies with lower costs, higher quality, and greater speed than their competitors, which could be utilized to contrive the core products. The core competence is the effective integration of technologies, specialized knowledge, skills, techniques, and experiences, and the core capability is the unique management ability of core competencies to develop core products and new business and then enter the new markets. Eventually, the firms will heighten their competitive advantage and result in business benefits and long-term development, which will exceed their competitors' in [35, 25].

In this period, Motorola encountered intense competition from their global competitors, especially from the Japanese competitors. The threats caused Motorola to execute the benchmarking from the Japanese electronics industry and found out that many Japanese electric products were with 6σ quality level, but Motorola's products were with 4σ quality level only. The weakness in quality led Motorola to initiate the Six-Sigma improvement programs. The aim was to achieve 6σ quality level in a 5-year period. The Six Sigma architects at Motorola focused on making improvements in all operations within a process—thus producing results far more rapidly and effectively.

The successful implementation of the Six Sigma program in Motorola resulted in huge benefits. Motorola recorded a significant reduction in defects and manufacturing time and also began to reap financial rewards. Within four years, the Six Sigma programs saved the company $2.2 billion in [36]. The crowning achievement for Motorola occurred when it was the winner of the Malcolm Baldrige National Quality Award in 1988 in [37]. IBM, Sony, and Allied Signal followed Motorola in successfully implementing the Six Sigma program. Allied Signal began its Six Sigma activities in the early 1990s and attained savings of US$2 billion during a five-year period in [37]. Such impressive results induced General Electric (GE) to undertake a thorough implementation of the Six-Sigma program (GE-6σ) since 1995 in [17].

GE implemented 6σ programs and reaped huge financial benefits. The 1999 annual report of GE showed that the implementation of GE-6σ produced more than US$2 billion in benefits in that year in [38]. The impressive benefits of implementing a Six Sigma program in Motorola, Allied Signal, and GE led to the Six Sigma methodology being widely adopted by industries and non-profit organizations throughout the world. Within a short time, the Six Sigma program thus became one of the world's most important tools in quality management in the last two decades.

5.2. The implementation system of Six-Sigma program

The huge contribution of the implementation of the Six-Sigma program is due to the realization of better practices and operation systems. In the initiative stage, Motorola and GE designed a complete implementation system. The main features of the system are discussed below under the following headings:

1. Implementation steps;
2. The support from organization;
3. Investment in training.

1. Implementation steps

There have been many improvement models for process improvement or re-engineering. Most of their implementations are based on the steps introduced by W. Edwards Deming, which can be characterized as 'Plan,' 'Do,' 'Study,' and 'Act' (PDSA) in [39]. The Six-Sigma program has a five-phase cycle: 'Define,' 'Measure,' 'Analyze,'

'Improve,' and 'Control' (DMAIC) for process improvement that has become increasingly popular in Six Sigma organizations. There is another cycle characterized as 'Define,' 'Measure,' 'Analyze,' 'Design,' and 'Verify' (DMADV) for process design (and redesign) in [17]. Like other improvement models, the DMAIC (or DMADV) model is grounded in the original Deming PDCA cycle. Table 1 describes the specific tasks in each step, and the tools and techniques used in the DMAIC steps. The tasks and tools used in the DNADV steps are similar to those used in the DMAIC steps.

Table 1. DMAIC steps and tools usage

Step	Specific tasks	Tools and techniques employed
Define	• Analyze voice of customers (VOC) • Identify improvement issues • Organize project team • Set-up improvement goal • Estimate financial benefit	• Customer complaint analysis • Cost of poor quality (COPQ) • Brainstorming • Run charts, control charts • Benchmarking
Measure	• Map process and identify inputs and outputs • Establish measurement system for inputs and outputs • Understand the existing capability of process	• Process map (SIPOC) • Cause and effect matrix • Gauge R&R • Control charts • Process capability analysis • Failure models and effects analysis (FMEA)
Analyze	• Identify sources of variation in process • Identify potential critical inputs • Discover the root causes • Determine tools used in the improvement step	• Cause-and-effect diagram • Pareto diagram • Scatter diagram • Brainstorming • Analysis of variance (ANOVA)
Improve	• Create the strategic actions to eliminate the root causes • Conduct improvement actions • Use experiments • Optimize critical inputs	• Design of experiment (DOE) • Quality function deployment (QFD) • Process capability analysis • Control charts
Control	• Standardize the process • Maintain critical inputs in the optimal area • Verify long-term capability • Evaluate the results of improvement projects	• Standard operation procedure • Process capability analysis • Fool-proofing (Poka Yoke) • Run charts • Failure models and effects analysis (FMEA)

2. The supports from organization

Along with the systematic implementation steps described above, the design of specific roles and their effective operations are important factors of the Six-Sigma program. Top management is ultimately responsible for the success of the projects through the provision of sufficient support, resources, and strong leadership. The implementation of the Six-Sigma program is thus top–down. The chief executive officer (CEO) is usually the driving force who sets up the vision, develops the strategies, and drives the changes. Apart from the critical role of the CEO, other players also have their specific roles: (i) the senior managers are the 'Champions,' who are the sponsors of the projects and responsible for success of Six-Sigma efforts; (ii) the 'Master Black Belts' (MBBs) are the full-time teachers and consultants; (iii) the 'Black Belts' (BBs) have the key operational role in the program as full-time Six Sigma players. They are the leaders of the Six-Sigma improvement projects, and therefore they need to show the best leadership; and (iv) the 'Green Belts' (GBs) are the part-time participants who, led by the BBs, work on Six Sigma projects while holding down their original job functions in the company in [40]. Additionally, other departments need to support the Six-Sigma teams as requested.

3. Investment in training

In Japan, employee education and training is a key ingredient in achieving success through QCC (quality control cycle) improvement. In the implementation of Six-Sigma, education and training is also an important success factor, thus Motorola, Allied Signal, and GE have invested heavily in employee training for the Six-Sigma programs in [17]. For example, GE has designed a complete training plan for the various roles described above—from the CEO, to the 'Champions,' 'MBBs,' 'BBs,' and 'GBs.' In addition, the training program extends to all other employees in the organization. The training courses are comprehensive and cover team leadership skills, the method of project management, measurement and analytical tools, improvement tools, planning and implementation skills, and so on. For example,

1. Champions have one week of champion training related to Six-Sigma development, leadership, and the implementation plan.
2. MBBs take over the responsibility of the training for all the BBs and GBs. They are promoted from BBs based on the successful leaders of at least ten Six-Sigma projects.
3. BBs spend about four to five weeks to receive the intensive, highly quantitative training, roughly corresponding to the five steps of the implementation of Six-Sigma project. Thus, the length of training is approximately 16-20 weeks.
4. GBs receive training for six to ten days. The courses include the statistical tools and the use of statistical software, the detailed modules of five steps, the innovative and improvement tools, and project management skills.

5.3. The features and CSF of the Six-Sigma program

In order to successfully implement the Six-Sigma program, the firms need to possess the related critical success factors (CSFs). The CSFs are dependent on the features of the Six-Sigma program.

1. The features of the Six-Sigma program.

Based on the above descriptions of the implementation of Six-Sigma, and several researches related to Six-Sigma issues in [41, 17, 42], we can summarize the major features of GE-6σ program as follows:

- ❖ GE-6σ projects are integrated with the company's visions and strategies;
- ❖ Most GE-6σ projects are created from the 'voice of customers';
- ❖ All GE-6σ projects are rigorously evaluated for financial results;
- ❖ All employees, from top management to the workers, participate in the progress of Six-Sigma program;
- ❖ GE-6σ is a top-down program, top managers are the sponsors of the projects, and major managers are responsible for success of Six-Sigma efforts;
- ❖ GE invested heavily in the employee education and training for the Six-Sigma program;
- ❖ The five implementation steps (DMAIC, or DMADV) are rigorously followed and result in significant benefits;
- ❖ Everyone who contributes to the success of the program receives significant rewards, especially in terms of staff promotion;
- ❖ Significant financial incentives (representing 40% of all bonuses received by employees) are tied to results of the GE-6σ projects;
- ❖ Many management, analysis, and improvement tools, especially the advanced statistical methods, are used in the implementation of GE-6σ projects;
- ❖ Almost all projects are completed rapidly (usually within 3–4 months); and
- ❖ The bottom-line results are expected and delivered.
- ❖ The critical success factors of the Six-Sigma program

Though the Six-Sigma program has been widely adopted by manufacturing and service industries, as well as non-profit organizations and government institutes in [43, 15], the failure rate of the implementation is very high. There are several obstacles that cause the high failure rate. For example, top management provides insufficient support to the Six-Sigma projects, lack of sufficient training for the employees, the financial incentives tied to the results of the Six-Sigma projects are deficient, etc.. Thus, it is worthy to investigate the critical factors for the successful implementation of Six-Sigma projects.

There are several researchers who have studied the critical success factors (CSFs) for the implementation of Six-Sigma in [40, 37, 43-46]. Yang et al. also investigated the CSFs for the Six-Sigma implementation in Taiwan using an empirical study. In this section, we integrate these studies in [47].

- Top management commitment and involvement.
- Full support from the organization.
- Cultural change—customer-orientation and quality-first.
- Communication with all employees to achieve congruence on the Six-Sigma program.
- Employee education and training in Six-Sigma.
- Linking Six Sigma to the corporate vision and business strategy.
- Linking Six Sigma to customers' needs (focused on the voice of customers).
- Familiarizing and implementing the methods, tools and techniques within Six Sigma.
- Complete evaluation system of project performance.
- Project prioritization and selection, and successful usage of project management.
- Organization infrastructure—the design of Champions, MBBs, BBs, and GBs.
- Employees' promotion and incentive compensation tied to the results of Six Sigma projects.

6. Integrated model of TQM and Six Sigma

In the last two decades, the public interest in TQM has declined. In contrast, the Six Sigma improvement method, especially in its form implemented by General Electric (GE-6σ), has become a popular management tool in the world. As a result, some researchers and practitioners assert that firms should implement Six Sigma in preference to TQM. Why have these kinds of contentions appeared? The literature contains reports of several cases in which the implementation of TQM has failed. Hubiak & O'Donnell, for example, have asserted that approximately two-thirds of the companies in the United States have either failed or stalled in their attempts to implement TQM in [48]. Many of these TQM programs have been cancelled, or are in the process of being cancelled, as a result of the negative impact on profits. The failure implementation of TQM is due to several factors. Besides the difficult achievement of TQM practices, one of them is that TQM has been a rather diffuse concept with many vague descriptions but few more graspable definitions, and the management does not have a complete picture of what TQM really means in [49]. Another one is that organizations around the world do not realize that implementation of TQM means a cultural change in [50]. In fact, academic discussion of TQM and its implementation has suffered a similar decline in recent years.

Is this trend really due to poor corporate business performance as a result of the implementation of TQM, with a consequent decline in the implementation of TQM, as has been asserted? Yang asserted that this is not an accurate reflection of the current status of TQM in [15]. Reports of instances of failed TQM implementation are only part of the explanation for the apparent declining trend in TQM. In reality, TQM has been so prominent for about twenty years that many firms and institutions have incorporated TQM into daily management activities. The result is that a well-established model of TQM has been so much a part of the routine business activities, that the 'decline' in discussion and implementation of TQM is apparent, rather than real.

6.1. The contentions related to the relations between TQM and Six-Sigma

Actually, the conspicuous success of the Six-Sigma program by GE (as GE-6σ) has gained great popularly in recent years in [38, 51]. It has even been suggested that TQM will be replaced by Six Sigma as the main strategy for successful business management. However, such assertions reveal a fundamental misunderstanding of the nature of TQM and its relationship with GE-6σ. For example, Pande et al. have asserted that TQM is less visible in many businesses than it was in the early 1990s, pointing to several major TQM gaffes as reasons for this apparent decline in [17]. According to Pande et al., these problems include a lack of integration, leadership apathy, a fuzzy concept, an unclear quality goal, failure to break down internal barriers, inadequate improvements in performance, and so on. They conclude that Six Sigma can overcome many of the pitfalls encountered in the implementation of TQM and, hence, that Six Sigma's expansion heralds a 'rebirth' of the quality movement in [17]. However, Klefsjö et al. and Lucas have a different perspective. Klefsjö et al. assert that Six Sigma is a methodology within – not alternative to – TQM in [37]. Lucas asserts that Six Sigma is essentially a methodology for disciplined quality improvement in [51]. Because this quality improvement is a prime ingredient of TQM, many firms have found that adding a Six Sigma program to their current business system gives them all, or almost all, of the elements of a TQM program.

It can be concluded that the approach of Lucas is correct, and that the TQM pitfalls noted by Pande et al. are not essential features of TQM in [17]. Rather, they are caused by incorrect practices adopted by firms, especially the lack of proper endeavour shown by management in the implementation of TQM. As a result, several assertions related to the relationship between TQM and GE-6σ have appeared, especially the treatise that TQM will be replaced by GE-6σ. However, there are very few studies in the literature that directly compare TQM with GE-6σ completely, and in the limited studies that do exist, conclusions on the relationship between TQM and GE-6σ have differed significantly.

Harry has claimed that Six Sigma represents a new, holistic, multidimensional systems approach to quality that replaces the "form, fit and function specification" of the past in [52]. However, it is not readily apparent from Harry which aspects of this multidimensional systems approach are presumed to be absent from TQM in [52]. Breyfogle et al. have stated that Six Sigma is more than a simple repacking of the best from other TQM programs in [41]. In view of a lack of consensus on the relationship between TQM and GE-6σ, Yang (2004) compared TQM and GE-6σ by using complete perspectives in [15]. The author reviewed several studies in [31, 53, 54], and selected the appropriate criteria used in these studies and then integrated them into 12 dimensions. They are: (i) development; (ii) principles; (iii) features; (iv) operation; (v) focus; (vi) practices; (vii) techniques; (viii) leadership; (ix) rewards; (x) training; (xi) change; and (xii) culture in [15].

6.2. Integration of TQM and GE-6σ

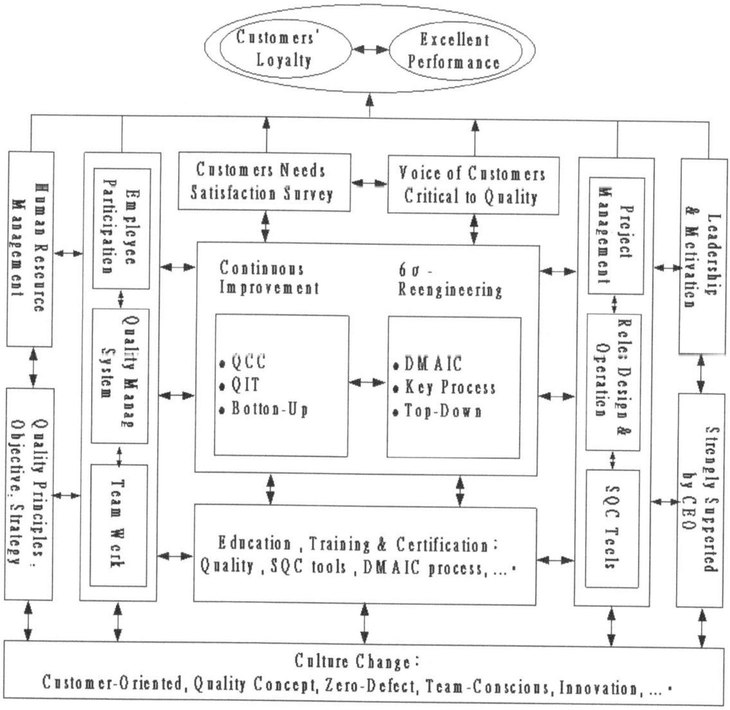

Figure 3. The integrated model of TQM and Six-Sigma program

Based on the comparison between TQM and Six-Sigma conducted by Yang in [15], it can be concluded that there is congruence among the quality principles, techniques, and cultural aspects of TQM and GE-6σ, and only a little difference between their management principles. As a result, the integration of TQM and GE-6σ is not as difficult as it might seem. The critical task is to combine the best aspects of TQM's continuous improvement with those of GE-6σ's re-engineering. Although the activities of a quality control cycle (QCC) and quality improvement team (QIT) cannot achieve significant effects in themselves, they can cultivate quality concepts and team awareness among employees, and hence the quality culture. Therefore, QCC and QIT can be performed by the operators and junior staff members to progress continuous improvements while focusing on daily operations and processes. GE-6σ projects can be applied by engineers and senior staff members to the key processes and systems that are related to customer requirements and the provision of performance in products and services. For GE-6σ projects, some aggressive goals can be set in conjunction with rapid project completion times. The target performances can be set according to the criteria of the critical-to-quality (CTQ) of key process—which are, in turn, determined according to the voice of customers (VOC). In TQM, the improvements are based on a customer satisfaction survey and an understanding of customers' requirements in [55]. In this fashion, these two ways of understanding customers' needs and expectations can be combined. See Figure 3 for a depiction of the model.

It has been suggested that the implementation of TQM results in an over-emphasis on customer satisfaction, with a relative neglect of the pursuit of profits. Indeed, several empirical studies have asserted that implementing TQM might not achieve any significant positive effect on profitability in [56, 52, 41]. Furthermore, Harry (2000a) has noted that "What's good for the customer is not always good for the company" in [57]. In contrast, it is argued that GE-6σ achieves both customer satisfaction and excellent financial performance. The major problem with TQM is that there is a disconnection between management systems designed to measure customer satisfaction and those designed to measure business profitability, and this has often led to unwise investments in quality in [41]. It should be recognized that the objective of TQM is to achieve customer satisfaction in order to increase customer loyalty. To sustain competitiveness and long-term profitability, companies not only devote themselves to attracting new customers, but also to retaining old customers in a continuous business relationship with incremental additional purchasing. For these reasons, increasing customer loyalty should be one of the main concerns of all companies in [58]. Any assessment of the effectiveness of TQM thus requires a system to measure customer loyalty.

If a management system cannot raise business performance and profitability, it will obviously be abandoned by firms. It is therefore apparent that indicators of customer loyalty and business performance should be added to TQM measurement systems. It is well known that GE-6σ pursues both customer satisfaction and high profits. If an integrated model of TQM and GE-6σ were developed, synergistic effects could be anticipated. In the integrated model proposed here, two major indicators are included—customer loyalty and high profit performance.

7. The development of a business excellence system

In section 3 we discuss the evolution of quality management, and state that now is an age of pursuing business excellence. In this section, we will develop a more comprehensive model, called a 'Business Excellence System,' based on an integrated model of the TQM and Six-Sigma programs developed in the above section and referring to several related researches. We also provide an example case, which is a good company that won the Deming Award in 2011.

7.1. Malcolm Baldrige National Quality Award and European Quality Award

Several studies in [59-62, 29] have suggested their own holistically strategic management system or business excellence system. These holistically integrated models can be used in association with the frameworks of the Malcolm Baldrige National Quality Award (MBNQA) or the model of European Quality Award (EQA), see Figure 4 and Figure 5. MBNQA was initiated by the USA in 1987 and is a framework of seven constructs: leadership, strategic planning, customer and market focus, information and analysis, human resource development and management, process management, and business results in [9]. The first six constructs are the critical management systems; the successful implementation of these systems will result in excellent business performances. Thus, MBNQA can be used to assess the performance of an organization, based on the realization of TQM and strategic management in [9].

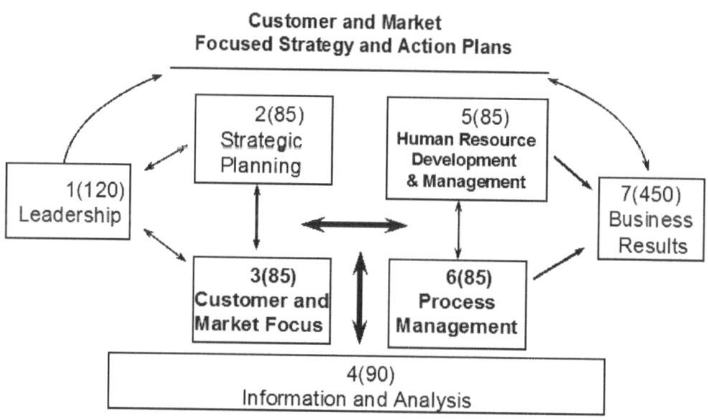

Figure 4. Framework of MBNQA

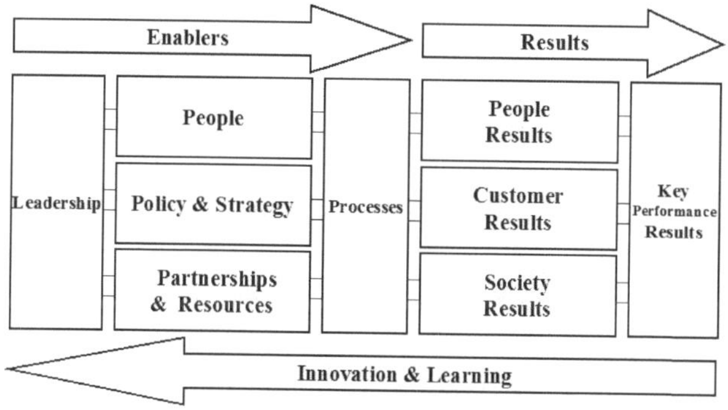

Figure 5. Framework of EQA

In 1992 European Foundation for Quality management (EFQM) launched the European Quality Award (EQA). EQA is a framework constituted by two parts: enablers and results in [9]. The enablers include the operation processes of leadership, people, police and strategy, and partnerships and resources, which are the means by which an organization can achieve the key performance results: people results, customer results, and society results. It is recognized that the frameworks of the MBNQA and the EQA are based on the 'holistic TQM system' and the enablers, especially the strategic management systems in [63, 21], and that the key metrics of the MBNQA and EQA models can thus be used to assess how well a firm is implementing the TQM system and the total business performances. As a result, many countries developed their National Quality Awards based on the Japanese Deming Award, MBNQA and EQA models before developing their business excellence model.

7.2. Strategic map of enterprise's long-term development

In order to pursue long-term profitability and successful development, a firm needs to develop core competencies and capabilities and possess core competitiveness. Therefore we suggest a 'strategic map of enterprise's long-term development,' which describes how a firm operates its core competencies and capabilities to achieve its 'vision: customer loyalty, successful development, and long-term profitability.' It consists of four constructs, and each construct includes several key essentials. They are stated in the following.

1. Growth force

It includes the business performances that will result in huge contributions to the firm, For example, increasing market share, entering new markets, new business development, and raising profits. Therefore, the firms need to successfully implement an integrated performance management system which is constituted of strategy management, Hoshin management, and a balanced scorecard.

2. Core competitiveness

This construct consists of the business model, management systems, or strategic actions which will form the core competitiveness for the firm, such as leader of core (innovative) products, capturing the customers' needs, high quality customer service, development of specialized technologies, and core business development.

3. Critical drivers

How to heighten the core competitiveness? The firm needs to effectively execute the critical drivers to attain the competitive advantage. The critical drivers are top management leadership and support, human resource management, total quality management, customer relationship management, the development of core competencies and capabilities, implementation of an IT and knowledge management

(KM) systems. The drivers are almost always included in the constructs of MBNQA or in the enablers of EQA.

4. Fundamental field

Fundamental field is the imperative resource which causes the firm to create the drivers. There are several critical ingredients of the fundamental field, which are realization of mission and value, innovative environment, investment in R&D, sufficient supporting systems, and a good organization culture.

These constructs and their involving items have a cause-and-effect relationship. The items of the 'fundamental field' construct will affect the development of the items in the 'drivers' construct. Effective implementation of the systems in the 'drivers' construct can result in advantage on the items in the 'core competitiveness' construct. As a result, the items in the 'growth force' construct will have the best performance. These relationships are manifested in Figure 6.

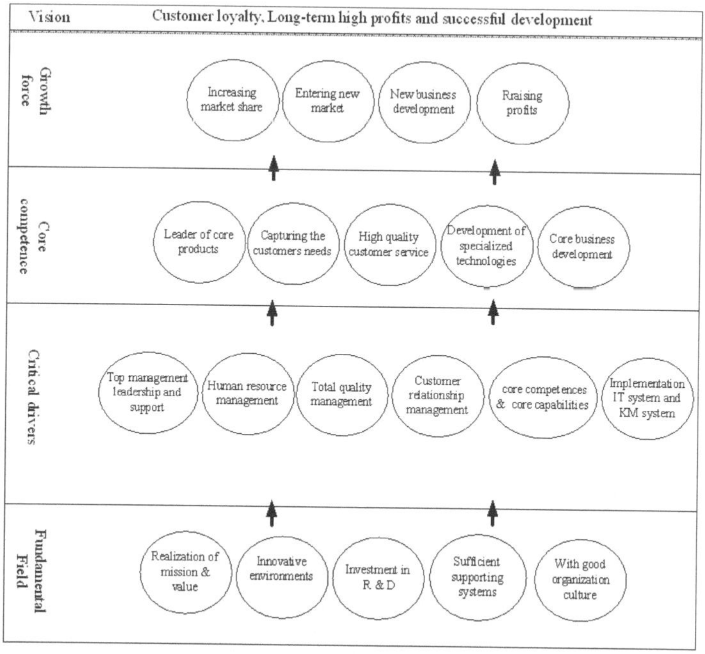

Figure 6. Strategic map of enterprise's long-term development

7.3. The integrated business excellence model

Basically, the business excellence system can be developed by combining 'the integrated model of TQM and the Six-Sigma program' and the 'strategic map of enterprise's long-term development' in [30]. In the integrated model of TQM and the Six-Sigma program, the critical activities are the improvement activities: QIT and QCC, and the Six-Sigma program, which are created by considering the voice of customers and their needs. In the integrated business excellence model, besides the improvement activities, development of core products is also the critical activity; its aim is to achieve the customers' latent needs and then delight the customers. In order to develop attractive and innovative products, the employees must have innovative concepts and lean thinking, then the products/services with attractive quality can be created; see the center part of Figure 7. The success of these activities is based on the implementation of TQM and the application of IT system and KM system.

Besides these two critical drivers (see Figure 6), the implementation of HRM will train the employees with specialized talents and realize the management of 'empowerment.' The top management leadership and support will lead the realization of 'full participation' and 'team-work'; see the left part of Figure 7. The firms possessing the core competencies and capabilities will develop the core products/services, and then create the related core businesses. Successful implementation of customer relationship management can result in customer service with good quality; see the right part of Figure 7.

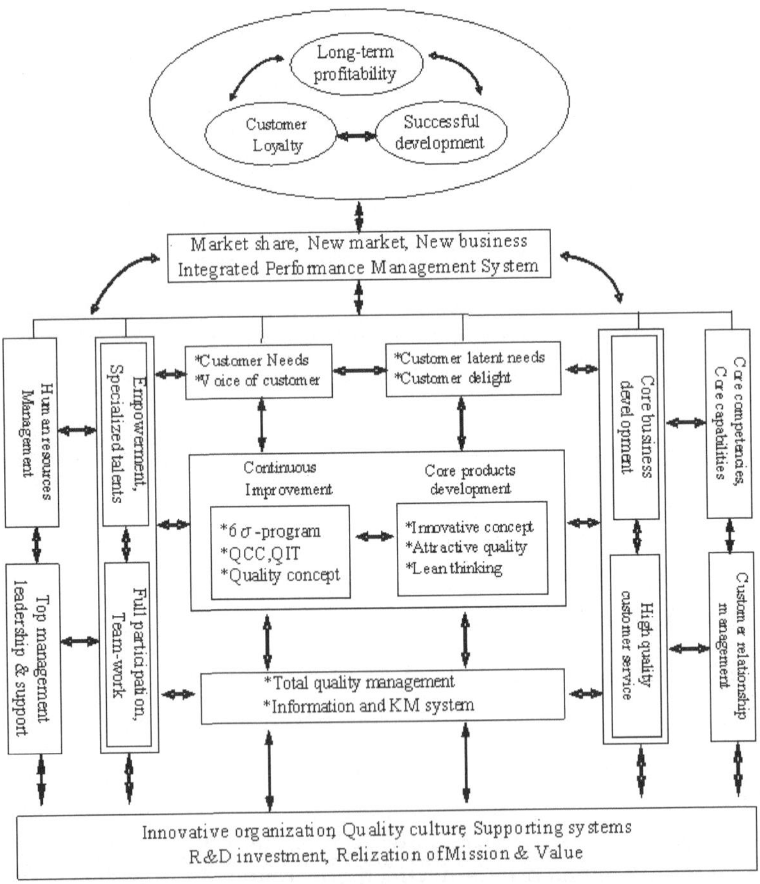

Figure 7. The integrated business excellence model

Drucker stated that the starting point both in theory and in practice may have to be "managing for performance" in [64]. The goal of an integrated business-excellence model is to go beyond mere 'customer satisfaction' to achieve customer loyalty and excellent performance, which is represented as long-term development and profitability through the strategies of creating new business, raising market share, and entering the new market (see Figure 7). The management systems, programs, and practices of this integrated model are the tools that can be used to achieve this goal. However, an appropriate performance management system is needed to monitor and evaluate the performance generated by the implementation of this integrated business excellence system.

The performance management system was developed by integrating Hoshin management in [65] with strategic planning and balanced scorecard (BSC) in [66]. We first take the implementation model of strategic planning as its starting-point. Firms commonly perform a SWOT analysis and develop a vision, objectives, and strategies of the whole organization. Having established its vision, objectives, and strategies, the firm can then develop a strategy map and the key performance indicators (KPIs) according to the four perspectives of BSC. The firm can then use the methods of Hoshin management to deploy the organization's objectives and strategies and its resulting performance indicators to the departments or units. During the implementation process, they commonly conduct a quality audit according to Hoshin management to produce progress reviews and an annual review. The organizations thus use an integrated model of performance management to evaluate the performance of TQM in [66].

The success of the implementation of this integrated business excellence model is dependent on the realization of the fundamental principles and conditions, including innovative environment, quality culture, compete supporting systems, R&D investment, and the realization of mission and vision. The top management must bear the responsibility of the cultivation of these fundamental principles and conditions.

8. Case study

Unimicron Technology Corporation, which is located in Taoyuan, Taiwan, was established in 1990 and is the heart of the printed circuit board (PCB) industry in Taiwan. This is currently the top-ranked industry in Taiwan and has been the fifth ranked worldwide since 2003. The company thus invests heavily in leading-edge technologies and its products are in high demand from customers.

The senior management of Unimicron strongly emphasizes the implementation of total quality management (TQM). Management introduced TQM in 1996, at which time the company established a TQM committee which currently has four subcommittees: a Six-Sigma/QIT subcommittee, an education and training subcommittee, a QCC (Quality control circle) subcommittee, and a quality &

standardization subcommittee. The company embarked on Hoshin management in 1998 and implemented Six-Sigma programs in 2001. In 2002, the company enhanced the element of strategic thinking in the Hoshin management system by introducing the management of strategic planning. With the increasing popularity of BSC around the world, Unimicron also initiated the implementation of BSC and a strategy map in 2003. Implementation of these systems simultaneously would have caused significant problems for both management and staff. The company therefore integrated the systems, as shown in Figure 8. Unimicron called this integrated model the 'Excellent Policy Management Model.'

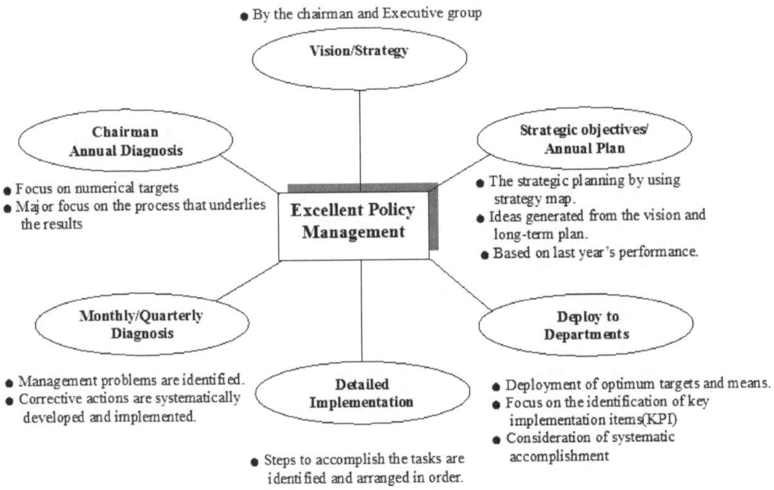

Figure 8. The integrated model 'Excellent Policy Management' adopted by Unimicron

The implementation principles of the 'excellent policy management' model were as follows:

1. PDCA cycle: integrating Deming's 'plan–do–check–act' language;
2. Focus: determining the direction and priorities of the organization's development;
3. Alignment: achieving consensus (regarding vision and strategy) with the employees who are likely to make a contribution;
4. Integration: integrating the 'excellent policy management system' with existing systems;
5. Review & diagnosis: using monthly/quarterly diagnosis to ensure that everyone is cooperating in the execution of strategic targets; and
6. Performance pursuit: ensuring desired performance through a focus on KPIs.

The company also developed a complete implementation model (see Figure 9).

Since Hoshin management was implemented in 1998, Unimicron has experienced strong growth in revenue, from US$0.18 billion in 1999 to US$ 33 billion in 2011. In the same period, profit increased from US$120 million in 1999 to US$3.5 billion in 2011. The company's worldwide ranking rose to No. 1 in 2009 (from No. 35 in 1999). These significant business successes have encouraged Unimicron to implement its 'excellent policy management' model even more comprehensively and thoroughly.

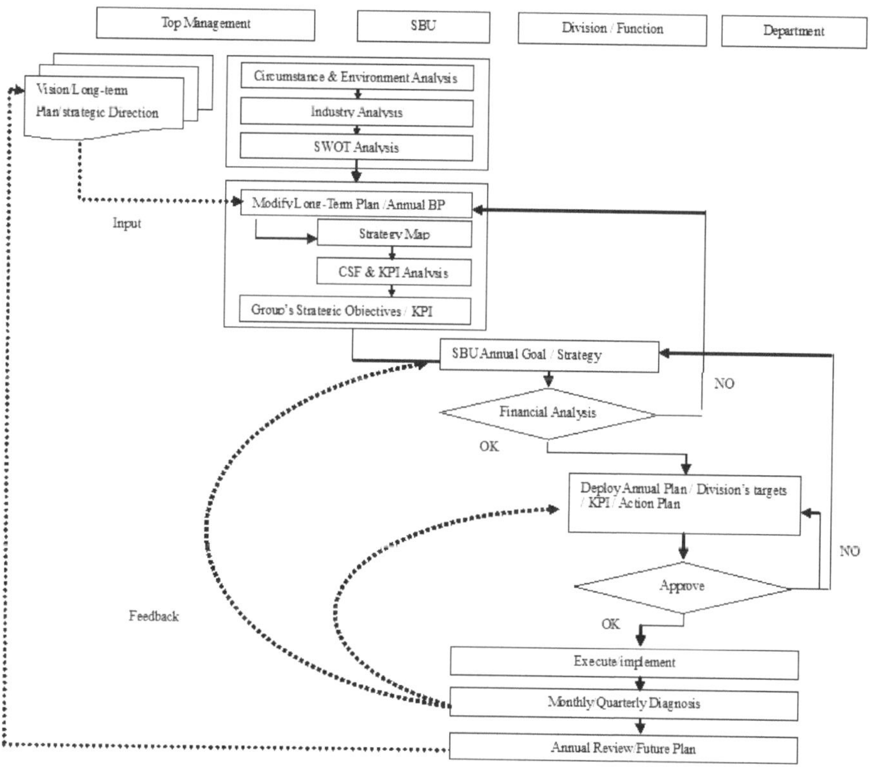

Figure 9. The implementation model of 'Excellent Policy Management' used by Unimicron

Since the business performances are very excellent in recent years, Unimicron applied for the National Quality Award in Taiwan for first time in 2006 and successfully won.

Therefore, Unimicron decided to apply for the Deming Award in 2007. It organized several committees and designed a complete schedule for the preparation. Unimicron aggressively utilized employee participation and team-work. All the employees paid more attention to the top objectives required to win the Deming Award. During the preparation period, they realized the 'Excellent Policy Management' system, effectively implemented quality audits, and took improvement actions immediately. Eventually, Unimicron won the Deming Award in October, 2011.

References

1. EdvardsC. D.Themeaning.ofquality.Quality Progress 19683639
2. JuranJ. M.QualityControl.Handbook 3rded.New York: McGraw-Hill; 1974
3. Crosby, P., Quality is Free: The Art of Making Quality Certain.New York: McGraw-Hill; 1979
4. ShettyY. K.RossJoel. E.Qualityitsmanagement.inservice.businessIndustrial Management 1985276712
5. Douthu, Naven, Quality control in services industry.Journal of Professional Services Marketing 1991713155
6. Edvardsson, B., Quality in new service development: Key concepts and a frame of reference.International Journal of Production Economics 19975231
7. Schvaneveldt, Shane J., Edkawa, Takan, & Miyakawa, Nasami, Customer evaluation perspectives and service quality: Evaluation factors and two-way model of quality. Total Quality Management 199122149161
8. DemingW. E.Outof.theCrisis. M. I. T.Centerfor.AdvancedEngineering.studyMA: Cambridge; 1986
9. Oakland, John S., Total Quality Management, Second edition. Jordan Hill, Oxford:Butterworth-Heinemann; 2000
10. RaoAshok.CarrLawrence. P.DambolenaIsmael.KoppRobert. J.MartinJohn. RafiiFarshad.SchlesingerPhyllis.FinemanTotal.QualityManagement. A.CrossFunctional.PerspectiveNew.YorkJohn.WileySons; 1996
11. Mitra, Amitava, Fundamentals of Quality Control and Improvement Second edition.London: Prentice-Hall International (UK) Limited; 1998
12. Dotchin, John A., & Oakland, John S., Theories and concepts in total quality management.Total Quality Management 199232133145
13. PowellT. C.TotalQuality.asManagementCompetitive.AdvantageA. ReviewEmpiricalStudy.Strategic Management Journal 19951615
14. Elshennawy, Ahmad K., Maytubby, Venetta J., & Aly, Nael A., Concepts and attributes of total quality management.Total Quality Management 1991217597
15. Yang, Ching-Chow, An integrated model of TQM and Six-Sigma.International Journal of Six Sigma and Competitive Advantage 20041197111
16. SilaI.EbrahimpourM.Aninvestigation.ofthe.totalquality.managementsurvey. basedon.researchpublished.between19.200Journal of Quality and Reliability International 2002
17. PandeP. S.NeumanR. P.CavanachR. R.TheSix.SigmaWay.New York: McGraw-Hill; 2000
18. Gates, Bill, Business @ the Speed of Thought: Using a Digital Nervous System. New York: Warner Books Inc.1109; 1999
19. Martin, Chuck, Get Future. New York: McGraw-Hill Inc.; 1999.

20. Ohmae, Kenichi, The Next Global Stage: Challenges and Opportunities in Our Borderless World.New Jersey: Wharton School Publishing; 2005
21. Tan, Kay C., Wong, Mei F., Mehta, Teena, & Khoo, Hsien H., Factors affecting the development of national quality awards.Measuring Business Excellence 2003733745
22. Petts, Nigel, Building growth on core competencies- a practical approach.Long Range Planning 1997304551561
23. PrahaladC. K.HamelGary.Thecore.competenceof.thecorporation.Harvard Business Review 1990May/June) 7991
24. Clark, Delwyn N., Implementation issues in core competence strategy making. Strategic Change 2000Mar-Apr) 115127
25. AlmerTamar.HashaiNiron.Thecompetitive.advantagestrategicconfiguration. ofknowledge-intensive.smallmedium-sizedmultinationals.modifiedA.resource-basedview.Journal of International Management 2004104479500
26. LaurieDonald. L.DozYves. L.SheerClaude. P.Creatingnew.growthplatforms. Harvard Business Review 2006May) 8090
27. Zook, Chris, Finding your next core business.Harvard Business Review 2007April) 6775
28. Kanji, Gopal L., Measurement of business excellence.Total Quality Management 199897633643
29. KanjiGopal. K.ParticiaS. Â.MouraE.Measuringleadership.excellenceTotal Quality Management 2001126701718
30. Yang, Ching-Chow, Development of an Integrated model of a business excellence system.Total Quality Management and Business Excellence 2011
31. BoadenR. J.Whatis.totalquality.management…anddoes.itmatter?.TotalQuality. Management199784153171
32. JuranJ. M.Grynafrank. M.QualityPlanning.AnalysisThird.editionNew York: Mcgraw-Hill Inc.; 1993
33. Yang, Ching Chow, Quality is the Best Strategy in Competition.Taiwan: APEX International Management Consulting Co.; 1997
34. SureshchandarG. S.RajendranC.AnantharamanR. N.ConceptualA.Modelfor. TotalQuality.Managementin.ServiceOrganization.Total Quality Management in Service Organization. Total Quality Management 2001123341363
35. Barney, Jay, Firm resources and sustained competitive advantage.Journal of Management 199117199120
36. HarryM. J.SchroederR.SixSigma.TheBreakthrough.ManagementStrategy. Revolutionizingthe.Worlds'Top.CorporationsNew York: Doubleday; 2000
37. Klefsjö, Bengt, Wiklund, Häkan, & Edgeman, Rick L., Six Sigma seen as a methodology for total quality management.Measuring Business Excellence 2001533135

38. Slater, Robert2001Get Better or Get Beaten! New York: McGraw-Hill; 2001.
39. Deming, W. E., The New Economics, for Industry, Government, Education. Cambridge, MA: Massachusetts Institute of Technology; 2001.
40. HendersonK. M.EvansJ. R.Successfulimplementation.ofSix.SigmaBenchmarking. GeneralElectric.companyBenchmarking: An International Journal 200074260281
41. Breyfogle III, Forrest W., Cupello, James M., & Meadows, Becki, Managing Six Sigma. New York: John Wiley & SonsInc; 2001
42. TreichlerDavid.CarmichaelRonald.KusmanoffAntone.LewisJohn. BerthiezGwendolyn.Designfor.SixSigma. .LessonsLearned.Quality Progress 2002January 3342
43. SandholmL.SorqvistL..requirementsfor.SixSigma.successSix Sigma success. Six Sigma Forum Magazine 2002211722
44. Coronado, Richardo Banuelas, & Antony, Jiju, Critical success factors for the successful implementation of Six Sigma projects in organizations.The TQM Magazine 20021429299
45. Antony, Jiju, & Banuelas, Richardo, Key ingredients for the effective implementation of Six Sigma program.Measuring Business Excellence 2002642027
46. Yun-YongJong.ChuaRichard. C. H.Samsunguses.SixSigma.tochange.itsimage.Six Sigma to change its image. Six Sigma Forum Magazine 2002211316
47. Yang, King-Jang, Yeh, Tsu-Ming, Pai, Fan-Yun, & Yang, Ching-Chow, The analysis of the implementation status of Six Sigma: an empirical study in Taiwan. International Journal of Six Sigma and Competitive Advantage 2008416080
48. HubiakW. A.O'DonnellS. J.DoAmericanshave.theirminds.setagainst. T. Q.M?Abstract.National Productivity Review 19961519
49. Hellsten, Ulrike & Klefsjö, Bengt, TQM as a management system consisting of values, techniques and tools.The TQM Magazine, 2000124238244
50. HanssonJohas.KlefsjöBengt. A.corevalue.modelfor.implementingtotal. qualitymanagement.insmall.organizationsThe TQM Magazine 20031527181
51. LucasJames. M.Theessential.SixSigma.Quality Progress 2002January 2731
52. HarryMikel. J.newA.definitionaims.toconnect.qualityperformance.withfinancial. performanceQuality Progress 20003316466
53. Hermel, Philippe, The new faces of total quality in Europe and the US.Total Quality Management 199784131143
54. GohT. N.strategicA.assessmentof.sixsigma.Quality and Reliability Engineering International 200218403
55. Yang, Ching Chow, Improvement actions based on the customers' satisfaction survey.TQM and Business Excellent 2003142919930
56. BergquistTimothy. M.RamsingKenneth. D.Measuringperformance.aftermeeting. awardcriteria.Studycompares.perceivedsuccess.tofinancial.dataof.awardwinners. applicantsQuality Progress 19993296672

57. Harry, Mikel J., The quality twilight zone.Quality Progress 20003326871

58. GorstJ.KanjiG.WallageW.Providingcustomer.satisfactionTotal Quality Management 1998

59. OrtizAngel.LarioFrancisco.RosLorenzo.Enterprise-businessintegration. processesintegrated.managementA.proposalfor. a.methodologyto. developenterprise.integrationprograms.Computers in Industry, 199940155

60. KaplanR. S.NortonD. P.Thebalanced.scorecard-measuresthat.driveperformance. Harvard Business Review 1992January-February 7179

61. KaplanR. S.NortonD. P.TheBalanced.ScorecardTranslating.strategyinto. actionBoston, MS: Harvard Business School Press; 1996

62. KaplanR. S.NortonD. P.Transformingthe.balancedscorecard.fromperformance. measurementto.strategicmanagement.PartI.Accounting Horizons 200115187104

63. PuayS. H.TanK. C.XieM.GohT. N.comparativeA.studyof.ninenational.qualityawards. The TQM Magazine 19981013039

64. Drucker, Peter F., Management Challenges for the 21st Century.New York: Harper Collins Publishers Inc.; 1999

65. Yang, Ching-Chow, Hoshin Management. Taiwan: Chinese Quality Management Society;1995

66. Yang, Ching-Chow, An integrated Implementation model of strategic planning, BSC, and Hoshin managemen.Total Quality Management and Business Excellence 2011

67. KondoYoshio.Hoshin-AKanri.participativeway.ofquality.managementin.JapanThe TQM Magazine 1998106425432

10

Qualitative and Quantitative Analysis of Six Sigma in Service Organizations

1. Introduction

Quality management has long been established as an important strategy for achieving competitive advantage. The aim of the businesses may differ, but the importance of customers is a matter of common interest. The ability of the organizations to adapt to new customer requirements in a globalized market is of vital importance for long-term success. Traditional quality initiatives such as statistical quality control, zero defects, and total quality management, have been key initiatives for many years. In last two decades, Six Sigma evolved as a new quality management initiative and now many organizations are working towards its implementation.

Six Sigma is a disciplined approach for improving manufacturing or service processes, based on defined metrics (Hahn et al., 1999). The strength of Six Sigma lies in its well defined framework involving methodology applying different tools and techniques (Goh, 2002). The Six Sigma journey started from Motorola in 1980s and spread its importance through adoptions by different high profile organizations such as General Electric (GE), Honeywell, Asea Brown Bovari (ABB), Lockheed-Martin, Polaroid, and Texas Instruments (Goh, 2002; Hahn et al., 1999). This initial success of Six Sigma has seen its implementation spreading in several other organizations mostly in mass-manufacturing sector (McAdam et al., 2005). These organizations adopted the systematic framework of Six Sigma through training and project management practices (Brady and Allen, 2006). The use of Six Sigma has been relatively high among many western organizations till now, see, for example, Inozu et al. (2006), Raisinghani et al. (2005), and Antony (2004b), but there exists a diversity of opinion among researchers regarding the actual benefits of Six

Sigma. Literature explaining about the positive effects on financial performance can be found in e.g. Jones Jr. (2004), Goh (2002), Caulcutt (2001), and Rucker (2000). However, McAdam and Lafferty (2004), Senapati (2004), and Paul (1999), for instance, express a more pessimistic view regarding the benefit of Six Sigma investments.

Similar to Six Sigma, services in the last two decades have become an important part in economies of developed as well as for developing nations. The importance of services has also increased as it became a major employment provider (Cook et al., 1999). This increased importance of service sector has various researchers contributing to the service literature. The service research from its beginning can be divided into stages such as an initial realization of the difference between goods and service, the development of conceptual frameworks, the empirical testing of these frameworks and the application of the tools and frameworks to improve service management (Johnston, 1999). The various stages of service research have gained by the major contributions from marketing and after that to some extent from operations management field.

But in case of definitions, services still lack a unified definition and similar can be said about the classification scheme. So, there is a need to provide a universally accepted classification scheme which can be done through empirical derivation and considering different dimensions of service organizations. This will facilitate in exploration of service quality and service strategy (Cook et al., 1999). As service quality is now the major focus of service organizations, so a better understanding of unique characteristics of services will be helpful. This in turn will help spreading new quality initiatives such as Six Sigma, in services.

This spread, however, is limited in service industries. A key argument here is that many service processes are unseen, intangible, and even immeasurable. As such, they are not amendable to improvement using a Six Sigma approach. This thinking has turned out to be rather presumptuous at least for the health care, banking, and call center services which have been able to apply Six Sigma (Hensley and Dobie, 2005). Other services such as education and hospitality are also beginning to see Six Sigma applications.

The question of Six Sigma and its implementation and performance in service organizations has not before been under investigation. The literature includes many papers describing Six Sigma implementation in a variety of business types; however, very few of these papers report empirical research and include single case studies (Hendry and Nonthaleerak, 2005). Examples of non-manufacturing contexts discussed in the literature include healthcare and financial services, as well as in non-production internal functions within a manufacturing organization (Nonthaleerak and Hendry, 2008). The paper by Wyper and Harrison (2000), discuss Six Sigma implementation in non-manufacturing context and highlights the difficulties specific to that context. Does et al. (2002), present a comparison of eight Six Sigma projects in non-manufacturing processes with a theoretical manufacturing application in a case study company in the Netherlands. This paper addressed various problems, typical of non-manufacturing and also identified difficulties in tools application. They conclude that Six Sigma can be applicable in non-manufacturing contexts with minor adaptations. Given that, the research is based in a single case study setting, there are limits on the degree to which the conclusions can be generalized (Nonthaleerak and Hendry, 2008). McAdam and Lafferty (2004) conducted a survey in a single company on Six Sigma implementation issue from process and people perspective. They found low success of Six Sigma in non-manufacturing areas.

The literature on discussion about Six Sigma in service organizations also concentrate about issues in implementation due to inherent differences between manufacturing and service. The possible reason being the manufacturing roots of Six Sigma like other quality management initiatives. Antony (2004a), Benedetto (2003), and Sehwall and De Yong (2003), argue for example that there are certain differences in Six Sigma implementation in services from manufacturing which acts as a barrier in Six Sigma implementation in service organizations.

Failed implementation initiatives, especially as extensive as Six Sigma implementation, result in financial losses and potential resistance towards change among the actors involved. It is therefore of importance that the implementation strategies used are well adapted, see, e.g. Biolos (2002). Hence, the literature has conflicting evidence regarding the applicability of Six Sigma to non-manufacturing settings and therefore there is a need to investigate further this issue (Nonthaleerak and Hendry, 2008).

The studies so far focused on Six Sigma implementation in non-manufacturing context at project level. The studies are mostly single case studies and descriptive in nature. The survey based studies are either pilot survey or focused on a single organization. Thus, there are a number of key research gaps in the literature, which our research aims to address.

- There is insufficient empirical evidence to verify and further explain the Six Sigma CSFs identified in service organizations.
- The existing difficulties in Six Sigma implementation in service organizations are not well understood.
- There is a scope to contribute to Six Sigma implementation in service organizations by enhancing the knowledge about tools and techniques usage.

Our research will focus on individual Six Sigma projects in service organizations to fill the identified research gaps.

2. Background

2.1. Six sigma definitions and philosophy

In 1924, Walter A. Shewart from Bell Telephone Laboratories, proposed the concept of using statistical charts to control the variables of products manufactured at Western Electric. This was the beginning of statistical quality control (Small, 1956). Dr. Shewart kept on with his efforts and applied the fundamentals of statistical quality control to industry. This lead to the modern attention to the use of statistical tools for the manufacture of products and process, originated prior to and during World War II, when the United States of America geared up to a massive build-up of machinery and arms to successfully conclude the war (Brady, 2005). The Western Electric manufacturing company is noteworthy during this time because it was the breeding ground for many quality leaders, not only Shewart but Joseph Juran, Edwards Deming and Kaoru Ishikawa all worked there at some time (Dimock, 1977). Two prominent individuals were Deming and Juran. Deming promoted the use of the plan-do-check-act (PDCA) cycle of continuous improvement. Later Juran introduced the concepts of project by project quality improvement. Any discussion on quality today will most likely cite at least one from the group of Deming, Juran, Crosby, Feigenbaum, and Ishikawa, if not all. They certainly represent the preponderance of information about quality. Adding to this group, Bill Smith, Motorola Vice President and Senior Quality Assurance Manager, is widely regarded as the father of Six Sigma, Shina (2002). Because Six Sigma was built on previous quality methodologies, a list of the pioneers of the quality and their contribution is included in Table 1.

According to Shina (2002) before, January 15, 1987, Six Sigma was solely a statistical term. Since then, the Six Sigma crusade, which began at Motorola, has spread to other companies which are continually striving for excellence. At Motorola, Six Sigma is defined as "A quality improvement program with a goal of reducing the number of defects to as low as 3.4 parts per million opportunities or 0.0003%". Six Sigma has a number of different meanings and interpretations (Henderson and Evans, 2000, pp 261). Its origin comes from statistics where sigma represents the amount of variation about a process average. From a business view of point, Six Sigma may be defined as "A business strategy used to improve business profitability, to improve the effectiveness and efficiency of all operations to meet or exceed customer's needs and expectations" (Kwak and Anbari, 2006, pp 709). Various other definitions include:

- Six Sigma is a formal methodology for measuring, analysing, improving, and then controlling or locking-in processes. This statistical approach reduces the occurrence of defects from a three sigma level or 66 800 defects per million to a Six Sigma level or less than four defects per million (Bolze, 1998).
- Six Sigma is a comprehensive, statistics-based methodology that aims to achieve nothing less than perfection in every single company process and product (Paul, 1999).

- Six Sigma is a disciplined method of rigorous data gathering and robust statistical analysis to pinpoint sources of error and ways of eliminating them (Harry and Schroeder, 1999).

- Six Sigma as an information-driven methodology for reducing waste, increasing customer satisfaction, and improving processes, with a focus on financially measurable results (As defined by Minitab in Goh, 2002).

Table 1. Pioneers of quality and their contribution to Six Sigma knowledge bank (adapted from Wortman, 2001)

Quality Gurus	Contribution
Philip B. Crosby	Senior manager involvement; four absolutes of quality management; quality costs measurements
W. Edwards Deming	Plan-do-study-act; top management involvement; concentration on system improvement; constancy of purpose
Armand V. Feigenbaum	Total quality control/management; top management involvement
Kauro Ishikawa	Cause and effect diagram; company-wide quality control;
Joseph M. Juran	Top management involvement; quality trilogy; quality cost measurement
Walter A. Shewart	Assignable cause versus chance cause; control charts; plan-do-check-act; use of statistics for improvement

The statistical focus of various Six Sigma definitions reflects its basic philosophy. Six Sigma is an operating philosophy that can be shared beneficially by everyone, including customers, shareholders, employees, and suppliers. Fundamentally, it is also a customer-focused methodology that drives out waste, raises levels of quality, and improves the financial performance of organizations to breakthrough levels (Chua, 2001).

Six Sigma's target for perfection is to achieve no more than 3.4 defects, errors, or mistakes per million opportunities whether this involves the design and production of a product or a customer-oriented service process. It is from this target that the name Six Sigma originated.

Compared to a process that has greater variation, a process with less variation will be able to fit more standard deviations or sigmas between the process centre and its specification limits. An increase in the number of sigmas between the specification limits means the acceptance of fewer defects. More sigmas imply a more consistent manufacturing or service delivery process (Chua, 2001).

2.2. Tools and techniques and six sigma methodologies

The concept of Six Sigma was introduced at and popularized by Motorola in 1987. Six Sigma is a logical extension of Statistical Process Control (SPC). The concept behind SPC is simple enough but powerful, indeed. Variation is present in every production/operations process and such variation is due either to common causes or special causes. The breakthrough made by Shewart was the statistical definition and measurement of variation, where variation within three-sigma limits was deemed to be random and produced by common causes, and variation outside of the three-sigma limits was produced by special causes, indicating a process problem (Shewart, 1931). The $\pm 3\sigma$ process limits mean a defect rate of 2.7/1000 or 2,700/1,000,000 opportunities, if one ignores lateral shifts in the process, and the capability of the process is thus defined as the range of natural variation, that is, $\pm 3\sigma$, or $Cpk = 6\sigma$. Six-Sigma doubles the range of normal variation to $\pm 6\sigma$, and allows for a 1.5σ lateral shift in the process average. The result is a dramatic tightening of acceptable defect rate target to 3.4/1,000,000 opportunities.

The basic elements of Six Sigma are not new. SPC, failure mode effect analysis, gage repeatability and reproducibility studies, and other tools and techniques, have been in use for some time. Six Sigma offers a framework that unites these basic quality tools and techniques with high-level management support.

There is much literature available on tools and techniques used in Six Sigma. Tools are mostly referred to as having a clearly defined role but narrow in focus, whereas techniques have wider application and require specific skills, creativity, and training (Antony, 2006). Similar to CSFs, CTQs, and KPIs; there is limited literature which discuss about STTs specific to service organizations. Discussion on STTs in the literature is mostly on its usage at different phases of DMAIC methodology. De Koning and De Mast (2006) used seven different literature sources and provided a summary of STTs used in DMAIC phases. Some other literature provide classification scheme for tools and techniques used. Henderson and Evans (2000) discussed about tool sets in three groups; team tools, process tools, and statistical tools. As for Six Sigma tools and techniques specific to service organizations, Antony (2006) provides a grid as a guideline for services.

A number of classification schemes for STTs exists, the majority of which are based on the DMAIC methodology. The classification schemes by the American Society for Quality (ASQ) and by Nancy Tague (1995) called the Tool Matrix provide an exhaustive list of tools and techniques which can be used during Six Sigma implementation. The ASQ classification scheme and the tool matrix have almost similar categories. The only difference being in the number of tools and techniques each category.

2.2.1. Classification scheme of tools and techniques

a. ASQ Classification

According to ASQ, tools and techniques that are utilized in different phases of DMAIC are classified according to their uses. There are 7 broad categories: Cause Analysis Tools, Data Collection and Analysis Tools, Evaluation and Decision

Making Tools, Idea Creations Tools, Process Analysis Tools, Project Planning and Implementation Tools, Seven Basic Quality Tools, and Seven New Management and Planning Tools.

Table 2. Classification of tools and techniques according to ASQ(Source: American Society for Quality Website)

Categories	Description	Tools
Cause analysis tools	Used to identify the cause of a problem.	Fishbone diagram, Pareto chart, Scatter diagram
Data collection and analysis tools	Used to collect or analysis data.	Check sheet, Control chart, Design of experiment, Histogram, Scatter diagram, Stratification, Survey
Evaluation and decision making tools	Used to select the best choices or to evaluate what is performance level of project so far.	Decision matrix, Multi-voting
Idea creations tools	Used to create ideas or organize ideas.	Affinity diagram, Benchmarking, Brainstorming, Nominal group technique
Process analysis tools	Used when an understanding of process flow is desired.	Flowchart, Failure mode effect analysis, Mistake-proofing
Seven basic quality tools	These tools are the most fundamental tools of quality control.	Cause and effect diagram/ Fishbone diagram, Check sheets, Control charts, Histogram, Pareto chart, Scatter diagram, Stratification
Seven new management and planning tools	Used to encourage innovation, communicate information and successful planning of key projects.	Affinity diagram, Relation diagram, Tree diagram, Matrix diagram, Arrow diagram, Process decision program chart

b. Tool Matrix

In Nancy R. Tague's The Quality Toolbox (1995), she developed a Tool Matrix that classifies quality tools according to what the tools can offer. It is quite similar to the categorization suggested by ASQ, but differs, as it encompasses more tools.

Table 3. Tool Matrix (Tague, 1995)

Categories	Tools
Ideas creation	Affinity diagram, Brainstorming. Brain writing Nominal group technique, Relation diagram
Process analysis	Cost of quality analysis, Critical-to-quality analysis, Deployment flowchart, Flowchart Matrix diagram, Relations diagram, Requirements matrix, Requirements-and-measure matrix, Storyboard, Top-down flowchart, Work-flow diagram

Categories	Tools
Cause analysis	Contingency diagram, Fishbone diagram, Force field diagram, Is-is not matrix, Matrix diagram Pareto chart, Scatter diagram, Stratification, Tree diagram, Why-why diagram
Planning	Activity chart, Arrow diagram, Contingency diagram, Deployment flowchart, Flowchart Force field analysis, Matrix analysis, Mission statement, Operational definitions, Plan-do-check-act cycle, Relations diagram, Storyboard, Top-down flowchart, Tree diagram, Work-flow diagram
Evaluation	ACORN test, Continuum of team goals, Decision matrix, Effective-achievable matrix, List reduction, Matrix diagram, Mission statement checklist, Multi-voting, Plan-results matrix, PMI
Data collection and analysis	Box plot, Check sheet, Control chart, Histograms, Importance-performance analysis, Kologorov-Smirnov test, Normal probability plot, Operational definitions, Pareto chart, Performance index, Process capability, Requirements-and-measures tree, Run chart, Scatter diagram, Stratification, Survey

c. Innovation Tools

The literature on service design and development talks about various tools which are effective in describing and analysing service problems. The tools are shown in Table 4.

Table 4. Innovation tools

S. No.	Tool	Description
1	Structured analysis and design technique	Used to model service system
2	Function analysis	Maps customer requirements to required functions and means
3	Service blueprinting	Analyse and represents the steps in a service process
4	Quality function deployment (QFD)	Translate customers' needs and expectations into specifications that are relevant to companies
5	Root cause analysis	Identify potential service failure points, service outcome or process problems in service recovery process
6	Theory of Inventive Problem Solving (TRIZ)	Algorithmic approach for solving technical and technological problems
7	Axiomatic design	Maintain the independence of the functional requirements and minimize the information content in a design

2.2.2. Six sigma methodologies

2.2.2.1. DMAIC methodology

Much information is available about the DMAIC (define, measure, analyze, improve, control) methodology. DMAIC is used mostly for existing processes. This approach not only makes use of various tools and techniques, it also incorporates other concepts such as financial analysis and project schedule development. The DMAIC methodology is excellent when dealing with an existing process in which reaching a defined level of performance will result in the benefits expected. There are number of articles and books providing details about DMAIC methodology. Table 5 provides the details about each phase taken from one of the literature.

Table 5. DMAIC methodology (Chua, 2001)

Phase	Description
Define	Identify, evaluate and select projects; prepare the mission; and select and launch the team
Measure	Measure the size of the problem, document the process, identify key customer requirements, determine key product characteristics and process parameters, document potential failure modes and effects; theorize on the cause or determinants of performance
Analyse	Plan for data collection; analyse the data and establish and confirm the "vital few" determinants of performance
Improve	Design and carry out experiments to determine the mathematical cause-effect relationships and optimize the process
Control	Design controls; make improvements, implement and monitor

2.2.2.2. Design for Six Sigma (DFSS): Overview

The emergence of Six Sigma since 1980s has been phenomenal. Initially, the major focus of the organizations was to improve from their existing three sigma limits to Six Sigma limit of product or service quality. The importance of innovation in products and services has changed the focus of organizations now more towards proactive approach rather than being reactive. The design for Six Sigma (DFSS) approach is relatively new compared to Six Sigma and is discussed in different ways in various literatures. Most of the literatures though agree that DFSS is a proactive approach and focuses on design by doing things right the first time. DFSS can be said as "A disciplined and rigorous approach to design that ensures that new designs meet customer requirements at launch" (El-Haik and Roy, 2005, pp 33). According to GE corporate research and development, the importance of DFSS is in the prediction of design quality up front and driving quality measurement and predictability improvement during the early design phases (Treichler et al, 2002). DFSS can also be explained as a data-driven methodology based on analytical tools which provide users with the ability to prevent and predict defects in the design of a product or service (De Feo and Bar-El, 2002). The major focus of DFSS approach is to look for inventive ways to satisfy and exceed the customer requirements. This can be achieved through optimization of product or service design function and then verifying that the product or service meets the requirements specified by the customer (Antony and Coronado, 2002).

The literatures also concentrate on the differences between DMAIC and DFSS approach. Though DFSS involves designing processes to reach Six Sigma levels and is considered as an aggressive approach, but it still lacks a single methodology unlike Six Sigma (Hoerl, 2004). The different methodologies used in DFSS are:

- IDOV (Identify, Design, Optimize, Validate)
- ICOV (Identify, Characterize, Optimize, Validate)
- DCOV (Define, Characterize, Optimize, Verify)
- DMADO (Define, Measure, Analyze, Design, Optimize)
- DMADV (Define, Measure, Analyze, Design, Verify)
- DMADOV (Define, Measure, Analyze, Design, Optimize, Verify)
- DCCDI (Define, Customer Concept, Design, Implement)
- DMEDI (Define, Measure, Explore, Develop, Implement)

Some of the other differences are:

- DFSS is a methodology that takes into account the issues highlighted by the end customers at the design stage while DMAIC solves operational issues (Ferryanto, 2005).
- Benefits in DFSS are difficult to quantify and are obtained in long term in comparison to Six Sigma, where the benefits are expressed mainly in financial terms and obtained rather quickly (www.ugs.com/products/nx/bpi).
- The DMAIC methodology tends to provide incremental improvements in comparison to DFSS where there can be radical improvements (El-Haik and Roy, 2005).
- The projects improved through DMAIC methodology are constrained by the assumptions made during the development and design stages, whereas DFSS builds quality into the design by implementing preventive thinking and tools in the product development process (Smith, 2001).

The tools and techniques involved in the DFSS methodology are also somewhat different from those of the DMAIC methodology. DFSS includes innovation tools such as the theory of inventive problem solving, axiomatic design, and quality function deployment, which DMAIC does not. Detailed information about the methodologies can be found in (Kwak and Anbari, 2006; Hendry and Nonthaleerak, 2005; El-Haik and Roy, 2005; Goel, et al., 2005; Raisinghani et al., 2005; Basu, 2004; Antony and Coronado, 2002; Stamatis, 2002 (a and b); Harry and Schroeder, 1999).

Though there are differences among Six Sigma and DFSS approaches but still these two complement each other. Different DFSS stages are shown in Figure 1. Problem definition is the first stage, where customer requirements are incorporated. This stage is followed by the characterization stage. The model of the problem in the process or engineering domain is developed at this stage, which is basically the translation of the voice of customer and the customer usage conditions into an engineering system (Ferryanto, 2005). As seen from Figure 1, improvements from the DMAIC are added to the model at the characterization stage. After model development, optimal and robust solutions are found out. At the last stage the solutions are verified for their usefulness to solve the real problem.

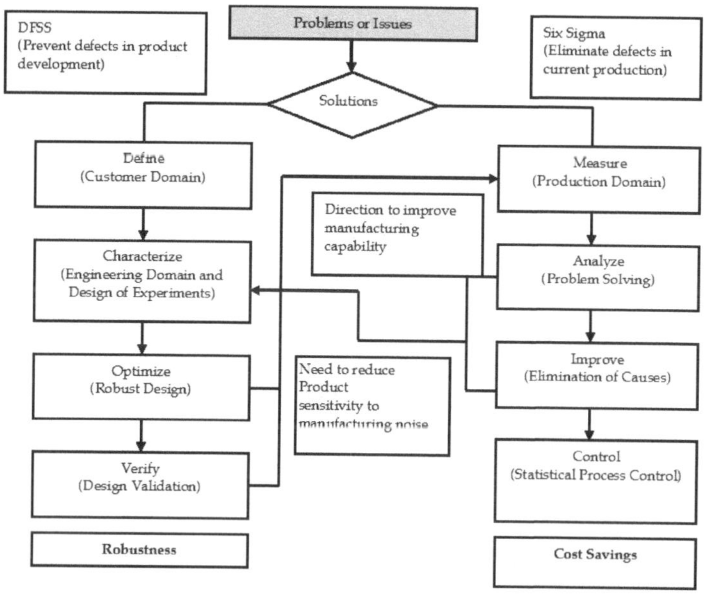

Figure 1. DFSS versus Six Sigma (Ferryanto, 2005)* The DFSS model illustrated is Ford Motor Co.'s DCOV approach

2.3. Critical Success Factors (CSFs)

CSFs are the essential ingredients required for success of Six Sigma projects in an organization (Coronado and Antony, 2002). There have been many studies on CSFs. One of the earliest is by Harry (2000), who discussed about six success factors involving management's leadership, belt system, etc. Later on Antony and Banuelas (2002) mentioned twelve success factors which include management involvement and commitment, linking Six Sigma to business strategy, etc. There are several other studies and all of them have at least one common CSF, i.e. top management commitment. The discussion on CSFs by Antony (2006) is the only one specific to service organizations. Some of the common CSFs are discussed below.

i. Top management commitment and involvement

Almost all the literature reviewed agrees that this factor is a must for successful Six Sigma implementation. And this has to be 'top-down' rather than initiated by a particular department or from the ground (Goh, 2002). Top management involvement helps to influence and restructure the business organization and the cultural change in attitudes of individual employees toward quality in a short implementation period (Henderson and Evans, 2000).

ii. Education and training

Another important feature of Six Sigma is the elaborate training and certification processes that result in Black Belts, Green Belts, etc (Goh, 2002). Education and training help people understand the fundamentals of Six Sigma along with the application of tools and techniques to different phases of DMAIC. Training is part of the communication process to make sure that manager and employees apply and implement the Six Sigma techniques effectively (Kwak and Anbari, 2006).

iii. Cultural change

Six Sigma is considered a breakthrough management strategy, and it involves the adjustment of a firm's values and culture. In some cases, substantial change to an organization's structure and infrastructure need to take place (Coronado and Antony, 2002). People facing cultural change and challenges due to the implementation of Six Sigma need to understand this requirement. Also needed are a clear communication plan and channels to motivate individuals to overcome resistance and to educate senior managers, employees, and customers on the benefits of Six Sigma (Kwak and Anbari, 2006).

iv. Customer focus

Customer focus is one of the major requirements in implementing Six Sigma. This is emphasized in terms of critical to quality characteristics. Six Sigma is highly much more sensitive to requirements for customer satisfaction (Goh, 2002).

v. Clear performance metrics

This is an important factor from a service point of view. Often the difficulty is with identifying what to measure (Sehwall and De Yong, 2003). Before starting any Six Sigma initiative it is better to have a clear idea and agreement on the performance metrics to be used.

vi. Attaching the success to financial benefits

Representing the success of Six Sigma projects in terms of financial benefits and measurement performance has made their selection and completion an important aspect for the organizations (Henderson and Evans, 2000). Financial benefits as a measure of achievement makes it easily understandable for the employees and helps them to relate to Six Sigma project outcome (Goh, 2002).

vii. Organizational understanding of work processes

The amount of effort that a service organization puts into measuring its work processes is important. Some organizations expend much time and effort in developing ways to measure the processes that ultimately impact customer satisfaction. Other organizations attempt this half-heartedly and measure only part of what is important to the customer. Like in hospitals the focus may be only on a particular laboratory or facility where the interaction with customer tends to be relatively greater. Because Six Sigma programs rely on measurements from processes, organizations with robust measurement systems in place are more likely to be ready for a Six Sigma implementation (Hensley and Dobie, 2005).

The factors discussed above are equally applicable to services and manufacturing. Our literature review found that top management commitment, education and training, cultural change, and financial benefits are the most important CSFs. Figure 2 summarizes the importance of the CSFs as seen by each of the articles that were reviewed.

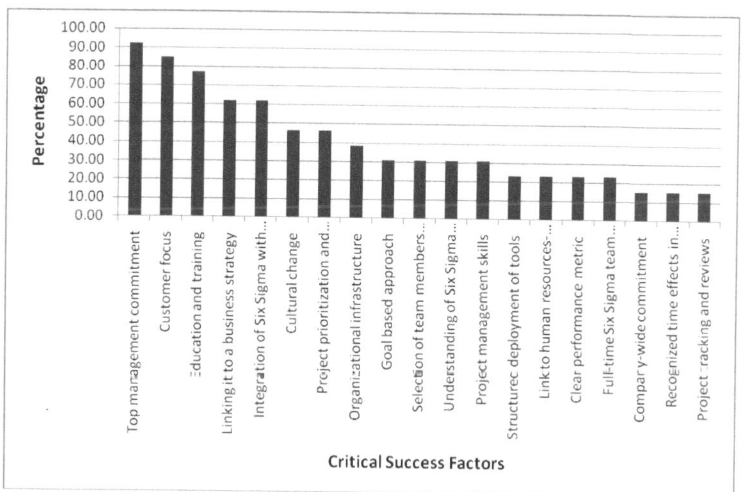

Figure 2. Percentage of articles mentioning each of 19 CSFs

2.4. Critical-to-Quality (CTQ) characteristics

In case of CTQs, we focused on its definitions mentioned in the literature. CTQ is defined in different ways in the literature but mostly they agree that it is a quality characteristic of product or service which is required to be improved from customer point of view. In other words, CTQ is generated from critical customer requirements derived from voice of customer (refer Figure 3).

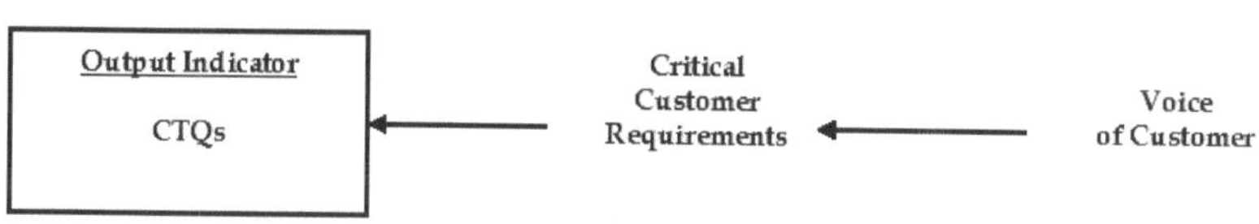

Figure 3. Understanding critical-to-quality (adapted from Muir, 2006)

CTQs are the key measurable indicators of a product or process whose performance standards or specification limits must be met in order to satisfy the customer. CTQs align improvement or design efforts with customer requirements. In a layman term, CTQs are what customers expect of a product or service. They are the spoken needs of the customer (isixsigma/dictionary). Six Sigma focuses on process improvement, and improving the service process is a major determinant of customer satisfaction.

The discussion on CTQs in the Six Sigma service literature is limited. Although services are widely different, the analysis from various literatures (Kwak and Anbari, 2006; Jones Jr., 2004; Sehwall and De Yong, 2003; Rucker, 2000) shows that some common CTQs exist across service. They are discussed below.

- **Time (service time, waiting time, cycle time):** In the case of services where the customer is involved in the process itself, time is an important consideration. The following three types of time should be considered:
    - Service time: The time required to serve a particular customer
    - Waiting time: The time customer waits in the system to get the work done
    - Cycle time: The total time including service and waiting time.
- **Cost:** Like time, cost is sometimes a critical factor from the customer's point of view. The two are in fact intertwined. Customers may at times be willing to pay more for a service that can be completed in a shorter time. The trade-off between cost and time is, thus, important for services.
- **Employee behaviour:** For services where there is high degree of customer contact, employee behavior may be an important consideration. An employee's attitude towards a customer's problem may well decide whether the customer wishes to continue being serviced by the organization.
- **Information (accurate information, timely information):** The growing importance of call center services shows the emergence of information needs. Getting the right information at the right time to one's customers is, thus, an important aspect from a customer point of view.

2.5. Key Performance Indicators (KPIs)

KPI is not well defined in the literature and there exist different interpretations of this term. Mostly the literature discuss about it as performance metrics, i.e., it is a measure of performance in terms of cost, quality, yield, and capacity (Basu and Wright, 2003; Hahn et al., 1999). A few of the suggested definitions of KPI is provided below (refer Table 6).

KPIs show actual data of a particular outcome. The outcomes of Six Sigma projects are usually required to be expressed in financial terms. This leads to a direct measure of achievement which is easy to understand (Goh, 2002). The majority of the KPI literature on Six Sigma in services talks about financial benefits. Other

KPIs include expressions in terms of customer satisfaction and efficiency. Similar to CTQs, some KPIs are common across services. Some of the common KPIs are discussed below.

Table 6. Key performance indicator definitions

Author(s)	KPI Definitions
Hahn et al. (1999)	Performance metrics are established that directly measure the improvement in cost, quality, yield, and capacity.
Basu and Wright (2003)	KPIs are measurements of a performance such as asset utilization, customer satisfaction, cycle time from order to delivery, inventory turnover, operations costs, productivity, and financial results.
Antony (2006)	KPIs can be termed as performance metrics of Six Sigma.
ASQ Glossary	KPI is a statistical measure of how well an organization is doing in a particular area. A KPI could measure a company's financial performance or how it is holding up against customer requirements.

- **Efficiency:** Efficiency in a service industry means the timely delivery of services at a reasonable cost.

- **Cost reduction:** Cost can be reduced by eliminating waste, such as reducing errors or mistakes in a process or reducing the time taken to complete a task. A concrete example is to reduce a patient's stay at a hospital (Heuvel et al., 2005), which can provide opportunity for more admissions.

- **Time-to-deliver:** Like in manufacturing, the time to deliver a service determines organizational performance. Examples may be the timely delivery of information or document as per customer requirement.

- **Quality of the service:** Quality of the service is a measure of the extent to which the service delivered, meets the customer's expectations. This depends on two aspects; one is the technical aspect and another is functional aspect. The technical aspect is the actual outcome of the service encounter. Functional aspect is the interaction between the service provider and customer i.e. the service process (Ghobadian et al., 1994).

- **Customer satisfaction:** This factor is difficult to measure as it varies from service to service. For example, for a call center service, customer satisfaction is measured by the receipt of timely information. For a hospital, the comfort and assurance that patient feels may be the all important criterion (Sehwall and De Yong, 2003). Overall customer satisfaction can also be indicated by the retention rate of one's customer.

- **Employee satisfaction:** This is another intangible measure of organizational performance. Employee retention rate can be an excellent indicator of employee satisfaction. Financial benefits due to Six Sigma can provide employees with a means to visualize their contribution. This may increase employee morale and satisfaction (Henderson and Evans, 2000).

- **Reduced variation:** Statistical process control and Six Sigma refer to the reduction of variation through improved standards and consistency. In the case of services, variation reduction may be in terms of, for example, the cycle time of processing statements, or the decision cycle of a process (such as credit process in a bank) or the inaccuracy of a billing process and incorrect laboratory test results (such as in a hospital) (Sehwall and De Yong, 2003; Rucker, 2000).

- **Financial benefits:** The impact of Six Sigma on the bottom line is huge (Henderson and Evans, 2000). In comparison to success and failure as a measure, financial bottom lines are a better indicator of the impact of improvements as well as a vivid calibration of progress (Goh, 2002).

2.6. Six sigma in manufacturing and service organizations

Although different terms may be used, scrap and rework exist in services just as they do in manufacturing. Inconsistent and out-of-specification processes cost money to rework. Such examples in services may include the need to re-contact a customer to verify an order, providing an incorrect service, providing a substandard service, or even over-servicing or providing more than what is required. Some widely publicized success stories due to implementation in services include GE Medical Systems, Mount Carmel Health System, Virtua Health, GE Capital Corp, Bank of America, and Citibank. Limited application can also be found in call centers, human resources such as DuPont de Nemours (Bott et al., 2000; Wyper and Harrison, 2000) and in product support services such as by Caterpillar (Schmidt and Aschkenase, 2004).

The literature analysis also revealed that applications are limited mostly to service organizations in North America and Europe. Benefits-wise, these are mostly expressed in financial terms and not much is published about the benefits in process improvement terms. The literatures (Brady and Allen, 2006; Inozu et al., 2006; Mortimer, 2006; Antony et al., 2005a; Dudman, 2005; Goel et al., 2005; Hensley and Dobie, 2005; Basu, 2004; McAdam and Evans, 2004; Schimdt and Aschkenase, 2004; Hill and Kearney, 2003; Sehwall and De Yong, 2003; Rucker, 2000; Hahn et al., 1999; Harry and Schroeder, 1999; Paul, 1999) on Six Sigma application in manufacturing or services discuss mainly about CSFs, CTQs, KPIs and STTs. The following section provides an overview of these factors. Table 7 presents the similarities and differences of these between manufacturing and services on the basis of observations from the literatures.

The above table provides some important insights regarding Six Sigma implementation aspects in manufacturing and services. There are similar CSFs in manufacturing and services but their order of preference differs between two. This difference in order of preference can also be observed within the literature involving Six Sigma implementation in services. The paper by Antony (2004b) shows that linking Six Sigma to business strategy is the most important of success factors whereas some other literatures discuss that top management commitment is the most important one, followed by education and training (Johnson and Swisher, 2003; Henderson and Evans, 2000).

Table 7. CSFs, CTQs, KPIs, and STTs (manufacturing versus service)

Dimensions		Manufacturing	Service
CSFs		Top management commitment, education and training, cultural change, linking Six Sigma to customers, linking Six Sigma to business strategy, effective communication	Top management commitment, cultural change, clear performance metrics, customer focus, education and training, attaching the success to financial benefits, organizational understanding of work processes
CTQs	Similarities	Cycle time, cost of quality, machine or human error	
	Differences	Product performance characteristics such as, strength; weight, defects, poor packaging, breakage, defects, inventory reduction, product travel distance, poor packaging, quantity of rework, time spent in rework	Service time, waiting time, employee behaviour, responding to customer complaints, providing accurate and timely information to customers
KPIs		Cost savings, customer satisfaction, reducing variation, employee satisfaction, increasing productivity, product quality improvement	Efficiency, cost reduction, time to deliver, quality of the service, customer satisfaction, employee satisfaction
STTs	Similarities	Histogram, Pareto analysis, cause and effect analysis, brainstorming, flowchart, project charter, process mapping, root cause analysis, control charts	
	Differences	FMEA, DOE, SPC, gauge repeatability and reproducibility, measurement system analysis, regression analysis, QFD	

CTQs show similarities in terms of cycle time and cost. The concentration in manufacturing is more on product specifications/characteristics, inventory reduction, and reducing variation whereas services focus more on service time, waiting time, responding to customer, employee behaviour, etc. The reason for this difference can be because of more customer contact in services.

KPIs for both manufacturing and services show much similarity and are not much discussed in literatures. The application of tools and techniques has similarities in usage of flowcharts, process map, histograms, Pareto analysis, etc. The use of statistical tools and techniques such as SPC and regression analysis is more prominent in manufacturing may be because of ease of data collection and continuity of the process. The tools and techniques such as gauge repeatability and reproducibility is commonly used in manufacturing but not so in services, the reason is non-repeatable nature of service processes (Does et al., 2002).

2.7. Review summary

First, although the industry has an increased interest in Six Sigma implementation and many companies have gained the profits and advantages from this disciplined approach, the literature is limited and the research impacts of Six Sigma implementation and factors contributing to its success remain unclear. Many articles on the impact analysis of operations performance do not mention the detailed improvements in the operating areas, but focus on the overall bottom line impact. Therefore, it is necessary to do a deeper and more detailed study in this area.

Second, only a few articles were found that dealt with factors in the area of success factor analysis to Six Sigma implementation. Existing studies are not well integrated and current concepts in the field of Six Sigma are largely based on case studies, anecdotal evidences and are prescriptive in nature. Consequently there is little consensus on which factors are critical to the success of the approach (Nonthaleerak and Hendry, 2008; Brady, 2005). Most of the articles concentrated on few success factors and reported that top management commitment is the main factor to Six Sigma success (Goh, 2002; Henderson and Evans, 2000). However, many other factors affecting Six Sigma's success are important and need to be better documented.

To fill this gap, Antony and Baneulas (2002) identified 10 typical CSFs from their review of literature. Several others also provided sets of CSFs which have similarities or differences among them. It could be argued that this list of CSFs is comprehensive and that many of the issues are in common with those found for any implementation process, and are thus not specific to Six Sigma. However, all of the papers that identify these issues are descriptive in nature and there is a need to verify them through rigorous empirical research.

Finally, some authors have called for theoretic research (Nonthaleerak and Hendry, 2008; Schroeder et al., 2008; Oke, 2007; Brady and Allen, 2006), as too much research is focused only on description of practice rather than on theory development that is of use to practitioners as well as academics.

3. Research methodology

Management research is mainly based on deductive theory testing and positivistic research methodologies (Alvesson and Willmott, 1996). These methodologies incorporate a more scientific approach with the formulation of theories and the use of large data samples to observe their validity. However, these approaches mostly fail to give deep insights and rich data in Six Sigma practice within service organizations. Schroeder et al. (2007) state the need for more theory grounded and contingency based research rather than be restrictive to deductive approaches. Antony et al. (2007) and Nonthaleerak and Hendry (2008) emphasize this point by saying there is a paucity of systematic and rigorous evaluation in many Six Sigma studies.

In this section we describe the three phase approach for this study. First phase involved literature review and exploratory case studies. A small-scale questionnaire survey and 15 case studies were done in the second phase. The third phase included a large-scale questionnaire survey and further case studies. Questionnaire structure and design for each phase is also discussed. Then details about the measures are provided. Finally, we explain how we test the sample bias, which population is targeted, and how to proceed for the data collection.

3.1. Phase I – Macro study

This phase focused on providing the necessary breadth to produce an understanding of the implementation of Six Sigma in service organizations and from which reliable patterns and theories can be formed. Next phase of this research focused on the issues uncovered by the first.

The phase is termed as macro study (Leonard and McAdam, 2001), and it provides an overview not only of Six Sigma implementation in services, but also a database of critical success factors (CSFs), critical-to-quality (CTQ) characteristics, key performance indicators (KPIs), and set of tools and techniques (STTs).

The study included two services one is library and the other one is a call center. During this phase, interviews were conducted with a black belt, who was considered by the organization as most knowledgeable and responsible for Six Sigma implementation. The study concentrated on the implementation aspect of Six Sigma which involves CSFs, CTQs, KPIs, STTs, and also the difficulties faced. The BB provided an essential insight and understanding of Six Sigma implementation in service organizations. The other methods of data collection in this phase involved documentation and archival records.

Once the macro study is completed and insights are developed, preparation for next phase was done to focus on additional relevant questions that had arisen in phase one. This next phase involved a small-scale questionnaire survey and simultaneous case studies. The study included multiple respondents which overcame the problem of using single respondent in phase one. It also provided a degree of validation.

3.2. Phase II – Small-scale questionnaire survey and case studies

3.2.1. Small-scale questionnaire survey

At this phase a questionnaire survey of Singapore service organizations was conducted to understand the status of Six Sigma implementation. The survey was exploratory in nature as the objective was to gain insights about Six Sigma in service organizations. This kind of survey helps to uncover or provide preliminary evidence of association among concepts. Further, it can help to explore the valid boundary of a theory (Forza, 2002).

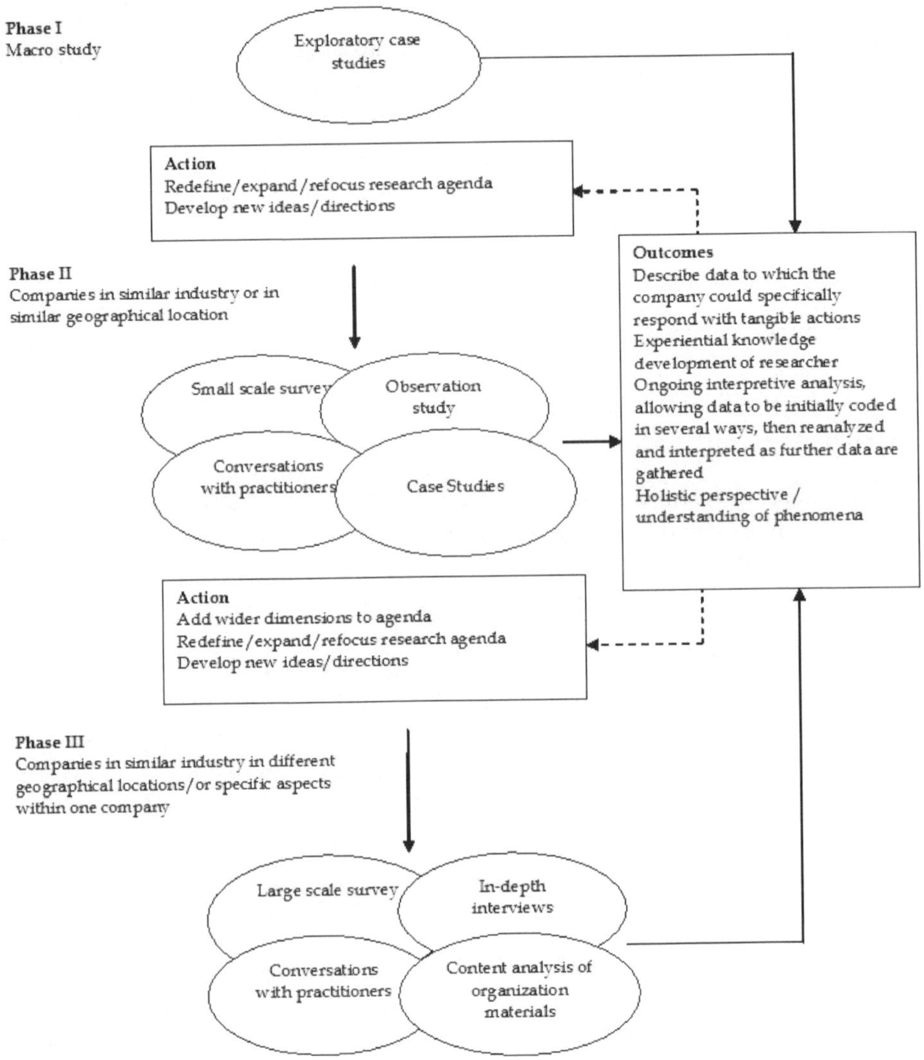

Figure 4. Three-phase approach (adapted from Leonard and McAdam, 2001; Gilmore and Carson, 1996)

Structure of the questionnaire

There are five parts in the questionnaire. The first part of the questionnaire is intended to get some general information of the respondent company, which includes the type of service organization, the size of the company, the type of company (local, multinational, or joint venture), whether they have quality department, there is a proper quality system in place, any business process improvement initiatives they are doing, and finally whether they have implemented Six Sigma. It is also designed as a filter to segregate the data based on service organizations which have or have not implemented Six Sigma.

The second part of the questionnaire attempts to identify the CSFs which are important while implementing Six Sigma in organization. The third part consists of two questions. First question is directed at identifying CTQs that are to be improved through Six Sigma implementation. Second question explores the tools and techniques used in Six Sigma DMAIC methodology and also in DFSS. The fourth part is focused on finding about KPIs while in the fifth part the objective is to identify the difficulties faced by service organizations in Six Sigma implementation. The fifth part is for those service organizations which have not implemented Six Sigma. There is one question in this part to explore about the reasons behind not implementing Six Sigma. The last part is designed to obtain background information on respondents including their name, job title, company, mailing address, phone/fax number, and e-mail. In order to share our findings with the respondents who are interested, we also left a space for them to tick whether they want to have the summary of our survey results.

Besides the six parts above, a cover letter with university letterhead explaining the aims and benefits of the research was designed.

Questionnaire design

The response format of the questionnaire is a major design consideration since this will alter the type and wording of the questions as well as focus on the type of analysis that the researcher wants to perform (Antony et al., 2007; Fowler, 2002; Kidder, 1986). For our research close-ended question format was considered since the data would be in a quantifiable form ensuring that statistical analysis can be used. Moreover, it is fast and easy to complete, enables automated data entry, and facilitates data analysis and summary of data (Antony et al., 2007; Fowler, 2002). The rating scale (Likert scale) and ranking used within this format is to obtain the answers from the respondents. The questionnaire focused on CSFs, CTQs, STTs, and KPIs as observed from the literature. CSFs, STTs, and KPIs are measured by a 5-point Likert-type scale (CSFs and KPIs: 1 = not important, 5 = very important; STTs: 1 = never, 5 = frequently). The Likert scale used provide a more precise measure than yes/no or true/false items and it is fast and easy to complete (Neuman, 2006). The rating scale used for few questions allows the respondents to indicate the relative importance of choices that facilitates the researchers in identifying the critical issues or factors (Antony et al., 2007).

Further, in the question content we intended to assure the respondents will be willing to answer honestly. To achieve this, personal information was not required across all the questions. The respondent profile which needed personal information was optional.

3.2.3. Case studies

The case studies in this phase focused on the critical issues that emerged from phase one. The case study was chosen as the research method primarily due to the nature of the research questions. Yin (1994) recommends this method as the most appropriate when contextual conditions are believed to be highly pertinent to the phenomenon of study. The case study method is also recommended when research questions embodies an explanatory component, such as in this study (how CSFs impact Six Sigma implementation in services?) (Yin, 1994).

Sample selection

We opted for an intricate sample design (Harrigan, 1983). This is a design where the sample is selected to coincide with sites that possess observable traits that are key factors in the propositions to be examined (Sousa and Voss, 2001).

The process for selecting individual service organizations was based on publicly available information and the respondents of small-scale questionnaire survey. From publicly available information, an initial list of 20 service organizations was compiled that were likely to comply with our research objective. 8 organizations which declined to participate in small-scale survey were removed from the initial list of 20, as they clearly mentioned about not revealing any data. We then started by contacting remaining 12 firms for participation in case study. At the end there were 3 firms which agreed to participate in the study. These 3 organizations were included for case study in next phase.

For second phase case studies we searched in public domain about the information available related to 17 organizations which were not interested for direct participation in the case study. We found Six Sigma implementation information related to 15 service organizations through different sources available. Finally second phase comprised 15 organizations which have completed around 29 projects between them for a period of 5 years, i.e. from 2003 to 2007.

Data collection procedure

A case study protocol was developed comprising a list of the research variables to address, and the respective questions, potential sources of information, and field procedures. Although data collection focused on the formal research variables, we also addressed other issues enabling us to understand the observed pattern of use of Six Sigma implementation, such as the history of use of its implementation and the difficulties experienced by the service organizations in using Six Sigma. Several data collection methods were used in both the phases including semi-structured interviews, direct observation, and secondary data.

Documentation

Evidence for case studies can be obtained from various sources such as documents, archival records, interviews, direct or participant observation, and physical artefacts. In this phase of the study the data collection is based on documentary evidences, which helps in providing specific details to corroborate information and also inferences can be made from documents (Yin, 1994). The documents considered for this study is in the form of articles, interviews, and speeches published in journals, magazines, newspapers, and websites.

The first two phases of case studies were devised to provide a breadth of data and understanding of the Six Sigma implementation in service organizations. Along with the small-scale survey the second phase further enhanced the database on different aspects of Six Sigma implementation. The third phase will provide more rich and deep data, which involve a large scale questionnaire survey and case studies with in-depth interviews.

Phases one and two are specific and as detailed and rich in data as the third phase, but they are limited in time, access, and practitioner involvement. Thus, specific areas of inquiry could be examined, but a true behind the scenes, and multi-faceted picture and understanding could not be provided. To provide such an understanding, in-depth case studies and large-scale survey were needed that would allow a significant access to different managerial levels and inputs from the use of different research techniques. This constituted the third phase where case studies ran parallel with the second phase.

Unit of Analysis

Except for single case versus multiple-case design possibilities, one can also distinguish a case design separating and choosing between a single unit of analysis and multiple unit of analysis, see Yin (1994). In the literature, unit of analysis refers to a great variety of objects of study, for example, a person, a program, an organisation, a classroom or a clinic (Mertens, 1998), or a community, state or nation (Patton, 1987). Other authors have considered the unit of analysis as interviews or diaries in their entity, and the amount of space allocated to a topic or an interaction under study (Downe-Wamboldt, 1992).

For case studies, the overarching unit of analysis was the Six Sigma projects but there are sub-units that were investigated in order to reveal the main unit as realistically as possible. These sub-units are the experiences from different expertise (belt levels), the difficulties faced, the tools and techniques used, which also counts for the opinions among the actors involved in the implementation work. The reason behind choosing different expertise is because of different roles of black belt (BB) and green belt (GB) in Six Sigma project. BBs are the project leaders who are responsible towards project management while GBs are involved in data collection and analysis process. Following the experiences from different expertise will help in understanding the concerns from different levels, about Six Sigma projects.

Therefore, during the case studies different expertise in Six Sigma are chosen, which indicates that the chosen research design is an embedded multiple-case design. The replication does not necessarily mean that each case study needs to be either holistic or embedded (Hansson, 2003). The individual cases, within a multiple case study design may be either. When an embedded design is used, each individual case may include the collection and analysis of high quantitative data including the use of surveys within each case (Yin, 1994). During this study, each individual case in the multiple-case design represented an embedded design. This unity between the individual cases was chosen in order to discover possible differences between the respondent groups, different levels of expertise and experience in Six Sigma implementation, as they may not share the same experience from an implementation process.

3.3. Phase III – Large-scale questionnaire survey and case studies

The small-scale questionnaire survey helped in understanding the status of Six Sigma in service organizations. It also highlighted certain issues which are required to be studied further in order to develop the theory. The next step is to conduct a large scale questionnaire survey and further case studies.

3.3.1. Case studies

The case study organizations were selected based on the idea of theoretical sampling. In case of theory building, theoretical sampling is preferable in comparison to generalizability concept in statistical studies. So, the cases are chosen for theoretical rather than statistical reasons (Schroeder et al., 2007; Eisenhardt, 1989).

The case studies in this phase involved three different service organizations and provided an opportunity for detailed understanding of Six Sigma implementation. After the second phase overall 8 organizations were contacted, 3 of which agreed to participate in the study. Similar to second phase, a range of data collection methods which include participant observation (e.g., organization tours), formal interviews, and review of company documents and archives, were used in this phase. This allowed a multi-perspective view on Six Sigma implementation in service organization.

Data collection procedure

Interviews

We conducted structured interviews with all the informants. In case of the four Staff Nurses for the healthcare service organization the questions were mainly towards their experience about the current projects, because of their limited knowledge of Six Sigma. The structured questionnaire involved questions on the Six Sigma initiative, project selection, Six Sigma implementation process, and the learning experience. As a part of Six Sigma initiative, we asked the informants about the reason they prefer Six Sigma over other initiatives, how the preparations were done to implement Six Sigma, and what was their approach to training personnel for Six Sigma. In project selection the informants were asked about the criteria of selection for the projects, factors involved in success of a project, and reasons behind unsuccessful projects. For the process of Six Sigma implementation, the questions are about their considerations on CTQs, tool and techniques used at different phases of DMAIC, selection criteria of STTs, and KPIs. We also asked the informants about their learning experience on the basis of Six Sigma's relevance to their organization, problems faced during the implementation process, and how they overcame those problems.

In total there were 10 interviews; 6 formal and 4 informal. All formal interviews were taped, transcribed, and coded. The list of interviewees is provided in Table 8.

Table 8. List of interviewees

Organization	Interviewee	Designation
Hospital	I1	Director, Human Resource
	I2	Head of Department
Construction and Related Engineering Service	I3	Assistant Director
	I4	Senior Development Officer
Consultancy	I5	Consultant
	I6	Building Manager

Having already established a database from the literature review, initial questionnaire survey, and exploratory case studies on different aspects of Six Sigma implementation in service organizations, phase three interviews were more focused and directed.

Participation in projects

In the case study of healthcare service organization, I was also involved in as a team member for two Six Sigma projects. This provided an opportunity to develop a partnership which lasted for six months. Combining retrospective and longitudinal study; as done in case of healthcare service organization for the study enhances construct, external, and internal validity (Barton, 1990). Moreover, this type of partnership in grounded theory research helps in observing phenomenon development and to develop framework from the collected data (Leonard and Mc Adam, 2001). Approaching the interviewees was not a problem, as I was visiting the organization on regular basis. This helped in getting completed answers on all questions and returning at a later date to seek clarification to questions that arose. For the other two organizations though there is no participant involvement, but the interviewees were approachable when required. Overall, interviewees represent different level in terms of experience and expertise with Six Sigma; this helps to avoid a bias or unqualified opinion which can be a problem in single respondent study (Nonthaleerak and Hendry, 2008; Voss et al., 2002).

My involvement in one of the case studies helped in observing changing attitudes towards Six Sigma and the development of the project. These observations included the challenges and issues involved in Six Sigma projects in service organizations. It allowed a more detailed history of Six Sigma implementation in the organization to be plotted, with wide access to documentation providing a clear picture of the reasoning for Six Sigma adoption, CSFs, CTQs, selection of tools and techniques, and the difficulties faced. Therefore, the case study research included an element of ethnography as what was being attempted was to learn the implementation of Six Sigma, and not only to accept or listen to the views articulated but also to actively use those views in discussion.

Documentation

The documentary evidences for information about these case studies is gathered through various sources which include websites such as Singapore Government website (PS21–Public Service for 21st Century), articles, interviews and speeches from newspapers, magazines, and journals. Other sources of data are the reports and presentations of the completed projects.

This third phase of research was being carried out at the same time as the studies in phase two. The issues emerging from the previous phases were brought over and examined in the three case studies. Though, the specific questions raised during previous phases could not be specifically answered by these three case studies but wider issues that were replicated throughout were examined. Thus, these three case studies allowed greater detail and more intricate issues to be dealt with.

This phase also involves a large-scale questionnaire survey by focusing on companies which operate in different geographical locations, following the integrative approach suggested by Gilmore and Carson (1996). Combining such compatible and complimentary methods provide depth, breadth, and subtlety of information to the study (Carson and Coviello, 1996). This survey is done concurrently with the case studies and the responses from it further strengthen the development of conceptual framework for Six Sigma implementation in service organizations.

3.3.2. Large-scale questionnaire survey

Structure of the questionnaire

There are six parts in the questionnaire. The first part of the questionnaire is intended to get some general information of the respondent company, which includes the type of service organization, the size of the company, the type of company (local, multinational, or joint venture), whether they have quality department, there is a proper quality system in place, any business process improvement initiatives they are doing, and finally whether they have implemented Six Sigma. It is also designed as a filter to segregate the data based on service organizations which have or have not implemented Six Sigma. The second part of the questionnaire attempts to identify the CSFs which are important while implementing Six Sigma in organization. The third part consists of four questions. First two questions are related to CTQs. One is asking about definition of CTQ and the other is to identifying CTQs that are to be improved through Six Sigma implementation. Third question explores the tools and techniques used in Six Sigma DMAIC methodology and last one is about their selection criteria. The fourth part focused on finding about KPIs and its definition while in the fifth part the objective is to identify the difficulties faced by service organizations in Six Sigma implementation. The last part is for those service organizations which have not implemented Six Sigma. There is one question in this part to explore about the reasons behind not implementing Six Sigma.

Besides the six parts above, the web-based respondents were sent an introductory letter and follow-up letter by e-mail. In each e-mail, the targeted person was directed to a specific web page address posted on the university internet server, where the survey was presented. After completing the survey and pressing a Submit button, the responses were automatically saved on the internet server with a date and time stamp.

Questionnaire design

Following Gilmore and Carson's (1996) integrative approach, we focused on survey of service organizations in different geographical locations. Web-based surveys are one of the most preferred methods when data collection is to be done from organizations spread world-wide. They have several advantages over other collection methods, such as low cost, broader distribution, potentially higher response rates, faster survey turnaround time times and high selectivity (Coderre and Mathieu, 2004; Boyer et al., 2002; Klassen and Jacobs, 2001).

The design of the survey web page was similar to hard-copy survey. Like a paper survey, the respondents can scroll through questions in a particular section and also browse through the questions in other sections without any restrictions. They could also answer questions in any order and could complete the survey in several sessions. In terms of appearance user friendly features was designed (e.g. radio buttons, check boxes, scrollable dialog boxes, etc. where appropriate, given the nature of the question) to speed completion of survey (Kalssen and Jacobs, 2001; Dillman, 1999).

Similar to small-scale questionnaire survey, here also in the question content we intended to assure the respondents will be willing to answer honestly. To achieve this, there was no requirement for personal information in any of the questions. The respondent profile which needed personal information was optional.

4. The framework consolidation

The framework involves three sections. First is CSFs, followed by Six Sigma implementation, and bottom-line result. The Six Sigma implementation section consists of CTQs or measurable process parameters, DMAIC methodology, and STTs.

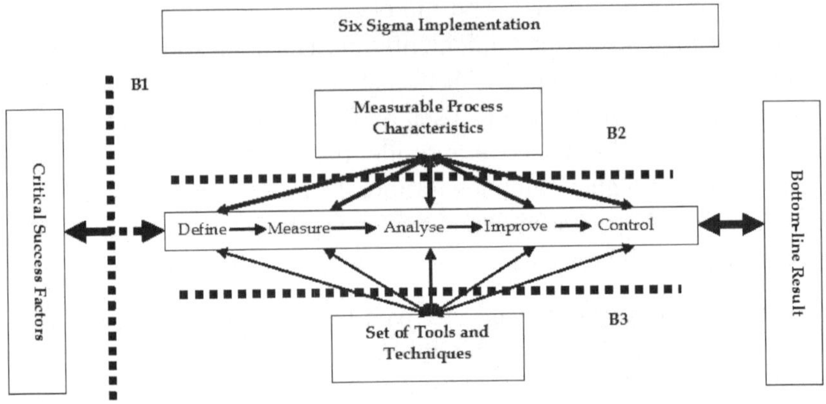

Figure 5. Conceptual framework for Six Sigma implementation in service organizations

The framework evolved on the basis of data collected through surveys, case studies, and continuous referrals with the literature. The initial framework developed was a kind of assessment model, i.e. in auditing role instead of an evaluative model to understand the dynamics of Six Sigma implementation in service organizations (Leonard and McAdam, 2001). There are works by Senapati (2004) and more recently by Parast et al. (2007) towards developing an assessment model for Six Sigma implementation. Though, both are descriptive in nature and lacks rigorous empirical research to support their applicability. We rather feel that a fuller understanding of dynamics of Six Sigma in service organizations, using suitable frameworks, will assist organizations to evaluate and predict the current and potential business benefits from Six Sigma implementation. Also, methods of reenergizing and directing Six Sigma efforts will be more specific and accurate. Furthermore there is a paucity of research literature in this area. Thus, our research study adds to the body of knowledge. The modified framework is presented in Figure 5.

4.1. Critical Success Factors (CSFs)

The idea of identifying CSFs as a basis for determining information needs of managers was popularized by Rockart (1979). In the context of Six Sigma implementation, CSFs represent the essential ingredients without which the initiative stands little chance of success (Antony et al., 2007). Based on this discussion and our findings from literature and data collected through surveys and case studies, CSFs are included in our framework. There are various CSFs identified but we feel only a few are essential as observed from the views of respondents from surveys and case studies.

First and foremost of CSF is top management commitment and involvement. Once top management buys in the decision to implement Six Sigma, they also have to involve themselves to ensure success of the program. This is highlighted by the respondents during interview sessions. They feel occasional involvement of top management during team meetings will motivate the team members and this will also help in solving certain problems which the team members cannot solve at their levels.

Next is support of team members. Since Six Sigma implementation requires project teams so, proper coordination and support between team members is an important aspect. Further in case of service organizations, the projects are done part-time so involvement of each member in team meetings becomes very important to keep everyone well versed about the project. This will also ensure timely completion of the project.

Linking Six Sigma to business strategy is another CSF which is mentioned both in surveys and case studies. This is important as there has to be alignment between the Six Sigma projects and company objectives as mentioned by one of the interviewee of consultancy service organization. This also ensures top management commitment towards Six Sigma program.

The surveys also highlighted two CSFs which are customer focus and education and training. Education and training on Six Sigma will be useful as it will help employees of service organization overcome fear on the use of rigorous statistical and quality tools and techniques (Nonthaleerak and Hendry, 2008).

4.2. Critical-to-Quality (CTQ) characteristics

We also propose definition for CTQ to have a clear distinction with KPI. CTQ can be defined as product or service process characteristic derived from critical customer requirements whereas KPI as mentioned is more specifically performance metric. The following example (adapted from Frings and Grant, 2005) will help in understanding about CTQ clearly.

In a call centre scenario: Customer quote: "I consistently wait too long to speak to a representative" CTQ definition: Representative responsiveness; CTQ measure: Time on hold (seconds)

So in order to reduce the ambiguity between terms we use only CTQ which include both CTQ definition and CTQ measure. Since CTQ is actually process characteristics so for a clear understanding, in the framework we mention it as measurable process characteristics.

The CTQs or measurable process characteristics which are important from service organization's perspective are time, cost, and quality. The study shows that most of the Six Sigma projects associated with service organizations are concerned with reduction in time. From our analysis of service strategy context we found that cycle time is an important CTQ for mass service organizations whereas waiting time is critical for professional service organizations. Reduction in cost is concerned with cost of transaction and quality is related to improved accuracy in information provided to customer or improved reliability of service systems, etc. The most important aspect related to measurable process characteristics is our finding that it is context dependant. As our research showed that importance of process parameters vary across service types. So to overcome the barrier of identification of process parameters it will be useful to position service organizations as professional service, service shop or mass service.

4.3. Key Performance Indicators (KPIs)

It is observed that there is ambiguity about KPI. It is often synonymously used with CTQ. The practitioners feel it more as key process input/output variable rather than key performance indicator. Key performance indicator is more like performance metric as mentioned in some literature and is strategic in nature. Table 9 provides the definitions identified from our study. Majority of these definitions are related to organization strategy.

It can be observed from the above table that KPI is significant when interpreted in terms of overall organization strategy rather than specific to Six Sigma strategy. Another interpretation of it being similarity with CTQ makes us to think of a uniform and clearly understandable term instead of two different terms. We feel in this scenario the term measurable process characteristic mentioned in our framework definition can overcome this problem.

Table 9. Definitions of KPI from large-scale survey

Strategic Oriented
KPI's is the chosen indicators to control a process and used to have data when a decision is required. KPIs are agreed in an organization as the measurement points without discussion.
The metrics that help guide the organization in the right directions. Tells you if you are succeeding in your goals.
We believe that excellence in service will be the nucleus for all actions and decisions. So the KPIs are a measure for our excellence in service and customer support. KPIs are based on SMART targets.
KPI is designed based on the results required and in line with the Company strategy.
We use KPIs to measure how we are doing on our strategic and financial goals.
KPIs are set up to indicate organization/department goals, set up dashboards and score cards w/ baseline metrics and monitor the performance based on key indicators.
Quantifiable indicators to measure the fit for purpose and efficiency of the organisation.
Process Oriented
Same as CTQs.
Key performance inputs.
KPI - key measure that are critical to evaluate the performance of a product/process.

In case of Six Sigma, financial benefits or bottom-line result is the most common performance metric (Goh, 2002). So, in the conceptual framework instead of key performance indicator, bottom-line result is included as the main outcome.

4.4. Set of Tools and Techniques (STTs)

In general Six Sigma projects utilize a number of tools and techniques in different phases of DMAIC methodology. The service organizations utilize specifically lesser number of tools and techniques compared to manufacturing (Antony et al., 2007). Our study observed that generally organizations mention a number of tools and techniques applicable in Six Sigma implementation but closer analysis showed that actually the number is quite small. We also found that organizations which have limited success and progress with Six Sigma so far use more number of tools and techniques in comparison to those service organizations which have moderate progress and success. Based on our findings we provide a set of tools and techniques which can act as a guide and provide better advice to those attempting to implement Six Sigma in service organizations. Table 10 provides STTs applicable to specific DMAIC phases for Six Sigma implementation in service organizations.

4.5. Difficulties or barriers in Six Sigma implementation

We also observed that literatures were mainly talking about difficulties or obstacles in Six Sigma implementation in service organizations. The reasons cited such as the inherent differences between services and manufacturing, or differences in application of tools and techniques (Antony et al., 2007; Hensley and Dobie, 2005; Antony, 2004b; Benedetto, 2003).

Table 10. STTs for service organizations

Define	Project charter, Brainstorming, Flowchart, Process map, Project management
Measure	Cause and effect diagram, Pareto analysis, Brainstorming, Check sheet, Histogram, Normal probability plot, Flowchart, Matrix diagram, Work flow diagram, Project Management, Process capability analysis
Analyse	Cause and effect diagram, Pareto analysis, Brainstorming, Histogram, Normal probability plot, Flowchart, Matrix diagram, Work flow diagram, Project management, Analysis of variance, Root cause analysis, Process capability analysis, Descriptive statistics
Improve	Brainstorming, Flowchart, Check sheet, Decision matrix, Project management
Control	Control chart, Project management

There are also views that Six Sigma will not work for every service processes, and adjustments may be required for it to suit even for those processes for which it does apply (Biolos, 2002). Like much of the literature in this area these obstacles discussed are descriptive and does not involve empirical studies for support. Building-up on this gap and based on our data collection we identified difficulties faced by service organizations during Six Sigma implementation and included them in our framework. These obstacles goes beyond the inherent differences between service and manufacturing, are practical problems faced by the organizations which may or may not be specific to a certain service organization.

One is between CSFs and Six Sigma implementation. This barrier includes lack of support from team members, resistance to change, long-term sustaining of Six Sigma, attaching incentives to successful Six Sigma projects, staff turnover, and lack of support from employees not involved in Six Sigma project. The second barrier is between measurable process characteristics and the methodology. This includes difficulty in identifying process parameters, difficulty in collecting data, and time consuming effort in collecting data. The third barrier is between STTs and the methodology. This barrier includes difficulty in identifying proper STTs, some tools and techniques are too complex to use and requires more time to learn.

4.6. Summary

The framework developed on the basis of grounded theory methodology, is an attempt to understand the aspects of Six Sigma implementation and performance in service organizations. The study contributes to Six Sigma knowledge through development of theory and building a prescriptive model to advice both managers and scholars attempting to implement or study Six Sigma in service organizations. The framework provides a set of CSFs, measurable process characteristics, and tools and techniques which will act as a guide and also overcome the difficulties or barriers in Six Sigma implementation in service organizations. The strength of our study is coming out of the service versus manufacturing differences paradigm and highlight the practical difficulties faced by service organizations in Six Sigma implementation.

5. Conclusion

Using qualitative analysis technique, we are able to find empirical support for critical success factors, measurable process parameters, tools and techniques, bottom line result and difficulties in Six Sigma implementation in service organizations. There exist different configurations of Six Sigma implementation for discriminating between high and low performance depending on the significance attributes to performance dimensions. This is in line with the systems approach to fit that upholds the criticality of the internal consistency of each design and match between the structural patterns of practices to the contingencies facing the organization. We are able to show which are the CSFs required for Six Sigma implementation at project level and how these CSFs can help in overcoming barriers observed at different phases of Six Sigma projects in service organizations. The research also highlights a set of tools and techniques used in Six Sigma projects and also explain the selection criteria for these STTs. One of the important developments is related to the interpretation of the term KPI related to Six Sigma at project level. KPI is better understood as key process input or output variables at Six Sigma project level and key performance indicator at strategic level in an organization. Last but not the least, understanding of the practical problems faced by service organizations during Six Sigma projects is a major contribution of our research since we feel this was one of the important missing links in existing literature.

5.1. Contribution to research process

We hope that this study encourages investigation of Six Sigma implementation in service organizations and promote rigorous development and explicit articulation of theories. It is necessary to increase theory development related to Six Sigma implementation that is grounded on relevant established theories and empirical evidence from related disciplines. So that empirical investigations of related phenomenon can be integrated into the building and modification of useful and interesting theories.

This study demonstrates the value of methodological triangulation in the development of framework and theory of Six Sigma implementation in service organizations using literature review, surveys, and case-based research. The use of different methods of investigation provides complimentary assessment of the same issues and brings out salient details that cannot be obtained by a single method of analysis.

The case-based research draws attention to the existence of contingencies and the need to further investigate the ambiguous role of contextual factors in affecting Six Sigma implementation in service organizations. Studies by Nonthaleerak and Hendry (2008); Schroeder et al. (2008); Antony (2004) prescribes that Six Sigma can be implemented in service organizations, our study suggests that the implementation and impact of Six Sigma can be affected by contextual factors such as service types.

In summary, this research contributes to theory-grounded empirical research. This is a worthwhile endeavour because contributions to valid and reliable measurements and explicit theory development help lay a foundation for future Six Sigma implementation studies. By identifying and testing theories we encourage the development of a stream of cumulative research.

5.2. Contribution to practice

This study offers conceptual clarity and specificity on Six Sigma implementation in service organizations, managers can use a guideline for choosing the fundamental practices that they can implement. We provide conceptual and empirical evidence on CSFs, measurable process characteristics, STTs, and difficulties faced by service organizations, encouraging managers to plan and implement Six Sigma with a systematic view of service environment. Furthermore, there is empirical evidence of the importance of committed and involved leadership in the implementation of Six Sigma in service organizations. We also find that a general emphasis on company-wide Six Sigma projects is significant in differentiating high and low performance.

5.3. Final discussion and future directions

The study has shown that there is a relationship between successful Six Sigma implementation and financial performance. The study also reveals that there are common features of the implementation of process of Six Sigma in service organization context. However there are still several areas that require further investigation related to these findings.

5.3.1. Six sigma implementation and success, progress, and service types

Six Sigma and organizational success, progress can be further studied based on the specific service types. The studies can involve the organizations included in the investigation, in order to study whether the advantageous financial performance of Six Sigma projects also holds in a longer perspective. Advantageous financial performance might be considered a major encouragement for commitment and motivation among employees and management. Since their commitment and involvement is vital areas for sustaining Six Sigma, see Goh (2002), maintained advantageous financial performance is vital for the future progress of Six Sigma. Furthermore one could include other organizations from individual service types, which have won quality awards or reached a certain level in the assessment, in order to enlarge the empirical foundation and further outline how different levels of Six Sigma implementation affect financial performance.

Additionally, an investigation aiming at exploring major Six Sigma achievements, e.g. increased customer satisfaction, reduction in cycle time or waiting time, among organizations that successfully implemented Six Sigma and studying their link to financial figures, could further explore the relationship between Six Sigma and success and progress. Also, a study of the major areas of costs when implementing Six Sigma compared with possible gain, and put in a relation to financial benefits after implementation, will possibly add supplementary information important for facilitating the understanding of the relationship between Six Sigma and its success and progress.

Furthermore, investigations based on individual service types and the effect on success and progress would further complement the findings presented within the framework of this thesis.

5.3.2. Six sigma implementation and service organizations

When considering the process of Six Sigma implementation in service organizations, several interesting opportunities could be mentioned for expanding the findings of the study. One appealing approach would be to do a longitudinal study in one or several service organizations that intend to start Six Sigma implementation in order to follow the implementation process in a more detailed manner and without being forced to totally rely on historical and personal information. A major problem with such a study might be that the outcome of the implementation efforts is not necessarily successful, i.e. the researcher will not know at the beginning that the study will investigate a successful Six Sigma implementation. If the studied organization(s) do not succeed in implementing, the findings may outline problems and reasons although not as reliable implications for successful implementation as the findings could have resulted otherwise.

Several core values of Six Sigma focus to a large extent on intangible factors related to e.g. support of team members, education and training, and top management issues. At the same time many of the concrete components, e.g. process parameters, tools and techniques, are more focused on tangible factors, of which some tools and techniques are statistical in nature and are not readily acceptable in service organizations. By making studies, with an increased focus on how service organizations which have implemented Six Sigma, address and develop intangible factors, and linking the findings to a further developed version of the implementation framework presented in chapter 7, an implementation framework even more adapted to service organizations could be created.

On the other hand, although the service organizations studied have implemented Six Sigma using CTQs, and STTs, with a focus on intangible factors, it is also very important to help service organizations to introduce and use different statistical tools and techniques to support and facilitate the handling and control of variation in process parameters in different ways. An interesting area is therefore, how to support service organizations' use of statistical tools and techniques. An approach might be to focus on a specific branch or sector in the service organization context. This will help to build a more specific background of the service organizations' characteristics within the chosen frame. By making such a study the specific characteristics of the included service organizations could be more accurately put in relation to their implementation process. Consequently, an increased consideration of contextual issues might be obtained.

Finally, the data collected for this study used the key informant approach (Bonner et al., 2002; Kumar et al., 1993). Therefore, all conclusions should be interpreted with this possible bias in mind. In addition although the reviewers in the pre-test did not find the survey questionnaires difficult to do, it was found that some of the questions are difficult to understand by some of the respondents. It was likely that the respondents who answered the surveys were more interested in Six Sigma than the non-respondents. Future studies with multiple respondents are recommended in which respondents come from different seniority and functional areas.

5.4. Concluding remarks

Six Sigma as a quality management practice is gaining importance in service organizations. Literature review shows that Six Sigma is mainly implemented in healthcare and banking service organizations. There is limited literature exploring Six Sigma implementation in service organizations and lacks rigorous empirical approach. Our study findings suggest Six Sigma implementation in different service organizations such as information technology, transportation, utilities, etc. Further most of these organizations are in moderate success and moderate progress category regarding Six Sigma implementation. On the basis of service types, it is found that most of the organizations are mass services.

Exploration of Six Sigma aspects in service organizations showed the importance of top management commitment and involvement along with some other CSFs. It is also observed that CSFs, CTQs, and STTs are to a certain extent depends on service types. There is some variation in CSFs, CTQs, and STTs across service types. The use of tools and techniques showed that successful organizations use limited number of tools in comparison to less successful organizations. One of the most significant finding is about KPI. The terms interpretation is best understood from strategic viewpoint. From the perspective of Six Sigma KPI is similar to CTQ and can be interpreted as process parameter. Another finding is about the difficulties faced in Six Sigma implementation by service organizations, which shows rather than the difficulty of data collection; part-time involvement, extension of project timeline, and staff turnover during projects or after training are the major difficulties. Unknown to us as a reason for not implementing Six Sigma prompts us to further understand the unique nature of service organizations and provide a customized approach to Six Sigma implementation in service organizations.

References

1. AlvessonM.WillmottH.1996Making sense of management, Sage Publications: London.
2. AntonyJ.2002Design for Six Sigma: A breakthrough business improvement strategy for achieving competitive advantage. Work Study, 51168
3. AntonyJ.2004aSome pros and cons of Six Sigma: An academic perspective. The TQM Magazine, 164303306
4. AntonyJ.2004bSix Sigma in the UK service organizations: Results from a pilot survey. Managerial Auditing Journal, 19810061013
5. AntonyJ.2006Six Sigma for service processes. Business Process Management Journal, 122234248
6. AntonyJ. .AntonyF. J. .KumarM.ChoB. R.2007Six Sigma in service organizations: Benefits, challenges and difficulties, common myths, observations and success factors. International Journal of Quality and Reliability Management, 243294311
7. AntonyJ.BanuelasR.2002Key ingredients for the effective implementation of Six Sigma program. Measuring Business Excellence, 642027
8. AntonyJ.CoronadoR. B.2002Design for Six Sigma. IEEE Manufacturing Engineer, Feb, 2426
9. AntonyJ. .KumarM.MaduC. N.2005Six Sigma in small and medium-sized UK manufacturing enterprises: Some empirical observations. International Journal of Quality and Reliability Management, 228860874

10. BartonD. L.1990A dual methodology for case studies: Synergistic use of a longitudinal single site with replicated multiple sites. Organization Science, 13248266
11. BasuR.2004Implementing quality: A practical guide to tools and techniques: Enabling the power of operational excellence, 1st Ed, Thomson Learning: London.
12. BasuR.WrightJ. N.2003Quality beyond Six Sigma, Butterworth Heinemann: Oxford.
13. BenedettoA. R.2003Adapting manufacturing-based Six Sigma methodology to the service environment of a radiology film library. Journal of Healthcare Management, 484263280
14. BiolosJ.2002Six Sigma meets service economy. Harvard Management Update, Nov, 35
15. BolzeS.1998A Six Sigma approach to competitiveness. Transmission and Distribution World, 50818
16. BonnerJ. M. .RuekertR. W.WalkerO. C.Jr 2002Upper-management control of new product development projects and project performance. Journal of Product Innovation Management, 193233245
17. BottC. .KeimE. .KimS.PalserL.2000Service quality Six Sigma case studies. ASQ's 54th Annual Quality Congress Proceedings, 225231
18. BoyerK. A. .OlsonJ. R. .CalantoneR. J.JacksonE. C.2002Print versus electronic surveys: A comparison of two data collection methodologies. Journal of Operations Management, 20357373
19. BradyJ. E.2005Six Sigma and the university: Teaching, research and meso-analysis. Ph.D Dissertation, Ohio State University: US.
20. BradyJ. E.AllenT. T.2006Six Sigma literature: A review and agenda for future research. Quality and Reliability Engineering International, 22335367
21. CarsonD.CovielloN.1996Qualitative research issues at the marketing/ entrepreneurship interface. Marketing Intelligence and Planning, 1465158
22. CaulcuttR.2001Why is Six Sigma so successful. Journal of Applied Statistics, 28Nos. 3 & 4, 301306
23. CoderreF.MathieuA.2004Comparison of the quality of qualitative data obtained through telephone, postal and e-mail surveys. International Journal of Market Research, 46Quarter 3, 347357
24. CookD. P. .GohC. H.ChungC. H.1999Service typologies: A state of the art survey. Production and Operations Management, 83318338
25. CoronadoR. B.AntonyJ.2002Critical success factors for the successful implementation of Six Sigma projects in organizations. The TQM Magazine, 1429299
26. De FeoJ. A.Bar ElZ.2002Creating strategic change more efficiently with a new Design for Six Sigma process. Journal of Change Management, 316080

27. De KoningH.De MastJ.2006A rational reconstruction of Six Sigma's breakthrough cookbook. International Journal of Quality and Reliability Management, 237766787

28. DillmanD. A.1999Mail and internet surveys: The tailored design method. 2nd Ed, J. Wiley: New York.

29. Dimock P.V.1977Engineering and Operations in the Bell System, Bell Telephone Laboratories, Inc.

30. DoesR. .HeuvelE. .MastJ.BisgaardS.2002Comparing non-manufacturing with traditional applications of Six Sigma. Quality Engineering, 151177182

31. Downe-WamboldtB.1992Content analysis: Method, applications, and issues. Health Care for Women International, 133313321

32. DudmanL.2005Big improvements for small parts, Quality Progress, Dec, 6772

33. EisenhardtK. M.1989Building theories from case study research. Academy of Management Review, 144532550

34. El -HaikB.RoyD. M.2005Service design for Six Sigma: A roadmap for excellence. John Wiley and Sons, Inc., Hoboken: New Jersey.

35. FerryantoL.2005DFSS: Lessons learned, ASQ Six Sigma Forum Magazine, Feb, 2428

36. ForzaC.2002Survey research in operations management: A process-based perspective. International Journal of Operations and Production Management, 222152194

37. FowlerF. J.2002Survey research methods. 3rd Ed., Thousand Oaks, Sage Publications: CA.

38 GhobadianA. .SpellerS.JonesM.1994Service quality: Concepts and models. International Journal of Quality and Reliability Management, 1194366

39. GilmoreA.CarsonD.1996Integrative qualitative methods in a services context. Marketing Intelligence and Planning, 1462126

40 GoelS. P. .GuptaP. .JainR.TyagiR. K.2005Six Sigma for transactions and service. McGraw-Hill: New York.

41. GohT. N.2002A strategic assessment of Six Sigma, Quality and Reliability Engineering International, 18403410

42 HahnG. J. .HillW. J. .HoerlR. W.ZinkgrafS. A.1999The impact of Six Sigma improvement- A glimpse into the future of statistics, The American Statistician, 533208215

43 HanssonJ.2003Total Quality Management- Aspects of implementation and performance: Investigations with a focus on small organizations. Ph. D Dissertation, Lulea University of Technology: Sweden.

44. HarriganK.1983Research methodologies for contingency approaches to business strategy. Academy of Management Review, 83398405

45. HarryM.SchroederR.1999Six Sigma: The breakthrough management strategy revolutionizing the world's top corporations. 1st Ed., Random House Inc.: New York.

46. HarryM. J.2000Six Sigma: A breakthrough strategy for probability, Quality Progress, May, 6064

47. HendersonK. H.EvansJ. R.2000Successful implementation of Six Sigma: Benchmarking General Electric company, Benchmarking: An International Journal, 74260281

48. HendryL.NonthaleerakP.2005Six Sigma: Literature review and key future research areas. LUMS Working Paper Series, Jun, 166

49. HensleyR. L.DubieK.2005Assessing readiness for Six Sigma in a service setting, Managing Service Quality, 15182101

50. HeuvelJ. .DoesR.BisgaardS.2005Dutch hospital implements Six Sigma, ASQ Six Sigma Forum Magazine, Feb, 1114

51. HillW. J.KearneyW.2003The Honeywell experience, ASQ Six Sigma Forum Magazine, Feb, 3437

52. HoerlR.2004The future of Six Sigma, ASQ Six Sigma Forum Magazine, Aug, 3843

53. InozuB. .NiccolaiM. J. .WhitcombC. A. .MacClaren. B. .RadovicI.BourgD.2006New horizons for ship building process improvement, Journal of Ship Production, 2228798

54. JohnsonA.SwisherB.2003How Six Sigma improves R&D, Research Technology Management, Mar, 1215

55. JohnsonK.2005Six Sigma delivers on-time service, Quality Progress, Dec, 5759

56. JohnstonR.1999Service operations management: return to roots, International Journal of Operations and Productions Management, 192104124

57. Jones JrM. H.2004Six Sigma: At a bank, ASQ Six Sigma Forum Magazine, 321317

58. KidderL. H.1986Research methods in social relations. 5th Ed., Holt, Rinehart and Winston: New York.

59. KlassenR. D.JacobsJ.2001Experimental comparison of web, electronic and mail survey technologies in operations management, Journal of Operations Management, 19713728

60. KumarM. .SternL. W.AndersonJ. C.1993Conducting inter-organizational research using key informants, Academy of Management Journal, 3616331651

61. KwakY. H.AnbariF. T.2006Benefits, obstacles, and future of Six Sigma approach, Technovation, 26708715

62. LeonardD.Mc AdamR.2001Grounded theory methodology and practitioner reflexivity in TQM research, International Journal of Quality and Reliability Management, 182180194

63. Mc AdamR.EvansA.2004Challenges to Six Sigma in a high technology mass-manufacturing environments, Total Quality Management, 155699706

64. Mc AdamR.LaffertyB.2004A multilevel case study critique of Six Sigma: Statistical control or strategic change, International Journal of Operations and Productions Management, 245530549

65. Mc AdamR. .HazlettS. A.HendersonJ.2005A critical review of Six Sigma: Exploring the dichotomies, The International Journal of Organizational Analysis, 132151174

66. MertensD. M.1998Research methods in education and psychology: Integrating diversity with quantitative and qualitative approaches, Sage Publications Inc., Thousand Oaks: London, New Delhi.

67. MortimerA. L.2006Six Sigma: A vital improvement approach when applied to the right problems, in the right environment, Assembly Automation, 2611017

68. MuirA. K.2006Lean Six Sigma statistics: Calculating process efficiencies in transactional projects. McGraw-Hill: New York: London.

69. NeumanW. L.2006Social research methods: Qualitative and quantitative approaches. 6th Ed., Allyn and Bacon: Boston.

70. NonthaleerakP.HendryL.2008Exploring the Six Sigma phenomenon using multiple case study evidence, International Journal of Operations and Production Management, 283279303

71. OkeS. A.2007Six Sigma: A literature review, South African Journal of Industrial Engineering, 182109129

72. ParastM. M.JonesE. C.AdamsS. G.2007Six Sigma and Baldrige: A quality alliance, Quality Progress, Sept, 4551

73. PattonQ. M.1987How to use Qualitative Methods in Evaluation, Sage Publications Inc.: Newsbury Park, London.

74. PaulL. G.1999Practice makes perfect", CIO Enterprise, 127Section 2.

75. RaisinghaniM. S.EtteH.PierceR.CannonG.DaripalyP.2005Six Sigma: Concepts, tools, and applications, Industrial Management and Data Systems, 1054491505

76. RockartJ. F.1979Chief executives define their own data needs, Harvard Business Review, Mar-Apr, 8193

77. RuckerR.2000Citibank increases customer loyalty with defect-free processes, Association for Quality and Participation, 3236

78. SchimdtM.AschkenaseS.2004The building blocks of service, Supply Chain Management Review, Jul/Aug, 3440

79. SchroederR. G.LindermanK.LiedtkeC.ChooA. S.2008Six Sigma: Definition and underlying theory, Journal of Operations Management, 26536554

80. SebastianelliR.TamimiN.2003Understanding the obstacles to TQM success, The Quality Management Journal, 1034556

81. SehwallL.De YongC.2003Six Sigma in health care, International Journal of Health Care Quality Assurance, 16615

82. SenapatiN. R.2004Six Sigma: Myths and realities, International Journal of Quality and Reliability Management, 216683690

83. ShewartW. A.1931Economic Control of Quality of Manufactured Products. Van Nostrand: New York, NY.

84. ShinaS. G.2002Six Sigma for Electronics Design and Manufacturing. McGraw-Hill.

85. SmallB. B.1956Statistical Quality Control Handbook, Western Electric Company. Mack Printing Company: Easton, PA.

86. SmithL. R.2001Six Sigma and the evolution of quality in product development, ASQ Six Sigma Forum Magazine, Nov, 2835

87. SousaR.VossC. A.2001Quality management: Universal or context dependent, Journal of Operations Management, 104383404

88. StamatisD. H.2002aGuidelines for Six Sigma design reviews- Part 1, Quality Digest, Apr, 2733

89. StamatisD. H.2002bGuidelines for Six Sigma design reviews- Part 2, Quality Digest, May, 4854

90. TagueN. R.1995The quality toolbox. ASQC Quality Press: Milwaukee, Wisconsin.

91. TreichlerD. .CarmichaelR. .KusmanoffA. .LewisJ.BerthiezG.2002Design for Six Sigma: 15 lessons learned, Quality Progress, Jan, 3342

92. VossC. .TsikriktsisN.FrohlichM.2002Case research in operations management", International Journal of Operations and Productions Management, 222195219

93. WortmanB.2001The certified Six Sigma black belt primer. 1st Ed, Quality Council of Indiana: West Terre Haute, IN.

94. WyperB.HarrisonA.2000Deployment of Six Sigma methodology in human resource function: A case study, Total Quality Management, 11Nos. 4-5, S720S727

95. YinR. K.1994Case study research: Design and methods. Sage Publications: California.

11

Lean Six Sigma - Making it 'Business as Usual'

1. Introduction

This chapter discusses a new approach to implementing Lean Six Sigma that is sustainable. All too often we hear of horror stories of failure that could be easily avoided. The basis of this approach is neither rocket science nor brand new thinking; indeed the individual elements have been around a long time; DMAIC, structured problem-solving, governance, coaching, dashboards etc.

What is unique and very relevant in today's fast moving e-business global market is the creation of a simple and effective process that brings all these elements together and then embeds them within the organisation as 'business as usual'. This process makes it easy to do the right things and difficult to fall back into old behaviours, and this is the difference that makes the difference.

This was one of the many challenges facing Oakland Consulting when engaged by several world leading large complex organisations. These companies were usually 'ahead of the game' and were already doing a lot of things right. They were typically expanding rapidly through acquisition or generic growth with a team of highly skilled, intellectual and creative staff. In many ways, an argument could be made for simply carrying on along this extremely successful path.

But in every case this was not the case as they recognised that to remain in the top slot also meant embracing even more competitive challenges to accelerate change, reduce costs, improve quality & delivery and protect reputation. This constitutes for many 'redefining quality' and was the basis on which we developed together a successful approach.

2. Lean Six Sigma – a brief step back in time

Some would argue that Lean Six Sigma is the hot topic of today, but this it is not an overnight phenomenon. Many of the tools & techniques have been around for a long time; Dr Walter A Shewhart [1] in 1924, working in Western Electric Company first introduced the idea of preventing defects in manufacturing rather than inspecting finished product, using the Control Chart to predict failure and manage processes economically. Dr Edwards Deming [2] later took his work to Japan and, together with Joseph Juran, transformed their thinking about achieving quality and reducing failure costs.

The idea of Six Sigma has been attributed to Bill Smith [3] who, when working with Motorola in the 1980's as Quality Assurance Manger, first applied the principles that led to Motorola winning the first Malcolm Baldridge National Quality Award in 1988. Others followed and there have been many success stories.

Similarly, although the term Lean is often attributed to James Womack [4], the basic tools of Lean have been used since the 1950's. At Toyota Motor Company, Taichii Ohno [5] and Shigeo Shingo [6], began to incorporate Ford production ideas and other techniques learnt from Deming, Juran and the Japanese 'guru' Ishikawa [7] into an approach known as the Toyota Production System or Just-In-Time. This was the precursor to Lean as we know it today.

3. Approach

Even with all this history, there are several quite unique differences in the way we have learned how to introduce Lean Six Sigma into many different types of companies. Firstly, we do not simply embark on a training programme. If the Lean Six Sigma process is to be effective in reducing cost, improving margin and delivery performance, it needs to become part of 'business as usual'.

This means there is a need to create a 'continuous improvement process' and a governance framework that firmly embeds into the company culture and structure. This is particularly important given the nature and characteristics of these businesses – usually global, highly innovative and responsive to rapid change with a highly skilled, technical and intellectual workforce.

The main areas of this chapter are presented under the following main headings:

- Continuous Improvement
- Training Materials and Workshops
- Coaching
- DMAIC Governance
- Talent Pool Utilisation

3.1. Continuous improvement

Right from the outset, it is critical that Lean Six Sigma is not a "here today, gone tomorrow" initiative. It has to become business as usual and part of a self sustaining process of continual improvement. With this is mind, it is important to create a continuous improvement process that embraces Lean Six Sigma and the DMAIC process in its entirety.

Lean Six Sigma projects need to be well led, managed and directed within the business, and a 9-step Continuous Improvement Process has been created for this purpose. This is a closed loop system that incorporates all the essential elements beginning with; Identified Projects, then Analysis & Rating, followed by developing the Mobilisation Charter and so on. This enables senior managers to steer effectively the Lean Six Sigma process within their strategic goals and is shown in Fig 1.

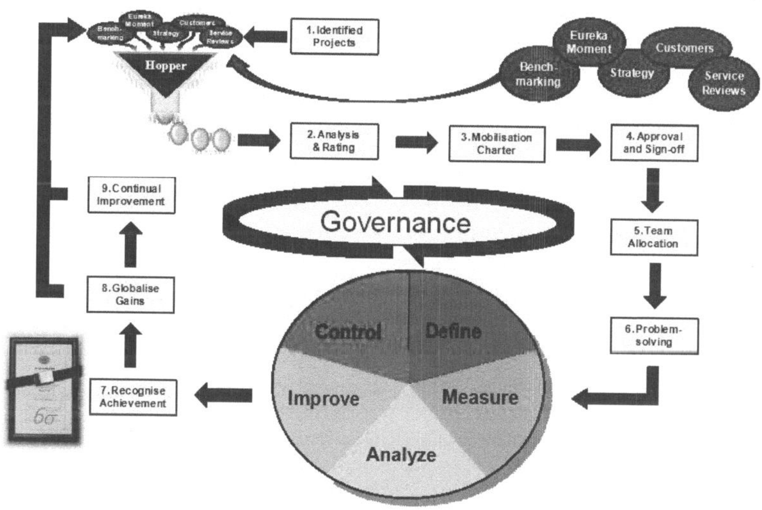

Figure 1. Continuous Improvement Process

For each of the 9-steps, a full description details explicitly what is required, when and who is responsible. Spreadsheet analysis tools are used as necessary. For example the Analysis & Rating step enables new project ideas to be assessed against their likely impact on criteria such as quality, margin and delivery as well as complexity and resource requirements.

In the Recognise Achievement step, a process is set up that embraces not only internal company policies, but also an external assessment by – say – the American Society for Quality [ASQ], or the British Quality Foundation [BQF] through which everyone who attends the training and completes their project[s] is entitled to register for an external accredited qualification. A final example, Globalise Gains step sets out how the Governance Body assesses the relevance and importance of completed projects in other parts of the business, in effect rolling out improvements already deemed to be successful.

The Governance Body comprises of senior managers, the Master Black Belt and Champion[s] who usually meet monthly to lead, manage and control the whole process. Their remit is summarised below;

Objectives

- To manage the 'Hopper' [quality and quantity of ideas], aligned to individual and strategic objectives captured within a Project Initiation Document [PID]
- To commit resources and ensure they are being well used
- To initiate good projects and ensure they deliver target savings within budget and timescale
- • To create a culture of continuous improvement

Process:

- Reviews PID's and Projects
- Follows the guidelines set out in the 9-Step Process
- Decides on critical mass of Lean Six Sigma talent
- Manages talent pool utilisation
- Sets Strategic Business Unit PID rating criteria
- Agrees project savings targets & trend profile
- Reviews and uses DMAIC metrics to drive the process that delivers tangible benefits.

Embedding the continuous improvement process within the organisation's culture means typically utilising the company intranet; this is used across the whole organisation daily to carry out its business. Easy access buttons are created enabling anyone to get an update on Lean Six Sigma progress, upload new project ideas, review the projects hopper, download training materials, and gain access to the GB/BB community, training waves, external accreditation and so on. There are both open and restricted areas created to allow for confidentiality to be honoured in certain cases.

3.2. Training materials and workshops

All materials are fully developed and delivered in-house to reflect company issues and challenges and to make them specific and relevant; four training workshops are created that are shown in Fig 2. The Black Belt is typically a 4 x 4 day course spanning a period of about 4 months with numerous practical examples and exercises. The latest statistical analysis software is used, such as MINTAB v16, as appropriate.

The Green Belt workshop is effectively the first four day training session of the Black Belt course. Combining the first session has enormous benefits in bringing together staff from across the different functions of engineering, quality, service, finance and purchasing etc., cross fertilising ideas and enabling team relationships to develop with a common understanding of the tools & techniques. A much simplified Excel based analysis tool is used, called a Toolbox Calculator, for analysis as the types of problems that Green Belts usually tackle are far less complex.

A Yellow Belt one day training course is created from the Green Belt materials. The purpose of this is to enable a large cross section of the organisation to learn and understand Lean Six Sigma so that they too can make a valuable contribution to improvement in their own work area.

Training Course	Objective	Roles & Responsibilities	Training Duration (days)
Green Belt	That key employees have the latest basic skill set and competence to improve performance significantly	▪ Support / Lead local improvement actions using DMAIC methodology ▪ Participate within BB improvement projects	4
Black Belt	That selected employees have a detailed understanding, skill & competence to address complex problems to make significant improvement	▪ Lead improvement projects using DMAIC ▪ Coach and disseminate DMAIC methodology across business units	16
Executive Black Belt	That senior managers have a good understanding enabling them to commit to the BB programme	▪ Support & deploy the BB improvement process across the business	0.5
Yellow Belt	That most of the organization are aware of the importance and relevance of Lean Six Sigma and therefore able to contribute to meaningful discussions	▪ Support GB and BB projects within their own work area ▪ Generate ideas for potential GB / BB projects ▪ Improve own processes	1

Figure 2. Training and Awareness modules

An Executive Black Belt workshop is designed and delivered for Senior Managers, so that they are aware of and understand the process, and therefore able to create the right environment for Lean Six Sigma to work effectively. After all, these people are responsible for sponsoring the projects, providing the resources and freeing up the barriers to change and improvement.

Additionally, a collection of ca 30 basic generic tools are delivered to the client as a Tool-Kit to accompany the training. These are problem-solving techniques structured under headings; What, Where, When, How and Why? This Tool-Kit is made available to everyone in the organisation to support the wider teams in contributing effectively to their projects, and includes tools such as; Affinity Diagrams, Brainstorming, Critical Path Analysis and Responsibility Charting.

3.3. Coaching

In our experience, coaching is one of the most misunderstood and undervalued elements of Lean Six Sigma initiatives. Time and resource is rarely budgeted for this at the outset, and is often seen as an unnecessary cost, with dire consequences. The need for coaching and recognising its importance in developing people skills and delivering great projects is paramount. To let loose newly trained Black Belts and Green Belts without this support can be a recipe for disaster.

Coaching is about harnessing the latent talent created during training to develop a confident and competent individual who can use effectively the new tools & techniques in developing the best solution. This is the "difference that makes the difference," enabling the right solution to be found, as there are always many.

The fundamental principle is to first understand each person's skill and competence and then to support them in their personal development in an agreed and structured way. This is such a worthwhile and vital part of applied learning. The company should set a minimum coaching requirement of 3 hours per Green / Black Belt per month as a high level figure for budgeting resource. Typically, coaching has been provided initially by Oakland consultants, to get the whole process moving swiftly and then we train a Master Black Belt[s], or a Black Belt[s] to impart the necessary coaching skills in close partnership with the HR department. A typical training and coaching 'wave' plan is given in Fig 3.

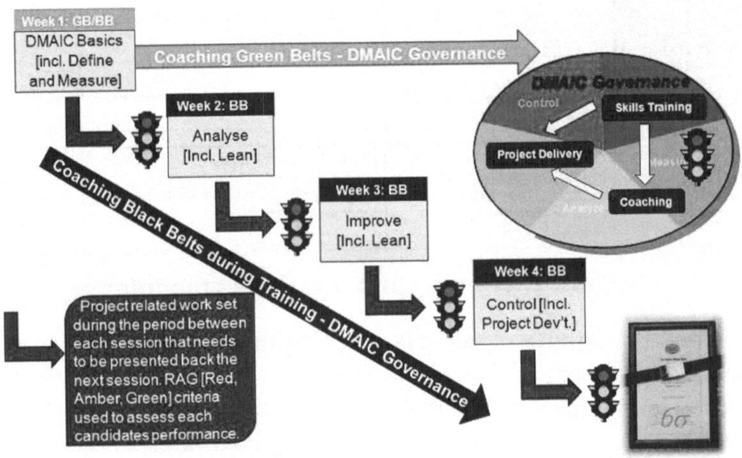

Figure 3. Overview of Wave Training and Coaching.

The Wave Training and Coaching plan depicts how the various elements are arranged. For those staff attending Black Belt training, their initial coaching is structured in between training sessions and is focused on them presenting back their work at the following sessions – in this way it is organised to follow the DMAIC pathway that leads them through to completing their projects. In contrast, the Green Belts receive coaching geared toward their individual needs, but of course based on the minimum requirement of 3 hours per month through to completion.

A RAG [red, amber and green] traffic light metric measures their achievement against agreed criteria and this is discussed more fully in the section headed DMAIC Governance.

To quote a practical example from one of our clients; the first two 'waves' saw 40 people being trained and then deployed part time on 25 projects [some in joint project teams]. Time allocated was budgeted for all 40 people x 3 hours or 120 hours per month, and not based on the 25 projects. This is because everyone working on joint projects must experience all elements of DMAIC [avoiding the; "I'll do Define, you do Measure" scenario]. This becomes critical in their eventual recognition as certified Green and Black Belts. The DMAIC process is given in Fig 4 described as a Lean Six Sigma Roadmap

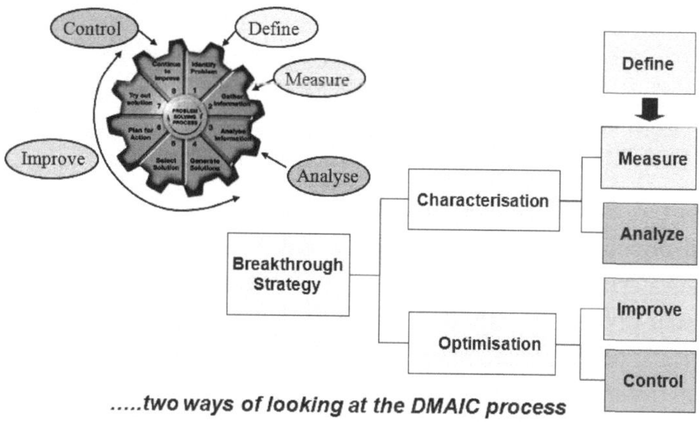

Figure 4. Lean Six Sigma Roadmap

The final sustainable element used within coaching is a unique Storyboard Workbook. This has enabled the Black Belts and Green Belts to work through the whole DMAIC process with a rigour and structure that 'gold-plates' and implements the best possible outcome. The Storyboard Workbook uses a worked-up template that gives clear instructions on what to do and when throughout the DMAIC process. Practical examples prompt formal sign-off at each stage, sample calculations of Sigma levels and financial savings etc enabling the coach and student to critique progress effectively.

The completed Storyboard Workbook is then be used as a record of the DMAIC project by other teams at a later date; for example, to see if these same improvements can also be made in their work area.

3.4. DMAIC governance

A usual question asked is; "How do we know how successfully we are leading and managing the Lean Sigma process?" To answer this question a set of metrics are created within Excel, which reports as a dashboard and becomes the 'heart beat' of the whole Lean Six Sigma process. This serves to report on and control how well desirable outcomes were being delivered. The spreadsheet model is then used as a framework to embed the dashboard within the companies own management information systems.

'What gets measured gets done' – is an all important axiom that is used as an underlying principle in building the dashboard. In all, 12 metrics are defined that measure the essential; 6 financial and 6 performance indices covering areas such as; aggregate and tracked savings, utilisation of talent pool, allocation of projects, coaching effectiveness and DMAIC progress. To ensure these metrics are easy to understand, RAG principles are used to measure achievement against agreed criteria and to report on trends.

The dashboard is updated real-time after each intervention / event by the Master Black Belt, whether a coaching or training session, or completion of a DMAIC element. This is then used by the Governance Body to lead and manage the whole process, as discussed more fully in the section on Continuous Improvement.

3.5. Talent pool utilisation

It is essential that the investment in training and coaching is harnessed and its impact on business performance optimised. Selecting the right staff for Black Belt training is a crucial element in creating a usable talent pool, as it is these same people who will predominantly lead projects and be key to achieving the savings. Aligning the right people with new project ideas is a top priority and guidelines for their selection are given in Fig 5.

Technical Skills	Interpersonal Skills	Managerial Skills	Others
▪ Basic knowledge & affinity with statistics ▪ Expertise in functional area ▪ Undertands relationship customer: supplier processes ▪ Is an expert in their specific field	▪ Good attitude and behaviour in working with people ▪ Natural team builder ▪ Effective communicator ▪ Offers and accepts helpful criticism ▪ Good leader ▪ Is objective ▪ Able to take risks ▪ Is an active listener ▪ Is adaptable	▪ Able to connect projects to business strategy ▪ Sense of urgency ▪ Able to manage meetings ▪ Understands how to manage change ▪ Can innovate ▪ Ability to spread knowledge ▪ A good planner ▪ Able to act under pressure	▪ Very credible and positive reputation ▪ Understands the 'big picture' ▪ A self-starter ▪ Desire to achieve excellence ▪ Promotes win-win solutions ▪ Effectively identifies priorities from a business point of view

......probably not everyone satisfies all criteria!

Figure 5. Criteria for Selecting Black Belts

An aspirational deployment and benefit strategy is developed that recognises the need to create a critical mass of Lean Six Sigma staff. Plans are discussed for between 2% to 5% of the organisation to become Black Belts – the wide variation reflects the different structures of the organisations we work with.

Similarly, the strategy articulates a need for circa 30% of the organisation to become Green Belts and over 75% Yellow Belt trained. Also, at any one time over 95% of the senior management team should have attended the Executive Black Belt workshop.

Guidelines are set for each Black Belt project to generate savings of at least 150k Euros and for a Green Belt project to achieve circa 50k Euros. In one 'mature' client whose improvement programme has been running for over five years, initial Black Belts projects are now averaging over 500k Euros, with total savings from the overall programme in hundreds of millions of Euros. Those Black Belts deployed full time would be expected to lead and deliver 8 projects over the first two years accruing target savings of around 1.5-2.0m Euros plus other non-financial / intangible benefits. Suitable HR policies are needed to support such programmes with back-filling of previous roles and/or clear career and personal development plans in place. By the same token, development planning through the Belt grades is also seen as an important element Yellow to Green to Black to Master Black Belt.

4. A final note

There is no easy route to success, but by using tried and tested principles and processes and introducing them in a structured and manageable way, sustainable change is possible provided that new behaviours and learning are well embedded into 'business as usual'. The bottom line is enabling the organisation to make it easier to do the new right things, rather than to fall back into old ways of working.

Einstein was accredited to have said; "Make things as simple as possible, but not simpler" and this is an important underlying principle in such programmes. Being able to recognise and develop new ways of working at the right level of detail, and then build the continuous improvement process that embraces DMAIC Governance, the training programme and coaching processes to fit for the organisation's culture is a crucial factor in its success.

References

1. ShewhartWalter. A.1931Economic Control of Quality of Manufactured Product. New York: D. Van Nostrand Company. 0-87389-076-0
2. Dr Edwards Deming2000Out of the Crisis. MIT Press, Centre for Advanced Engineering Study.
3. Mikel Harry and Richard Schroeder2000Six Sigma The Breakthrough Management Strategy Revolutionizing The World's Top Corporation.
4. JamesP.WomackDanielT.Jones1996Lean Thinking. Simon & Schuster. 0-68481-035-2
5. Taiichi Ohno1988Toyota Production System: Beyond Large-Scale Production, Productivity Press, 0-91529-914-3
6. Shigeo Shingo1990Modern Approaches to Manufacturing Improvement: The Shingo System. Productivity Press, 091529964
7. Kaoru Ishikawa1976Guide to Quality Control. Asian Productivity Organization, UNIPUB. 9-28331-036-5

Index

A

Actions 135, 177, 185, 189, 191, 223, 226
Activation 157, 159
Adaptation 98
Adoption 82
Airports 189
Analogy 4, 9-10, 12
Arguments 132
Argyris 28-29, 39

B

Budgeting 46
Buffer 145-146
Business 4-5, 11-12, 14-15, 18-19, 23, 65, 201, 213, 218, 221, 226, 230, 236, 238-241, 243, 245, 247-249, 254, 261, 286, 293, 295-296

C

Circuit 79
Community 76, 87-89, 96
Conformity 208, 210
Creation 75
Culminate 79
Currently 243
Customers 119, 134, 222-223, 227-228, 235, 238, 242, 264

D

Dashboard 299-300
Database 177
Designs 110-111, 117
Digest 291
Digital 89-90, 95-96
Documents 50, 52, 95-96, 207, 273
Domain 148
Dynamics 42, 278

E

Economy 218
Element 156, 166-168, 196, 300
Elementary 142, 166
Elements 144, 152, 166-168, 195, 205, 293
E-Mail 277
Emergency 31-32, 185-186
Employees 19, 84, 93, 124, 126, 213, 221, 225, 227-229, 234-235, 242, 262
Equipment 90, 177, 184, 186-187, 211
Ergonomic 18
Evolution 223
External 5, 224, 295
Extinction 183

F

Factors 47, 79, 84, 103, 107, 110-113, 116, 124, 126, 165, 182, 203, 218, 234, 261, 268, 283, 285
Fractional 110-111

Fractures 32

Fusion 77

G

Gender 122

Generation 89, 176

Geographic 93

Governance 294

H

Hardware 152

Hazards 98

Health 41, 69, 71-72, 84-86, 97-99, 119-120, 157, 266, 290

Healthcare 25, 31, 39, 41, 275

Henderson 262

History 202

Humidity 183

Hypothesis 125

I

Indicators 238, 243

Influence 153, 157, 159-160, 173, 192, 196

Inspection 207, 212, 223-224

Internet 70, 87-89, 91, 174, 277

L

Lifecycle 64

M

Magazine 248-249, 289

Manpower 121

Mapping 115

Materials 52, 83, 296

Mechanisms 26, 28, 35, 37-39

Metrics 135, 137, 166, 173, 262, 300

Metrology 205-206, 212

Mirror 104, 106, 112, 116

Moderate 286

Moments 149

Motivation 124

N

Networks 114, 118

Neurons 114

O

Operations 287-291

P

Parameters 115, 137, 164, 174, 182, 280, 282, 285

Passengers 188, 190, 192

Patients 31-32, 119-122, 124, 126

Permanent 159

Philosophy 95-96, 104, 217-218, 255

Planning 6, 44-45, 47, 80, 227, 243, 257

Poisson 145

Proactive 157, 182, 259

Protocol 87

R

Random 137, 149, 165

Regularity 147

Roadmap 37-38, 299

Robust 104

S

Safety 31, 69, 71-72, 79, 85-86, 91, 96-98, 100, 131, 135-137, 172, 184-185, 187-188, 190-191, 193, 197

Selection 274

Sessions 298

Signal 231

Simulation 48, 53-54, 56, 63

Software 8-9, 18, 22, 130, 132-135, 139, 141-142, 152, 160, 165, 169, 171, 174, 176, 188, 193, 197, 199

Solutions 33, 139, 261

Statistics 65, 117

Subsystems 144, 186

Systems 3-5, 9, 11-12, 18-23, 78, 85, 103, 114, 129, 131, 134-136, 139, 141, 145-146, 160, 165, 173, 185, 198-200, 208, 221, 238-241, 243-244

T

Terrorists 188-193

Tobacco 209

Tool-Kit 297

Topology 115

Tossing 165

Toyota 294

Training 12, 37, 114-115, 124, 177, 213-217, 219, 225, 227, 229, 233-234, 262, 279, 294, 296-298, 300

V

Validate 260

Validation 87, 89, 169

Violations 166-167

W

Warning 190

Workplace 72, 82-85, 98-99, 214

Workshops 296

9 789360 847630

Printed at Repro India Ltd.